Electro
D0764858
FOR
DUMMIES
A Wiley Brand
3rd Edition

by Cathleen Shamieh

Electronics For Dummies®, 3rd Edition

Published by: **John Wiley & Sons, Inc.,** 111 River Street, Hoboken, NJ 07030-5774, www.wiley.com

Library of Congress Control Number: 2015944208

ISBN 978-1-119-11797-1 (pbk); ISBN 978-1-119-11799-5 (ebk); ISBN 978-1-119-11798-8 (ebk)

Manufactured in the United States of America

10 9 8 7 6 5 4 3 2

Contents at a Glance

Table of Contents

Chapter 6: Obeying Ohm's Law 89

Chapter 7: Getting Charged Up about Capacitors 105

Introduction

Are you curious to know what makes your iPhone tick? How about your tablet, stereo system, GPS device, HDTV — well, just about every other electronic thing you use to entertain yourself and enrich your life?

Or have you wondered how resistors, diodes, transistors, capacitors, and other building blocks of electronics work? Been tempted to try building your own electronic devices? Well, you've come to the right place!

Electronics For Dummies, 3rd Edition, is your entrée into the exciting world of modern electronics. Loaded with illustrations and plain-English explanations, this book enables you to understand, create, and troubleshoot your own electronic devices.

About This Book

All too often, electronics seems like a mystery because it involves controlling something you can't see — electric current — which you've been warned repeatedly not to touch. That's enough to scare away most people. But as you continue to experience the daily benefits of electronics, you may begin to wonder how it's possible to make so many incredible things happen in such small spaces.

This book offers you a chance to satisfy your curiosity about electronics while having a lot of fun along the way. You get a basic understanding of exactly what electronics is, down-to-earth explanations (and gobs of illustrations) of how major electronic components — and the rules that govern them — work, and step-by-step instructions for building and testing working electronic circuits and projects. Although this book doesn't pretend to answer all your questions about electronics, it gives you a good grounding in the essentials and prepares you to dig deeper into the world of electronic circuits.

I assume that you may want to jump around this book a bit, diving deep into a topic that holds special interest for you and possibly skimming through other topics. For this reason, I provide loads of chapter cross-references to point you to information that can fill in any gaps or refresh your memory on a topic.

The table of contents at the front of this book provides an excellent resource that you can use to quickly locate exactly what you're looking for. You'll also find the glossary useful when you get stuck on a particular term and need to review its definition. Finally, the folks at Wiley have thoughtfully provided a thorough index at the back of the book to assist you in narrowing your reading to specific pages.

It is my hope that when you're finished with this book, you realize that electronics isn't as complicated as you may have once thought. And, it is my intent to arm you with the knowledge and confidence you need to charge ahead in the exciting field of electronics.

Foolish Assumptions

In writing this book, I made a few assumptions about the skill level and interests of you and other readers when it comes to the field of electronics. I tailored the book based on the following assumptions:

- You don't know much — if anything — about electronics.
- You aren't necessarily well versed in physics or mathematics, but you're at least moderately comfortable with introductory high school algebra.
- You want to find out how resistors, capacitors, diodes, transistors, and other electronic components actually work.
- You want to see for yourself — in simple circuits you can build — how each component does its job.
- You're interested in building — and understanding the operation of — circuits that actually accomplish something useful.
- You have a pioneering spirit — that is, a willingness to experiment, accept periodic setbacks, and tackle any problems that may crop up — tempered by an interest in your personal safety.

I start from scratch — explaining what electric current is and why circuits are necessary for current to flow — and build from there. You find easy-to-understand descriptions of how each electronic component works supported by lots of illustrative photographs and diagrams. In 9 of the first 11 chapters, you find one or more mini projects you can build in 15 minutes or so; each is designed to showcase how a particular component works.

Later in the book, I provide several fun projects you can build in an hour or less, and I explain the workings of each one in detail. By building these projects, you get to see firsthand how various components work together to make something cool — sometimes even useful — happen.

As you embark on this electronics tour, expect to make some mistakes along the way. Mistakes are fine; they help enhance your understanding of and appreciation for electronics. Keep in mind: no pain, no gain. (Or should I say, no transistor, no gain?)

Icons Used in This Book

Tips alert you to information that can truly save you time, headaches, or money (or all three!). You'll find that if you use my tips, your electronics experience will be that much more enjoyable.

This icon reminds you of important ideas or facts that you should keep in mind while exploring the fascinating world of electronics.

Even though this entire book is about technical stuff, I flag certain topics to alert you to deeper technical information that might require a little more brain power to digest. If you choose to skip this information, that's okay — you can still follow along just fine.

When you tinker with electronics, you're bound to encounter situations that call for extreme caution. Enter the Warning icon, a not-so-gentle reminder to take extra precautions to avoid personal injury or prevent damage to your tools, components, circuits — or your pocketbook.

Beyond the Book

I have written a lot of extra content that you won't find in this book. Go online to find the following:

- **Cheat sheet** (`www.dummies.com/cheatsheet/electronics`): Here you'll find important formulas and other information you may want to refer to quickly and easily when you're working with a circuit.
- **Online articles covering additional topics** (`www.dummies.com/extras/electronics`): Discover how semiconductors conduct, find out what an oscilloscope is and how it is used, and get more information designed to enhance your knowledge and use of electronics.
- **Updates** (`www.dummies.com/extras/electronics`): Go to this link to find any significate updates or corrections to the material in this book. (My editor made me add the part about corrections, but since I dun't maek mistacks, they're wont be any errata posted. By the weigh, do you want too by a bridge?)

Where to Go from Here

You can use this book in a number of ways. If you start at the beginning (a good place to start), you discover the basics of electronics, add to your knowledge one component at a time, and then put it all together by building projects in your fully outfitted electronics lab.

Or, if you've always been curious about, say, how transistors work, you can jump right into Chapter 10, find out about those amazing little three-legged components, and build a couple of transistor circuits. With a chapter each focused on resistors, capacitors, inductors, diodes, transistors, and integrated circuits (ICs), you can direct your energy to a single chapter to master the component of your choice.

This book also serves as a useful reference, so when you start creating your own circuits, you can go back into the book to refresh your memory on a particular component or rule that governs circuits.

Here are my recommendations for good places to start in this book:

- **Chapter 1:** Start here if you want to get introduced to three of the most important concepts in electronics: current, voltage, and power.
- **Chapter 3:** Jump straight to this chapter if you're anxious to build your first circuit, examine voltages and currents with your multimeter, and make power calculations.
- **Chapter 13:** If you know you'll be addicted to your electronics habit, start with Chapter 13 to find out how to set up your mad-scientist electronics lab, and then go back to the earlier chapters to find out how all the stuff you just bought works.

I hope you thoroughly enjoy the journey you are about to begin. Now, go forth, and explore!

Part I

Fathoming the Fundamentals of Electronics

Check out www.dummies.com/extras/electronics for more great content online.

In this part . . .

- Discovering what makes electronics so fascinating
- Shopping for circuit components and tools
- Experimenting with series and parallel circuits

Chapter 1

Introducing You to Electronics

In This Chapter

- Seeing electric current for what it really is
- Recognizing the power of electrons
- Using conductors to go with the flow (of electrons)
- Pushing electrons around with voltage
- Making the right connections with a circuit
- Controlling the destiny of electrons with electronic components
- Applying electrical energy to loads of things

If you're like most people, you probably have some idea about the topic of electronics. You've been up close and personal with lots of consumer electronics devices, such as smartphones, tablets, iPods, stereo equipment, personal computers, digital cameras, and televisions, but to you, they may seem like mysteriously magical boxes with buttons that respond to your every desire.

You know that underneath each sleek exterior lies an amazing assortment of tiny electronic parts connected in just the right way to make something happen. And now you want to understand how.

In this chapter, you find out that electrons moving in harmony through a conductor constitute electric current — and that controlling electric current is the basis of electronics. You discover what electric current really is and find out that you need voltage to keep the juice flowing. You also get an overview of some of the incredible things you can do with electronics.

Just What Is Electronics?

When you turn on a light in your home, you're connecting a source of electrical energy (usually supplied by your power company) to a light bulb in a complete path, known as an *electrical circuit.* If you add a dimmer or a timer to the light bulb circuit, you can *control* the operation of the light bulb in a more interesting way than just manually switching it on and off.

Electrical systems use electric current to power things such as light bulbs and kitchen appliances. *Electronic systems* take this a step further: They *control* the current, switching it on and off, changing its fluctuations, direction, and timing in various ways to accomplish a variety of functions, from dimming a light bulb (see Figure 1-1), to flashing your holiday light display in sync with your favorite holiday tune, to communicating via satellites — and lots of other things. This control distinguishes electronic systems from electrical systems.

The word *electronics* describes both the field of study that focuses on the control of electrical energy and the physical systems (including circuits, components, and interconnections) that implement this control of electrical energy.

To understand what it means to control electric current, first you need a good working sense of what electric current really is and how it powers things such as light bulbs, speakers, and motors.

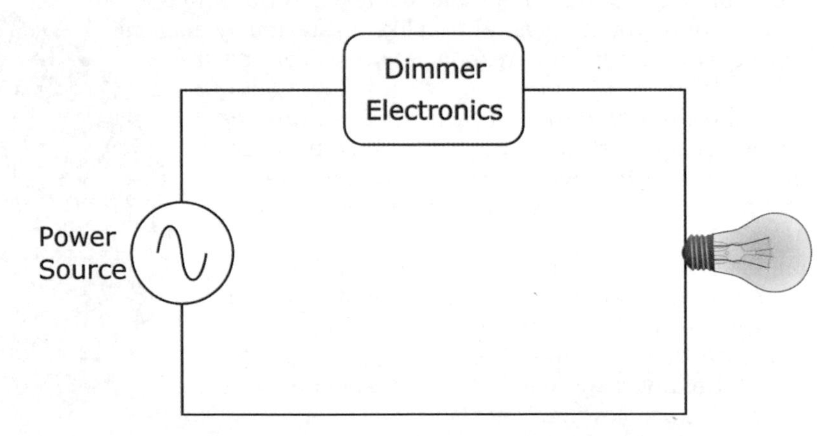

Figure 1-1: The dimmer electronics in this circuit control the flow of electric current to the light bulb.

What is electricity?

The term *electricity* is ambiguous, often contradictory, and can lead to confusion, even among scientists and teachers. Generally speaking, electricity has to do with how certain types of particles in nature interact with each other when in close proximity.

Rather than rely on the term electricity as you explore the field of electronics, you're better off using other, more precise, terminology to describe all things electric. Here are some of them:

- **Electric charge:** A fundamental property of certain particles that describes how they interact with each other. There are two types of electric charges: positive and negative. Particles of the same type (positive-positive or negative-negative) repel each other, and particles of the opposite type (positive-negative) attract each other.
- **Electrical energy:** A form of energy caused by the behavior of electrically charged particles. This is what you pay your electric company to supply.
- **Electric current:** The movement, or flow, of electrically charged particles. This connotation of electricity is probably the one you are most familiar with and the one I focus on in this book.

Checking Out Electric Current

Electric current, sometimes known as electricity (see the sidebar "What is electricity?"), is the movement in the same direction of microscopically small, electrically charged particles called *electrons.* So where exactly do you find electrons, and how do they move around? You'll find the answers by taking a peek inside the atom.

Exploring an atom

Atoms are the basic building blocks of everything in the universe, whether natural or manmade. They're so tiny that you'd find millions of them in a single speck of dust. Every atom contains the following types of subatomic particles:

- **Protons** carry a positive electric charge and exist inside the *nucleus,* or center, of the atom.
- **Neutrons** have no electric charge, and exist along with protons inside the nucleus.
- **Electrons** carry a negative electric charge and are located outside the nucleus in an *electron cloud.* Don't worry about exactly where the electrons of a particular atom are located. Just know that electrons whiz around outside the nucleus, and that some are closer to the nucleus than others.

The specific combination of protons, electrons, and neutrons in an atom defines the type of atom, and substances made up of just one type of atom are known as *elements*. (You may remember wrestling with the *Periodic Table of the Elements* way back in Chemistry class.) I show a simplistic representation of a helium atom in Figure 1-2 and one of a copper atom in Figure 1-3.

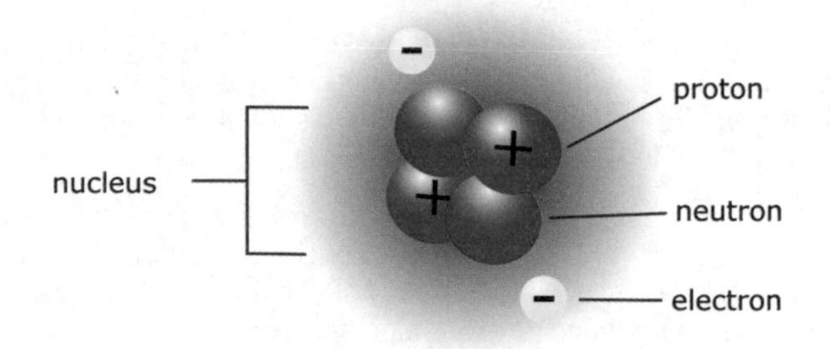

Figure 1-2: This helium atom consists of 2 protons and 2 neutrons in the nucleus with 2 electrons surrounding the nucleus.

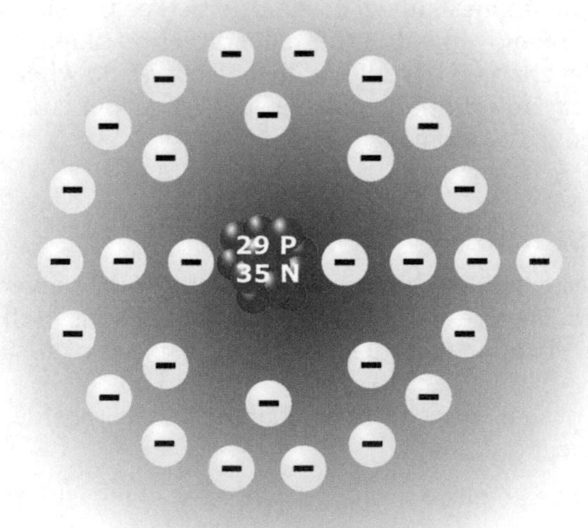

Figure 1-3: A copper atom consists of 29 protons, 35 neutrons, and 29 electrons.

Getting a charge out of protons and electrons

Electric charge is a property of certain particles, such as electrons, protons, and quarks (yes, quarks) that describes how they interact with each other. There are two different types of electric charge, somewhat arbitrarily named positive and negative (much like the four cardinal directions are named north, south, east, and west). In general, particles carrying the same type of charge repel each other, whereas particles carrying opposite charges attract each other. Within each atom, the protons inside the nucleus attract the electrons that are outside the nucleus.

You can experience a similar attraction/repulsion phenomenon with magnets. If you place the north pole of a bar magnet near the south pole of a second bar magnet, you'll find that the magnets attract each other. If, instead, you place the north pole of one magnet near the north pole of another magnet, you'll observe that the magnets repel each other. This mini-experiment gives you some idea of what happens with protons and electrons — without requiring you to split an atom!

Under normal circumstances, every atom has an equal number of protons and electrons, and the atom is said to be *electrically neutral.* (Note that the helium atom has 2 protons and 2 electrons and that the copper atom has 29 of each.) The attractive force between the protons and electrons acts like invisible glue, holding the atom together, in much the same way that the gravitational force of the Earth keeps the moon within sight.

The electrons closest to the nucleus are held to the atom with a stronger force than the electrons farther from the nucleus; some atoms hold on to their outer electrons with a vengeance, while others are a bit more lax. Just how tightly certain atoms hold on to their electrons turns out to be important when it comes to electricity.

Identifying conductors and insulators

Materials (such as copper, silver, aluminum, and other metals) containing loosely bound outer electrons are called *electrical conductors,* or simply *conductors.* Copper is a good conductor because it contains a single loosely bound electron in the outermost reaches of its electron cloud. Materials that tend to keep their electrons close to home are classified as *electrical insulators.* Air, glass, paper, and plastic are good insulators, as are the rubber-like polymers that are used to insulate electrical wires.

In conductors, the outer electrons of each atom are bound so loosely that many of them break free and jump around from atom to atom. These free electrons are like sheep grazing on a hillside: They drift around aimlessly but don't move very far or in any particular direction. But if you give these free electrons a bit of a push in one direction, they will quickly get organized and move together in the direction of the push.

Mobilizing electrons to create current

Electric current (often called electricity) is the displacement of a large number of electrons in the same direction through a conductor when an external force (or push) is applied. That external force is known as *voltage* (which I describe in the next section, "Understanding Voltage").

This flow of electric current appears to happen instantaneously. That's because each free electron — from one end of a conductor to the other — begins to move more or less immediately, jumping from one atom to the next. So each atom simultaneously *loses* one of its electrons to a neighboring atom and *gains* an electron from another neighbor. The result of this cascade of jumping electrons is what we observe as electric current.

Think of a bucket brigade: You have a line of people, each holding a bucket of water, with a person at one end filling an empty bucket with water, and a person at the other end dumping a full bucket out. On command, each person passes his bucket to his neighbor on the right, and accepts a bucket from his neighbor on the left. Although each bucket moves just a short distance (from one person to the next), it appears as if a bucket of water is being transported from one end of the line to the other. Likewise, with electric current, as each electron displaces the one in front of it along a conductive path, it appears as if the electrons are moving nearly instantaneously from one end of the conductor to the other. (See Figure 1-4.)

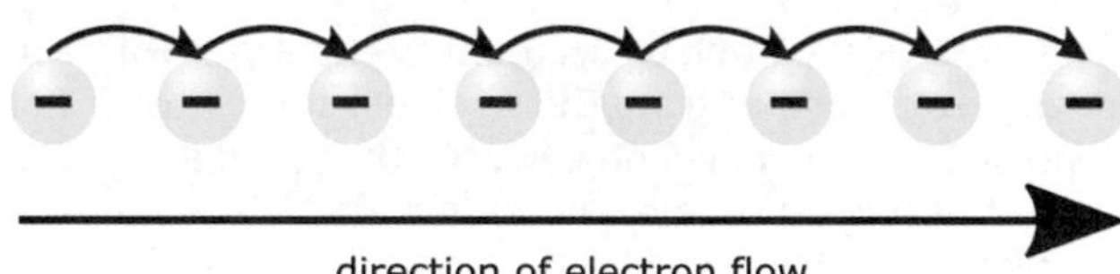

Figure 1-4: Electron flow through a conductor is analogous to a bucket brigade.

The strength of an electric current is defined by how many charge carriers (usually electrons) pass a fixed point in one second, and is measured in units called *amperes,* or *amps* (abbreviated as A). One ampere is defined to be 6,241,000,000,000,000,000 electrons per second. (A more concise way to express this quantity, using scientific notation, is 6.241×10^{18}.) Measuring electric current is analogous to measuring water flow in gallons per minute or liters per second, for instance. The symbol *I* is used to represent the strength of an electric current. (It may help to think of *I* as representing the intensity of the current.)

Experiencing electricity

You can personally experience the flow of electrons by shuffling your feet across a carpet on a dry day and touching a doorknob; that zap you feel (and the spark you may see) is the result of electrically charged particles jumping from your fingertip to the doorknob, a form of electricity known as static electricity. *Static electricity* is an accumulation of electrically charged particles that remain static (unmoving) until drawn to a bunch of oppositely charged particles.

Lightning is another example of static electricity (but not one you want to experience personally), with charged particles traveling from one cloud to another or from a cloud to the ground. The energy resulting from the movement of these charged particles causes the air surrounding the charges to rapidly heat up to nearly 20,000 Celsius — lighting the air and creating an audible shock wave better known as thunder.

If you can get enough charged particles to move around and can control their movement, you can use the resulting electrical energy to power light bulbs and other things.

You may hear the term *coulomb* (pronounced "cool-ome") used to describe the magnitude of the charge carried by 6,241,000,000,000,000,000 electrons. A coulomb is related to an amp in that one coulomb is the amount of charge carried by one amp of current in one second. Coulombs are nice to know about, but amps are what you really need to understand because moving charge, or current, is at the heart of electronics.

A typical refrigerator draws about 3–5 amps of current, and a toaster draws roughly 9 amps. That's a whole lot of electrons at once, much more than are typically found in electronic circuits, where you're more likely to see current measured in milliamps (abbreviated mA). A *milliamp* is one one-thousandth of an amp, or 0.001 amp. (In scientific notation, a milliamp is 1×10^{-3} amp.)

Understanding Voltage

Electric current is the flow of negatively charged electrons through a conductor when a force is applied. But just what is the force that provokes the electrons to move in harmony? What commands the electronic bucket brigade?

Let the force be with you

The force that pushes electrons along is technically called an *electromotive force* (abbreviated *EMF* or *E*), but it is more commonly known as *voltage* (abbreviated *V*). You measure voltage by using units called (conveniently) *volts* (abbreviated V). Apply enough voltage to a conductor, provide a complete path through which an electric charge can move, and the free electrons in the conductor's atoms will move in the same direction, like sheep being herded into a pen — only much faster.

Think of voltage as electric pressure. In much the way water pressure pushes water through pipes and valves, voltage pushes electrons through conductors. The higher the water pressure, the stronger the push. The higher the voltage, the stronger the electric current that flows through a conductor.

Why voltage needs to be different

A *voltage* is simply a difference in electrical charge between two points. In a battery, negatively charged atoms (atoms with an abundance of electrons) build up on one of two metal plates, and positively charged atoms (atoms with a dearth of electrons) build up on the other metal plate, creating a voltage across the plates. (See Figure 1-5.) If you provide a conductive path between the metal plates, you enable excess electrons to travel from one plate to the other, and current will flow in an effort to neutralize the charges. The electromotive force that compels current to flow when the circuit is completed is created by the difference between charges at the battery terminals. (You discover more about how batteries work in the later section "Getting direct current from a battery.")

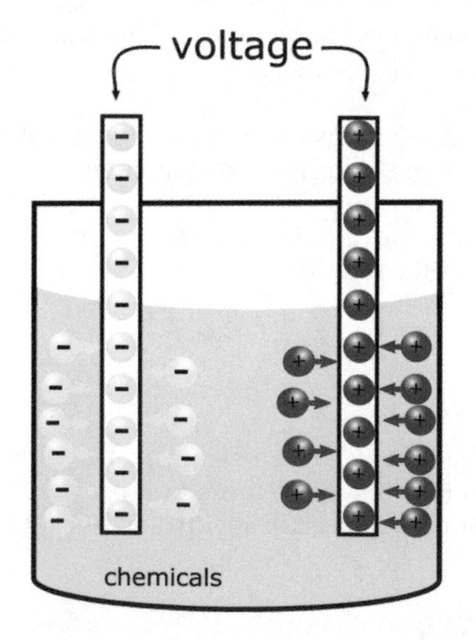

Figure 1-5: A difference in charge between metal plates in a battery creates a voltage.

You may also hear the terms *potential difference, voltage potential, potential drop,* or *voltage drop* used to describe voltage. The word *potential* refers to the possibility that a current may flow if you complete the circuit, and the words *drop* and *difference* both refer to the difference in charge that creates the voltage. You read more about this in Chapter 3.

Putting Electrical Energy to Work

Ben Franklin was one of the first people to observe and experiment with electricity, and he came up with many of the terms and concepts (for instance, *current*) we know and love today. Contrary to popular belief, Franklin didn't actually hold the key at the end of his kite string during that storm in 1752. (If he had, he wouldn't have been around for the American Revolution.) He may have performed that experiment, but not by holding the key.

Franklin knew that electricity was both dangerous and powerful, and his work had people wondering whether there was a way to use the power of electricity for practical applications. Scientists such as Michael Faraday, Thomas Edison, and others took Franklin's work further and figured out ways to harness electrical energy and put it to good use.

As you begin to get excited about harnessing electrical energy, remember that over 250 years ago, Ben Franklin knew enough to be careful around the electrical forces of nature — and so should you. Even tiny amounts of electric current can be dangerous — even fatal — if you're not careful. In Chapter 13, I explain more about the harm that current can inflict and the precautions you can (and must) take to stay safe when working with electronics.

In this section, I explain how electrons transport energy — and how that energy can be applied to make things, such as light bulbs and motors, work.

Tapping into electrical energy

As electrons travel through a conductor, they transport energy from one end of the conductor to the other. Because like charges repel, each electron exerts a noncontact repulsive force on the electron next to it, pushing that electron along through the conductor. As a result, electrical energy is propagated through the conductor.

If you can transport that energy to an object that allows work to be done on it, such as a light bulb, a motor, or a loudspeaker, you can put that energy to good use. The electrical energy carried by the electrons is absorbed by the object and transformed into another form of energy, such as light, heat, or motion. That's how you make the bulb glow, rotate the motor shaft, or cause the diaphragm of the speaker to vibrate and create sound.

Because you can't see gobs of flowing electrons, try thinking about water to help make sense out of harnessing electrical energy. A single drop of water can't do much on its own, but get a whole group of water drops to work in unison, funnel them through a conduit, direct the flow of water toward an object (for example, a waterwheel), and you can put the resulting energy to good use. Just as millions of drops of water moving in the same direction constitute a current, millions of electrons moving in the same direction make an electric current. In fact, Benjamin Franklin came up with the idea that electricity acts like a fluid and has similar properties, such as current and pressure.

But where does the original energy — the thing that starts the electrons moving in the first place — come from? It comes from a *source* of electrical energy, such as a battery. (I discuss electrical energy sources in the section "Supplying Electrical Energy," in this chapter.)

Working electrons deliver power

To electrons delivering energy to a light bulb or other device, the word *work* has real physical meaning. *Work* is a measure of the energy consumed by the device over some time when a force (voltage) is applied to a bunch of electrons in the device. The more electrons you push, and the harder you push them, the more electrical energy is available and the more work can be done (for instance, the brighter the light, or the faster the motor rotation).

Power (abbreviated *P*) is the total energy consumed in doing work over some period of time, and it is measured in *watts* (abbreviated *W*). Power is calculated by multiplying the force (voltage) by the strength of the electron flow (current):

$$power = voltage \times current$$

or

$$P = V \times I$$

The power equation is one of a handful of equations that you should really pay attention to because of its importance in keeping you from blowing things up. Every electronic part, or *component,* has its limits when it comes to how much power it can handle. If you energize too many electrons in a given component, you'll generate a lot of heat energy and you might fry that part. Many electronic components come with maximum power ratings so you can avoid getting into a heated situation. I remind you about the importance of power considerations in later chapters when I discuss specific components and their power ratings, as well as how to use the power equation to ensure that you protect your parts.

Using Circuits to Make Sure Electrons Arrive at Their Destination

Electric current doesn't flow just anywhere. (If it did, you'd be getting shocked all the time.) Electrons flow only if you provide a closed conductive path, known as an *electrical circuit,* or simply a *circuit,* for them to move through, and initiate the flow with a battery or other source of electrical energy.

As shown in Figure 1-6, every circuit needs at least three basic things to ensure that electrons get energized and deliver their energy to something that needs work done:

- **Source of electrical energy:** The *source* provides the voltage, or force, that nudges the electrons through the circuit. You may also hear the terms *electrical source, power source, voltage source,* and *energy source* used to describe a source of electrical energy.
- **Load:** The *load* is something that absorbs electrical energy in a circuit (for instance, a light bulb, a speaker, or a refrigerator). Think of the load as the destination for the electrical energy.
- **Path:** A conductive *path* provides a conduit for electrons to flow between the source and the load. Copper and other conductors are commonly formed into wire to provide this path.

An electric current starts with a push from the source and flows through the wire path to the load, where electrical energy makes something happen (such as light being emitted) and then back to the other side of the source.

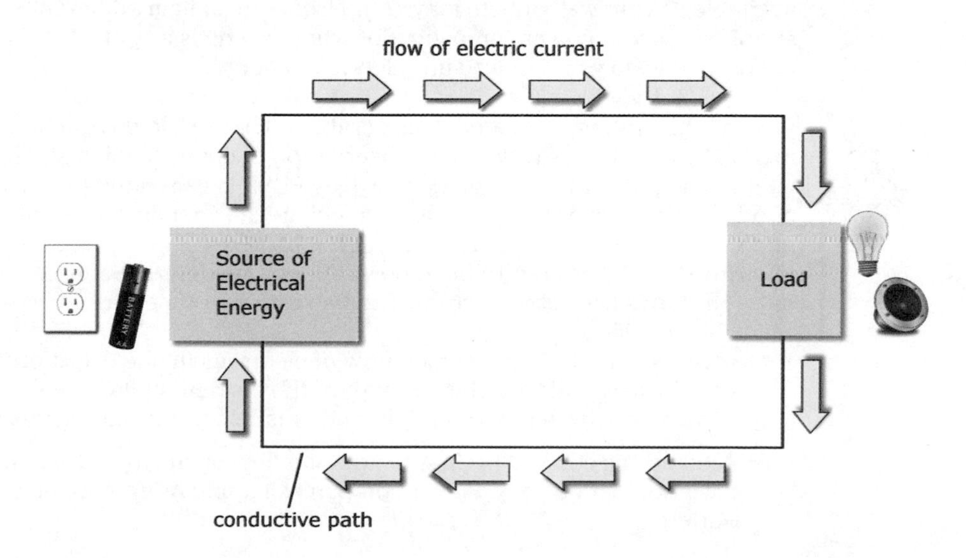

Figure 1-6: A simple circuit consisting of a power source, a load, and a path for electric current.

Most often, other electronic parts are also connected throughout the circuit to control the flow of current.

If you simply provide a conductive path in a closed loop that contains a power source but no light bulb, speaker, or other external load, you still have a circuit and current will flow. In this case, the role of the load is played by the resistance of the wire and the internal resistance of the battery, which transfer the electrical energy into heat energy. (You find out about resistance in Chapter 5.) Without an external load to absorb some of the electrical energy, the heat energy can melt the insulation around a wire or cause an explosion or release of dangerous chemicals from a battery. In Chapter 3, I explain more about this type of circuit, which is known as a *short circuit*.

Supplying Electrical Energy

If you take a copper wire and arrange it in a closed loop by twisting the ends together, do you think the free electrons will flow? Well, the electrons might dance around a bit, because they're so easy to move. But unless a force pushes the electrons one way or another, you won't get current to flow.

Think about the motion of water that is just sitting in a closed pipe: The water isn't going to go whooshing through the pipe on its own. You need to introduce a force, a pressure differential, to deliver the energy needed to get a current flowing through the pipe.

Likewise, every circuit needs a source of electrical energy to get the electrons flowing. Batteries and solar cells are common sources; the electrical energy available at your wall outlets may come from one of many different sources supplied by your power company. But what exactly is a source of electrical energy? How do you "conjure up" electrical energy?

Electrical energy isn't created from scratch. (That would go against a fundamental law of physics called the conservation of energy, which states that energy can neither be created nor destroyed.) It is generated by converting another form of energy (for instance, mechanical, chemical, heat, or light) into electrical energy. Exactly how electrical energy is generated by your favorite source turns out to be important because different sources produce different types of electric current. The two different types are

- **Direct current (DC):** A steady flow of electrons in one direction, with very little variation in the strength of the current. Cells (commonly known as batteries) produce DC and most electronic circuits use DC.
- **Alternating current (AC):** A fluctuating flow of electrons that changes direction periodically. Power companies supply AC to your electrical outlets.

Getting direct current from a battery

A battery converts chemical energy into electrical energy through a process called an *electrochemical reaction.* When two different metals are immersed in certain chemicals, the metal atoms react with the chemical atoms to produce charged atoms, known as *ions*. As you see in Figure 1-7, negative ions build up on one metal plate, known as an *electrode,* while positive ions build up on the other electrode. The difference in charge across the two electrodes creates a voltage. That voltage is the force that electrons need to push them around a circuit.

You might think that the oppositely charged ions would move towards each other inside the battery, because opposite charges attract, but the chemicals inside the battery act as a barrier to prevent this from happening.

To use a battery in a circuit, you connect one side of your load — for instance, a light bulb — to the negative terminal and the other side of your load to the positive terminal. (A *terminal* is just a piece of metal connected to an electrode to which you can hook up wires.) You've created a path that allows the charges to move, and electrons flow from the negative terminal, through the circuit, to the positive terminal. As they pass through the wire filament of the light bulb, some of the electrical energy supplied by the battery is converted to light and heat, causing the filament to glow and get warm.

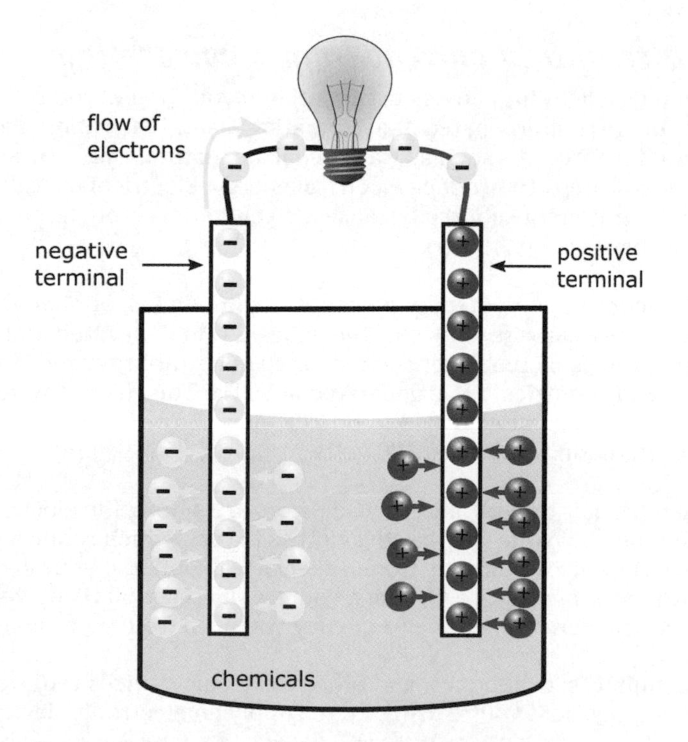

Figure 1-7: Direct current (DC) generated by a battery.

The electrons keep flowing as long as the battery is connected in a circuit and the electrochemical reactions continue to take place. As the chemicals become depleted, fewer reactions take place, and the battery's voltage starts to drop. Eventually, the battery can no longer generate electrical energy, and we say that the battery is flat or dead.

Because the electrons move in only one direction (from the negative terminal, through the circuit, to the positive terminal), the electric current generated by a battery is DC. The AAA-, AA-, C-, and D-size batteries you can buy almost anywhere each generate about 1.5 volts — regardless of size. The difference in size among those batteries has to do with how much current can be drawn from them. The larger the battery, the more current can be drawn, and the longer it will last. Larger batteries can handle heavier loads, which is just a way of saying they can produce more power (remember, power = voltage × current), so they can do more work.

Technically speaking, an individual battery isn't really a battery (that is, a group of units working together); it's a *cell* (one of those units). If you connect several cells together, as you often do in many types of flashlights and children's toys, *then* you've created a battery. The battery in your car is made up of six cells, each generating 2 to 2.1 volts, connected together to produce 12 to 12.6 volts total.

Using alternating current from a power plant

When you plug a light into an electrical outlet in your home, you're using electrical energy that originated at a generating plant. Generating plants process natural resources — such as water, coal, oil, natural gas, or uranium — through several steps to produce electrical energy. Electrical energy is said to be a *secondary* energy source because it's generated through the conversion of a primary energy source.

The electric current generated by power plants fluctuates, or changes direction, at a regular rate known as the *frequency*. In the United States and Canada, that rate is 60 times per second, or 60 hertz (abbreviated Hz), but in most European countries, AC is generated at 50 Hz. The electricity supplied by your average wall outlet is said to be 120 volts AC (or 120 VAC), which just means it's alternating current at 120 volts.

Heaters, lamps, hair dryers, and electric razors are among the electrical devices that use 120 volts AC directly; clothes dryers, which require more power, use 240 volts AC directly from a special wall outlet. If your hair dryer uses 60 Hz power, and you're visiting a country that uses 50 Hz power, you'll need a *power converter* to get the frequency you need from your host country.

Tablets, computers, cellphones, and other electronic devices require a steady DC supply, so if you're using AC to supply an electronic device or circuit, you'll need to convert AC to DC. *Regulated power supplies,* also known as *AC-to-DC adapters,* or *AC adapters,* don't actually *supply* power: They convert AC to DC and are commonly included with electronic devices when purchased. Think of your cellphone charger; this little device essentially converts AC power into DC power that the battery in your cellphone uses to charge itself back up.

Transforming light into electricity

Solar cells, also known as *photovoltaic cells,* produce a small voltage when you shine light on them. They are made from *semiconductors,* which are materials that are somewhere between conductors and insulators in terms of their willingness to give up their electrons. (I discuss semiconductors in detail in Chapter 9.) The amount of voltage produced by a solar cell is fairly constant, no matter how much light you shine on it — but the *strength* of the current you can draw depends on the intensity of the light: The brighter the light, the higher the strength of the available current (that is, until you reach the solar cell's maximum output, at which point no more current can be drawn).

Solar cells have wires attached to two terminals for conducting electrons through circuits, so you can power your calculator or the garden lights that frame your walkway. You may have seen arrays of solar cells used to power calculators (see Figure 1-8), emergency road signs, call boxes, or lights in parking lots, but you probably haven't seen the large solar-cell arrays used to power satellites (not from close up, anyway).

Solar panels are becoming increasingly popular for supplying electrical power to homes and businesses. If you scour the Internet, you'll find lots of information on how you can make your own solar panels — for just a couple of hundred dollars and a willingness to try. You can read more about this topic in *Solar Power Your Home For Dummies,* 2nd Edition, by Rik DeGunther (Wiley Publishing, Inc.).

Figure 1-8: This calculator is powered by photovoltaic cells.

Using symbols to represent energy sources

Figure 1-9 shows the symbols commonly used to represent different energy sources in circuit diagrams, or *schematics.*

In the battery symbol (see Figure 1-9, left), the plus sign signifies the positive terminal (sometimes called the *cathode*); the minus sign signifies the negative terminal (or the *anode*). Usually the battery's voltage is shown alongside the symbol. The sine wave in the symbol for an AC voltage source (see Figure 1-9, center) is a reminder that the voltage varies up and down. In the symbol for a photovoltaic cell (see Figure 1-9, right), the two arrows pointing towards the battery symbol signify light energy.

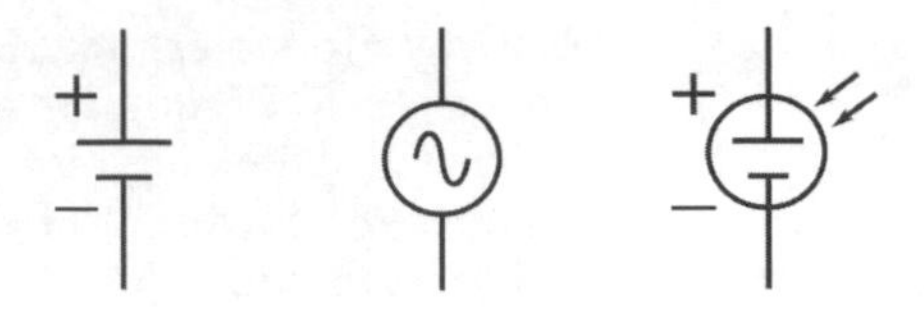

Figure 1-9: Circuit symbols for a battery (left), AC power source (center), and photovoltaic cell (right).

Marveling at What Electrons Can Do

Imagine applying a constant electric current to a pair of speakers without using anything to control, or "shape," the current. What would you hear? Guaranteed it wouldn't be music! By using the proper combination of electronics assembled in just the right way, you can control the way each speaker diaphragm vibrates, producing recognizable sounds such as speech or music. There's so much more you can do with electric current when you know how to control the flow of electrons.

Electronics is all about using specialized devices known as *electronic components* (for example, switches, resistors, capacitors, inductors, and transistors) to control current (also known as the flow of electrons) in such a way that a specific function is performed.

The nice thing is that you after you understand how a few individual electronic components work and how to apply some basic principles, you can begin to understand and build interesting electronic circuits.

This section provides just a sampling of the sorts of things you can do by controlling electric current with electronic circuits.

Creating good vibrations

Electronic components in your iPod, car stereo, and other audio systems convert electrical energy into sound energy. In each case, the system's speakers are the load, or destination, for electrical energy. The job of the electronic components in the system is to "shape" the current flowing to the speakers so that the diaphragm within each speaker moves in such a way as to reproduce the original sound.

Seeing is believing

In visual systems, electronic components control the timing and intensity of light emissions. Many remote-control devices, such as your TV remote, emit infrared light (which is not visible) when you press a button, and the specific pattern of the emitted light acts as a sort of code that is understood by the device you are controlling. A circuit in your TV detects the infrared light and, in effect, decodes the instructions sent by the remote.

A flat-screen liquid crystal display (LCD) or plasma TV consists of millions of tiny picture elements, or *pixels,* each of which is a red, blue, or green light that can be switched on or off electronically. The electronic circuits in the TV control the timing and on/off state of each pixel, thus controlling the pattern painted across the TV screen, which is the image you see.

Sensing and alarming

Electronics can be used also to make something happen in response to a specific level or absence of light, heat, sound, or motion. Electronic *sensors* generate or change an electrical current in response to a stimulus. Microphones, motion detectors, temperature sensors, humidity sensors, and light sensors can be used to trigger other electronic components to perform some action, such as activating an automatic door opener, sounding an alarm, or switching a sprinkler system on or off.

Controlling motion

A common use of electronics is to control the on/off activity and speed of motors. By connecting various objects — for instance, wheels, airplane flaps, or fan blades — to motors, you can use electronics to control their motion. Such electronics can be found in robotic systems, aircraft, spacecraft, elevators, and lots of other places.

Computing

In much the same way that the ancients used the abacus to perform arithmetic operations, so you use electronic calculators and computers to perform computations. With the abacus, beads were used to represent numbers, and calculations were performed by manipulating those beads. In computing systems, patterns of stored electrical energy are used to represent numbers, letters, and other information, and computations are performed by manipulating those patterns using electronic components. (Of course, the worker-bee electrons inside have no idea that they are crunching numbers!) The result of a computation is stored as a new pattern of electrical energy and often directed to special circuits designed to display the result on a monitor or other screen.

Voice, video, and data communications

Electronic circuits in your cellphone work together to convert the sound of your voice into an electrical pattern, manipulate the pattern (to compress and encode it for efficient, secure transmission), transform it into a radio signal, and send it out through the air to a communication tower. Other electronic circuits in your handset detect incoming messages from the tower, decode the messages, and convert an electrical pattern in the message into sound (through a speaker) or a text or video message (through your phone's display).

Data communication systems use electronics to transmit information encoded in electrical patterns between two or more endpoints. When you shop online, your order is transmitted by sending an electrical pattern from your data communication device (such as a laptop, smartphone, or tablet) over the Internet to a communication system operated by a vendor. With a little help from electronic components, you can get electrons to convert your materialistic desires into shopping orders — and charge the order to your credit card.

2

Gearing Up to Explore Electronics

In This Chapter

- Appreciating the many different ways to control electric current
- Acquiring the tools and components you need to start building circuits
- Getting a briefing on how to use a solderless breadboard

Controlling electrical current is similar in many ways to controlling H_2O current. How many different ways can you control the flow of water using various plumbing devices and other components? Some of the things you can do are restrict the flow, cut off the flow completely, adjust the pressure, allow water to flow in one direction only, and store water. (This water analogy may help but it isn't 100% valid; you don't need a closed system for water to flow — and you *do* need a closed system to make electric current flow.)

Many electronic components can help control the electrical energy in circuits. Among the most popular components are *resistors,* which restrict current flow, and *capacitors*, which store electrical energy. *Inductors* and *transformers* are devices that store electrical energy in magnetic fields. *Diodes* are used to restrict current flow in one direction, much like valves, while *transistors* are versatile components that can be used to switch circuits on and off, or amplify current. *Integrated circuits (ICs)* contain multiple discrete (that is, individual) components in a single package and are able to control current in many ways, depending on the particular IC. Sensors, switches, and other parts also play important roles in circuits.

In Chapters 3 through 12, you find out how these different electronic components manipulate current and work together to make useful things happen. Most of those chapters include simple experiments designed to show you firsthand what each component can do. Chapter 17 contains more involved projects, each of which involves many components working together as a team to make something useful (or just fun) happen. In this chapter, you find out what you need to build these experimental circuits and projects.

Getting the Tools You Need

To complete the experiments and projects in this book, you'll need a few tools that may cost you $100 to $250 total, depending on where you shop. I list the essential tools here, and provide a more detailed list of tools and supplies for the serious electronics hobbyist in Chapter 13.

In the following list, I provide some model numbers (identified by #) and prices, but feel free to shop around online or at yard sales to seek better deals.

- **Multimeter:** This tool enables you to measure voltage, resistance, and current, and is essential for understanding what's going on (or not) in the circuits you build. Buy RadioShack #22-813 ($40), shown in Figure 2-1, or similar. Purchase a set of spring-loaded test clips, too, such as RadioShack #270-334 ($3.49). Chapter 16 provides detailed information about how to use a multimeter.

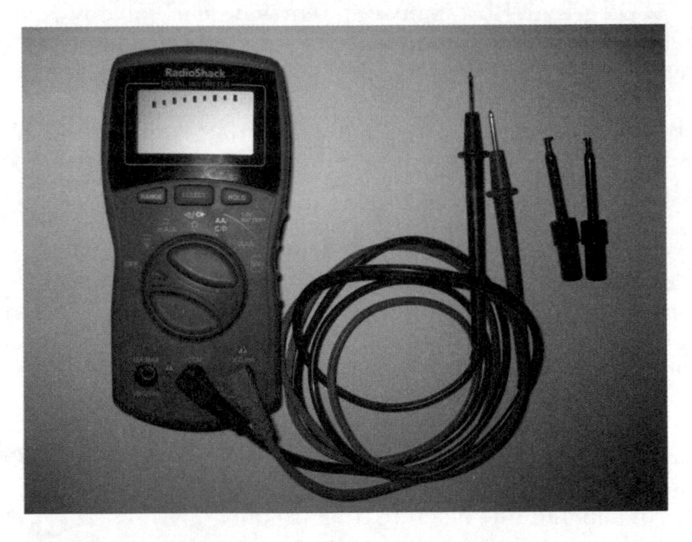

Figure 2-1: A multimeter and spring-loaded test clips.

- **Solderless breadboard:** You use a breadboard to build, explore, update, tear down, and rebuild circuits. I recommend you purchase a larger model, such as the Elenco #9425 830-contact breadboard (roughly $14 at various online suppliers), which is shown in Figure 2-2.

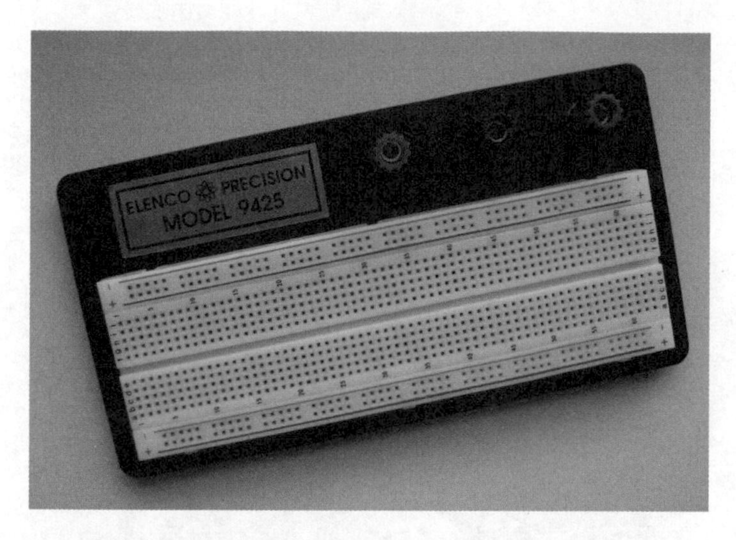

Figure 2-2: This solderless breadboard has 830 contact holes.

- **Soldering iron:** This tool (shown in Figure 2-3) enables you to create a conductive joint between parts such as wires, component leads, and circuit boards. You will need it to attach leads to a few potentiometers (variable resistors). Models range from a low-end Weller SP25NKUS ($20 at Home Depot) to the mid-range Weller WLC-100 ($44 at Home Depot) to the top-notch Weller WES51 ($129 at Mouser.com). You need 60/40 rosin-core solder in either 0.031-inch or 0.062-inch diameters, such as top-of-the-line Kester 44 ($30 for a 1-lb. spool).

Figure 2-3: The Weller WES51 soldering station includes a temperature-adjustable soldering iron and a stand.

- **Hand tools:** Must-have hand tools include needle-nose pliers for bending leads and wire and a multipurpose wire stripper/cutter (see Figure 2-4). The pliers also come in handy for inserting and removing components from your solderless breadboard. Expect to spend at least $10 each on these items at your local hardware store or online electronics supplier.

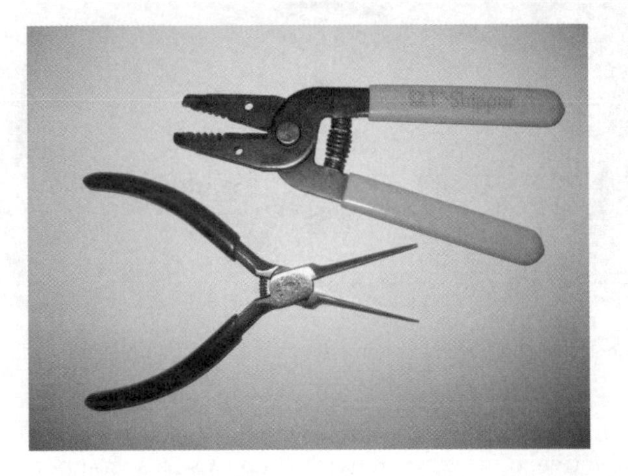

Figure 2-4: A gauged wire stripper/cutter and needle-nose pliers.

- **Antistatic wrist strap:** You need to use a strap like the one in Figure 2-5 to prevent the charges that build up on your body from zapping — and potentially damaging — static-sensitive integrated circuits (ICs) during handling. Purchase a Zitrades #S-W-S-1 ($10) or similar.
- **Calculator:** You use a little math when choosing certain components for your circuits and to help you understand circuit operation. Even if you're a rock star at math, it's still a good idea to use a calculator.

Figure 2-5: An antistatic wrist strap can prevent you from zapping sensitive components.

Stocking Up on Essential Supplies

In this section, I provide a comprehensive list of the electronic components, power supplies, interconnections, and other parts you need to complete the experiments in Chapters 3–11 and the projects in Chapter 17. You can find most of these products at RadioShack stores (many of which are in the process of closing as of this writing). If you plan, you can find great deals online at Amazon.com, eBay.com, Parts-Express.com, and other websites. Check consumer reviews of products, shipping costs, and delivery-time windows before ordering online. See Chapter 19 for more parts sources.

In the list that follows, I sometimes specify a product code (identified by #) and price (as of this writing, in mid-2015). I do this just to give you an idea of what to look for and roughly how much you should expect to pay. Because multiple options are usually available, feel free to shop around. Here's your shopping list of electronic parts, most of which are shown in Figure 2-6:

- **Batteries and accessories**
  - One (minimum) fresh 9-volt disposable (not rechargeable) battery.
  - Four (minimum) fresh AA disposable batteries.
  - One four-battery (AA) holder with leads or terminals for a battery snap connector. Parts Express #140-972 ($1.49) or similar.
  - One 9-volt battery clip (sometimes called a snap connector). Buy two if your four-battery holder has terminals for a clip rather than leads. Parts Express #090-805 ($0.65 each) or similar.

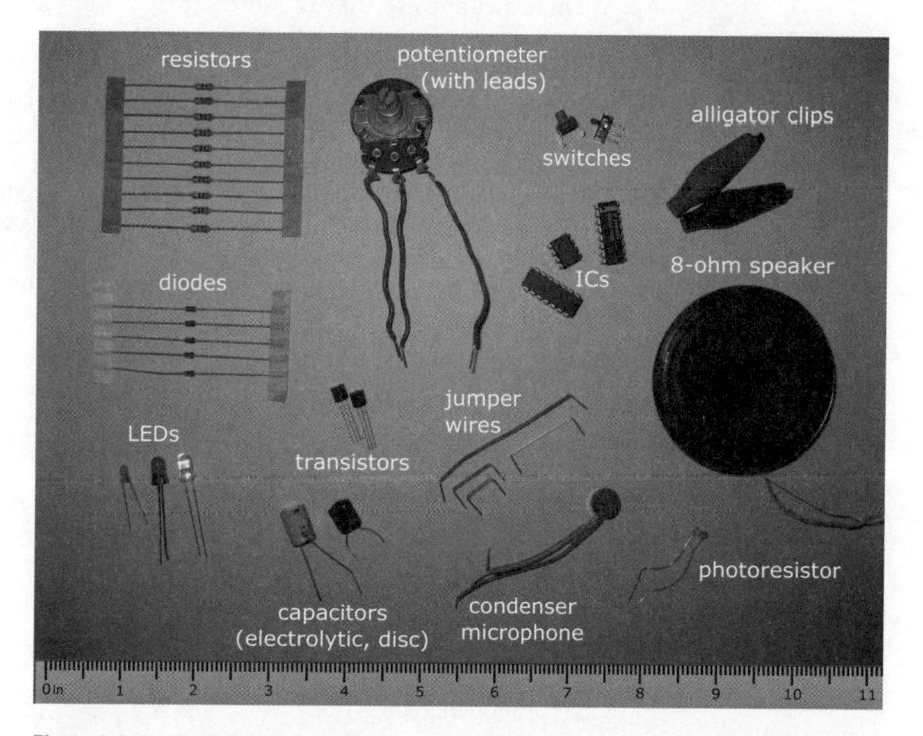

Figure 2-6: A sampling of the electronic components used in the experiments and projects in this book.

- **Wire, alligator clips, and switches**
  - 22-gauge solid wire, insulated, at least 4 feet total (multiple colors are preferable but not necessary). Elenco #884420 (red), #884440 (yellow), and #884410 (black) cost less than $3 each at various online suppliers. Each model provides 25 feet of wire on a spool.
  - Assorted precut, prestripped jumper wires (optional, but highly recommended). Purchase RadioShack #276-173 ($7) or similar.
  - Alligator clips, fully insulated. Get one set of 10, preferably in assorted colors. Purchase RadioShack #270-378 (1-1/4-inch mini clips) or #270-356 (2-inch clips) or similar ($2.50-3.50 per set).
  - Five (minimum) single-pole, double-throw (SPDT) slide switches. Make sure these switches are breadboard friendly, with pins spaced 0.1 inch (2.54mm) apart. Mouser #123-09.03201.02 ($1.15 each), Banana Robotics #BR010115 (5-pack for $0.99), or similar.
  - Eight mini pushbutton (momentary on, normally open) switches. Purchase SparkFun Electronics #COM-00097 ($0.35 each), Amico #a12011500ux0302 (100-pack for $3.90 on Amazon.com), or similar. You may want to use your pliers to straighten the curved legs on these mini switches so they fit more snugly in your solderless breadboard. Banana Robotics sells a 10-pack of 2-pin pushbutton switches for $0.99 (#BR010084).
- **Resistors:** You'll need an assortment of resistor values. Many suppliers sell resistors in packs of 5 or 10 for $1 or less. Resistors rated at 1/4 watt with 10% or 20% tolerance are fine. You can order a multipack for each resistance value, or you can buy a variety pack, such as RadioShack #271-312 ($14.49), which contains 500 assorted 1/4 W resistors with 5% tolerance and includes all the values listed next. Here are the resistor values, the color codes used to identify them, and the minimum quantities you need:
  - One 330 Ω (orange-orange-brown)
  - Three 470 Ω (yellow-violet-brown)
  - One 820 Ω (grey-red-brown)
  - Two 1 kΩ (black-brown-red)
  - One 1.2 kΩ (brown-red-red)
  - Two 1.8 kΩ (brown-grey-red)
  - Two 2.2 kΩ (red-red-red)
  - One 2.7 kΩ (red-violet-red)
  - One 3 kΩ (orange-black-red)
  - One 3.9 kΩ (orange-white-red)
  - One 4.7 kΩ (yellow-violet-red)

- Four 10 kΩ (brown-black-orange)
- One 12 kΩ (brown-red-orange)
- One 15 kΩ (brown-green-orange)
- One 22 kΩ (red-red-orange)
- One 47 kΩ (yellow-violet-orange)
- One 100 kΩ (brown-black-yellow)

- **Potentiometers (variable resistors)**
  - One 10 kΩ; Parts Express #023-628 ($1.55) or similar
  - One 50 kΩ; Parts Express #023-632 ($1.55) or similar
  - One 100 kΩ; Parts Express #023-634 ($1.55) or similar
  - One 1 MΩ; Parts Express #023-640 ($1.60) or similar
- **Capacitors:** For the capacitors in the following list, a voltage rating of 16 V or higher will do. Prices range from roughly $0.10 to $1.49 each, depending on the size and supplier (online is cheaper).
  - Two 0.01 μF disc
  - One 0.047 μF disc
  - One 0.1 μF disc
  - One 4.7 μF electrolytic
  - Three 10 μF electrolytic
  - One 47 μF electrolytic
  - One 100 μF electrolytic
  - One 220 μF electrolytic
  - One 470 μF electrolytic
- **Diodes:** Minimum quantities are specified in the following list, but I recommend you purchase at least a few more of each (they're cheap — and they're fryable).
  - Ten 1N4148 diodes. These diodes cost a few pennies each online, or you can purchase a 10-pack at RadioShack for about $2.
  - Ten diffused light-emitting diodes (LEDs), any size (3mm or 5mm recommended), any color. You may want to buy at least one red, one yellow, and one green for the traffic light circuit in Chapter 17. These LEDs cost $0.08–$0.25 each online (for instance, Parts Express #070-020).
  - Eight ultrabright LEDs, 5mm, any color. You may want to purchase red ones, such as Parts Express #070-501 ($0.58 each), if you're serious about using the LED Bike Flasher you build in Chapter 17.

- **Transistors:** Buy one or two more than the minimum specified quantity of each type, just in case you fry one. They cost about $0.30 each online, or $1.49 each in RadioShack stores.
  - Two 2N3904, 2N2222, BC548, or any general-purpose NPN bipolar transistors
  - One 2N3906, 2N2907, or any general-purpose PNP bipolar transistor
- **Integrated circuits (ICs)**
  - One 74HC00 CMOS quad 2-input NAND gate, 14-pin dual-in-line package (DIP). Get two because they can be easily damaged by static discharge. Purchase Jameco #906339 ($0.79) or similar.
  - Two 555 timers (8-pin DIP). I recommend you purchase one or two extra chips. This IC costs about $0.25-$1 online or $2 in RadioShack stores.
  - One LM386 audio power amplifier (8-pin DIP). Expect to pay $1 to $2 online or at RadioShack.
  - One 4017 CMOS decade counter. I recommend you get at least one extra chip, due to its sensitivity to static. Per chip costs range from $0.35 to $2 online, depending on quantity.
- **Miscellaneous**
  - One 8 Ω, 0.5 W speaker. Purchase RadioShack #273-092 ($3.99) or similar.
  - One or more photoresistors (any value will do). RadioShack sells a 5-pack (#276-1657) for $3.99, but you can find larger quantities for less money online.
  - Condenser microphone (optional). Purchase RadioShack #270-092 ($3.99) or similar.
  - One wooden pencil or small-diameter wooden dowel.
  - One relatively strong bar magnet, approximately 2 inches long.

Getting Ready to Rumble

After you've purchased all your supplies, tools, and components, you need to do a few things before you can start building circuits:

- **Attach a battery clip to a 9-volt battery.** The clip provides leads so you can connect the 9-volt battery to your solderless breadboard. The leads are color-coded: Red indicates the positive battery terminal and black indicates the negative battery terminal. (See Figure 2-7.)

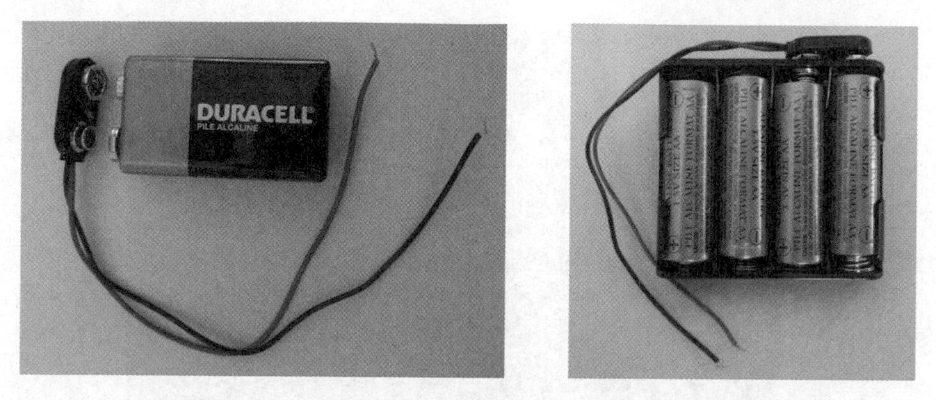

Figure 2-7: Prepare your batteries for use in a solderless breadboard.

- **Insert the four AA batteries into the four-battery holder, observing the polarity markers.** The battery holder is wired to connect the four batteries end-to-end, creating a battery pack that supplies $4 \times 1.5 = 6$ volts. If your four-battery holder does not have leads, attach a battery clip to the snap connectors on the holder. (See Figure 2-7.)
- **Attach leads to the potentiometers.** This step involves cutting three short lengths (2–3 inches each) of 22-gauge solid wire for each potentiometer, stripping both ends of each wire, and soldering the stripped leads to the potentiometer terminals (refer to the top row in Figure 2-6). See Chapter 15 for detailed instructions on soldering.

Using a Solderless Breadboard

This section provides a brief overview of how to use a solderless breadboard. I explain much more about solderless breadboards in Chapter 15, and I strongly encourage you to read that chapter before you get too deep into building circuits because you need to know about the limitations of these handy circuit-building platforms.

A *solderless breadboard* is a reusable rectangular plastic board that contains several hundred square *sockets,* or contact holes, into which you plug components such as resistors, capacitors, diodes, transistors, and integrated circuits. Groups of contact holes are electrically connected by flexible metal strips running below the surface. The photo in Figure 2-8 shows part of an 830-contact solderless breadboard with yellow lines added to help you visualize the underlying connections between contact holes.

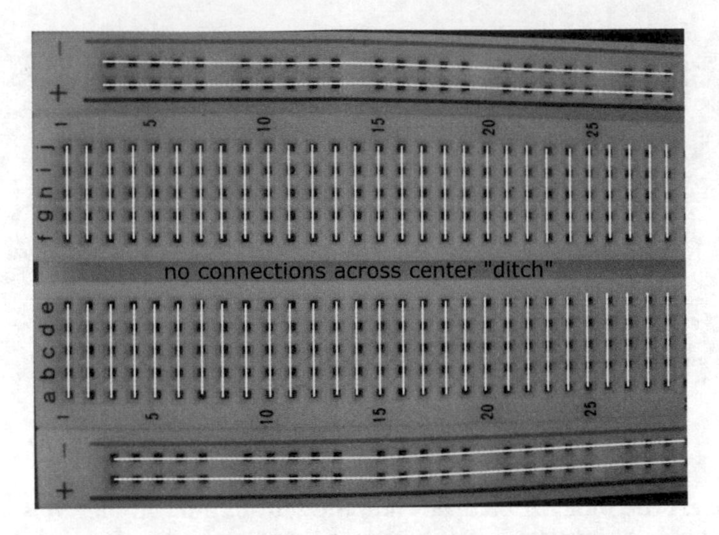

Figure 2-8: The contact holes in a solderless breadboard are arranged in rows and columns that are electrically connected in small groups below the surface.

Say you poke one lead of a resistor into hole b5 in the breadboard in Figure 2-8. You're connecting that lead to the underlying metal strip that connects the five holes in column 5, rows a through e. By plugging, say, one lead of a capacitor into hole d5, you make a connection between the resistor and the capacitor because holes b5 and d5 are electrically connected. You can build a working circuit — without permanently bonding components — by plugging in components to make the connections you need, and then running wires from your breadboard to your power supply (say, a 9-volt battery).

Solderless breadboards enable you to test a circuit easily by swapping components in and out. The not-so-nice thing about them is that it is easy to make a mistake. Common mistakes include plugging both leads of a component into holes in the same row (that is, creating an unintended connection) and plugging a lead into a hole in the row next to the row you want (that is, leaving an intended connection unconnected).

3

Running Around in Circuits

In This Chapter

- Making the right connections with a circuit
- Being positive about the direction of current flow
- Shedding light on a circuit in action
- Seeing how voltages and currents measure up
- Finding out how much electrical energy is used

Electric current doesn't flow just anywhere. (If it did, you'd be getting shocked all the time.) Electrons only flow if you provide a closed conductive path, known as an *electrical circuit,* or simply a *circuit,* for them to move through, and initiate the flow with a battery or other source of electrical energy.

In this chapter, you discover how electric current flows through a circuit and why conventional current can be thought of as electrons moving in reverse. You also explore the depths of a simple electronic circuit that you can build yourself. Finally, you find out how to measure voltages and currents in that circuit and how to figure out the amount of power supplied and used in a circuit.

Comparing Closed, Open, and Short Circuits

You need a closed path, or *closed circuit,* to get electric current to flow. If there's a break anywhere in the path, you have an *open circuit,* and the current stops flowing — and the metal atoms in the wire quickly settle down to a peaceful, electrically neutral existence. (See Figure 3-1.)

Picture a gallon of water flowing through an open pipe. The water will flow for a short time but then stop when all the water exits the pipe. If you pump water through a closed pipe system, the water will continue to flow as long as you keep forcing it to move.

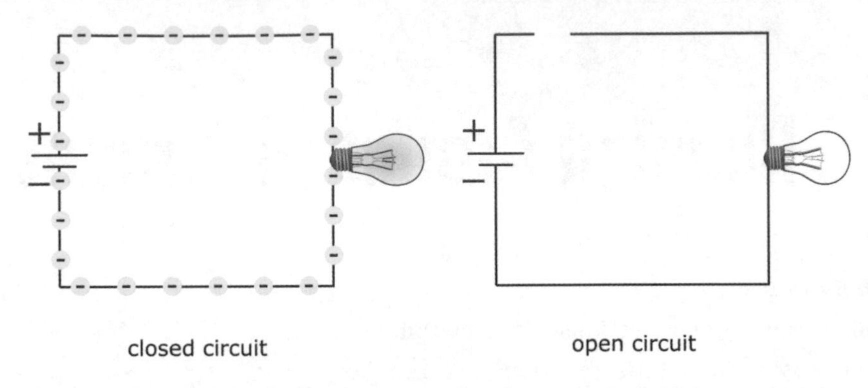

Figure 3-1: A closed circuit allows current to flow, but an open circuit leaves electrons stranded.

Open circuits are often created by design. For instance, a simple light switch opens and closes the circuit that connects a light to a power source. When you build a circuit, it's a good idea to disconnect the battery or other power source when the circuit is not in use. Technically, that's creating an open circuit.

A flashlight that is off is an open circuit. In the flashlight in Figure 3-2, the flat black button in the lower left controls the switch inside. The switch is nothing more than two flexible pieces of metal in close proximity to each other. With the black button slid all the way to the right, the switch is in an open position and the flashlight is off.

Figure 3-2: A switch in the open position disconnects the light bulb from the battery, creating an open circuit.

Turning the flashlight on by sliding the black button to the left pushes the two pieces of metal together — or closes the switch — and completes the circuit so that current can flow. (See Figure 3-3.)

Figure 3-3: Closing the switch completes the conductive path in this flashlight, allowing electrons to flow.

Sometimes open circuits are created by accident. You forget to connect a battery, for instance, or there's a break in a wire somewhere in your circuit. When you build a circuit using a solderless breadboard (which I discuss in Chapters 2 and 15), you may mistakenly plug one side of a component into the wrong hole in the breadboard, leaving that component unconnected and creating an open circuit. Accidental open circuits are usually harmless but can be the source of much frustration when you're trying to figure out why your circuit isn't working the way you think it should.

Short circuits are another matter entirely. A *short circuit* is a direct connection between two points in a circuit that aren't supposed to be directly connected, such as the two terminals of a power supply. (See Figure 3-4.) As you discover in Chapter 5, electric current takes the path of least resistance, so in a short circuit, the current will bypass other parallel paths and travel through the direct connection. (Think of the current as being lazy and taking the path through which it doesn't have to do much work.)

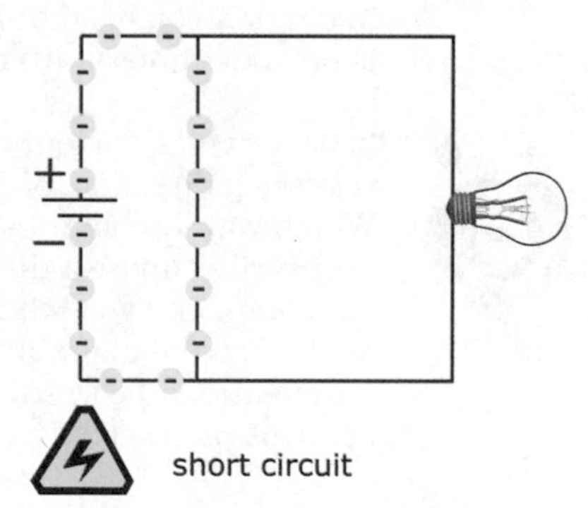

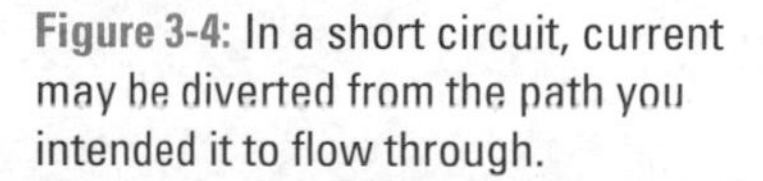

Figure 3-4: In a short circuit, current may be diverted from the path you intended it to flow through.

If you short out a power supply, you send large amounts of electrical energy from one side of the power supply to the other. With nothing in the circuit to limit the current and absorb the electrical energy, heat builds up quickly in the wire and in the power supply. A short circuit can melt the insulation around a wire and may cause a fire, an explosion, or a release of harmful chemicals from certain power supplies, such as a rechargeable battery or a car battery.

Understanding Conventional Current Flow

Early experimenters believed that electric current was the flow of positive charges, so they described electric current as the flow of a positive charge from a positive terminal to a negative terminal. Much later, experimenters discovered electrons and determined that they flow from a negative terminal to a positive terminal. That original convention is still with us today — so the standard is to depict the direction of electric current in diagrams with an arrow that points opposite the direction of actual electron flow.

Conventional current is the flow of a positive charge from positive to negative and is the reverse of real electron flow. (See Figure 3-5.) All descriptions of electronic circuits use conventional current, so if you see an arrow depicting current flow in a circuit diagram, you know it is showing the direction of conventional current flow. In electronics, the symbol *I* represents conventional current, measured in amperes (or amps, abbreviated *A*). You're more likely to encounter *milliamps* (*mA*) in circuits you build at home. A milliamp is one one-thousandth of an amp.

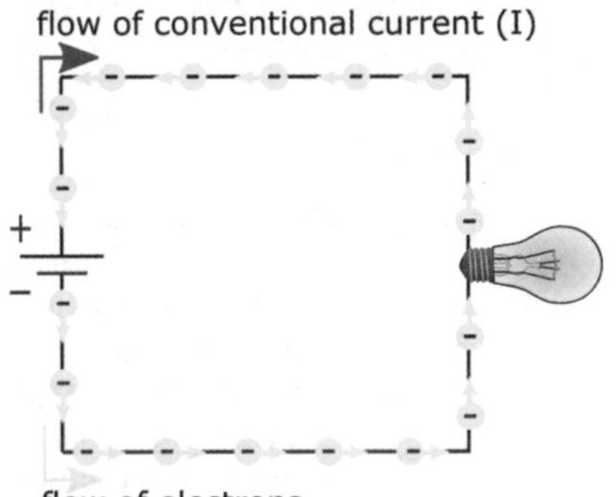

Figure 3-5: Conventional current flows one way; electrons flow the other way.

In AC circuits, current is constantly reversing direction. So how do you show current flow in a circuit diagram? Which way should the arrow point? The answer is that it doesn't matter. You arbitrarily choose a direction for the current flow (known as the *reference direction*), and you label that current *I*. The value of *I* fluctuates up and down as the current alternates. If the value of *I* is negative, that just means that the (conventional) current is flowing in the direction opposite to the way the arrow is pointing.

Examining a Basic Circuit

The diagram in Figure 3-6 depicts a battery-operated circuit that powers a light-emitting diode (LED), much like what you might find in a mini LED flashlight. What you see in the figure is a circuit diagram, or *schematic,* that shows all the components of the circuit and how they are connected. (I discuss schematics in detail in Chapter 14.)

The battery is supplying 6 volts DC (that is, a steady 6 volts) to the circuit. The plus sign near the battery symbol indicates the positive terminal of the battery, from which current flows (conventional current, of course).

The negative sign near the battery symbol indicates the negative terminal of the battery, to which the current flows after it makes its way around the circuit. The arrow in the circuit indicates the reference direction of current flow, and because it's pointing away from the positive terminal of the battery in a DC circuit, you should expect the value of the current to be positive all the time.

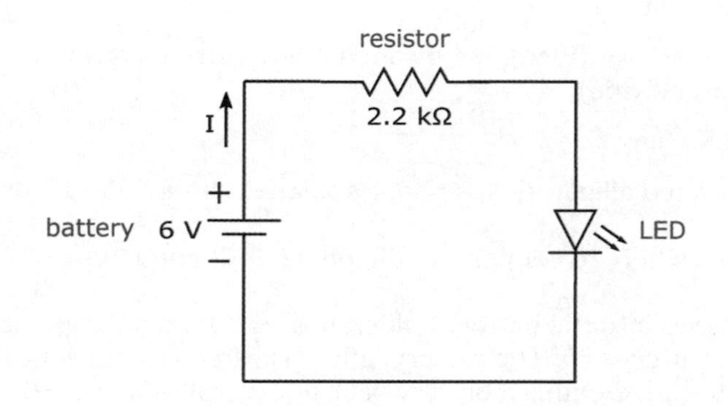

Figure 3-6: Current delivers electrical energy from the battery to the resistor and the LED.

The lines in the circuit diagram show how the circuit components are connected, using wire or other connectors. (I discuss various kinds of wire and connectors in Chapter 12.) Electronic components are usually made with *leads* — protruding wires connected to the innards of the component that provide the means to connect the component to other circuit elements.

The zigzag symbol in the circuit diagram represents a resistor. The role of the *resistor* is to limit the amount of current that flows through the circuit, much like a kink in a garden hose restricts water flow. Chapter 5 gives you more information about resistors, but for now, just know that resistance is measured in units called ohms (symbolized as Ω) and that the resistor in this circuit is keeping the LED from being destroyed.

The LED is symbolized by a triangle with a line segment on one end and two arrows pointing outward. The triangular part of the symbol represents a *diode,* and the two arrows facing out represent the fact that this diode emits light (hence, it is a light-emitting diode). Diodes are part of a special class of electronic component known as *semiconductors,* which I describe in Chapter 9.

By building this circuit and taking some measurements of voltages and current, you can learn a lot about how circuits work. And voltage and current measurements are the key to finding out how the electrical energy generated by the battery is used in the circuit. So let's get started!

Building the basic LED circuit

Here are the parts you need to construct the LED circuit:

- Four 1.5-volt AA batteries (make sure they're fresh)
- One four-battery holder (for AA batteries)
- One battery clip
- One 2.2 kΩ resistor (identified by a red-red-red stripe pattern and then a gold or silver stripe)
- One red LED (any size)
- Three insulated alligator clips *or* one solderless breadboard

You can find out where to get parts in Chapter 2 or Chapter 19.

Insert the batteries into the battery holder, observing the polarity markers, and attach the battery clip. The battery holder is wired to connect the four batteries end-to-end, creating a battery pack that supplies $4 \times 1.5 = 6$ volts via the wires extending from the clip.

Before you build the circuit, you may want to use your multimeter to verify the voltage of your battery pack and the value of your resistor (especially if you're not sure of their values). For details on using a multimeter, see Chapter 16.

Set your multimeter to measure DC voltage, hold the black (negative) multimeter probe to the black lead coming out of the battery pack, and hold the red (positive) multimeter probe to the red lead coming out of the battery pack. You should get a reading of at least 6 volts, because fresh batteries supply a higher voltage than their rating. If the reading is much less than 6 volts, remove the batteries and check each one individually.

To check the resistor value, switch your multimeter selector to measure ohms, and touch one multimeter lead to each side of the resistor (it doesn't matter which way). Verify that the resistor value is roughly 2.2 kΩ (which is 2,200 Ω).

You can build the LED circuit by using alligator clips to connect the components or by using a solderless breadboard to make the connections. I walk you through both construction methods. Note that when I discuss taking voltage and current measurements, I provide pictures showing the breadboard circuit. You can read more about constructing circuits in Chapter 15.

Building the circuit with alligator clips

Use the alligator clips to make connections in the circuit, as shown in Figure 3-7. Note that the orientation of the resistor doesn't matter, but the orientation of the LED does matter. You connect the longer lead of the LED to the resistor, and the shorter lead of the LED to the negative side (black wire) of your battery pack. When you make your final connection, the LED should glow.

If you connect an LED the wrong way, it won't light and it might become damaged. You find out why in Chapter 9.

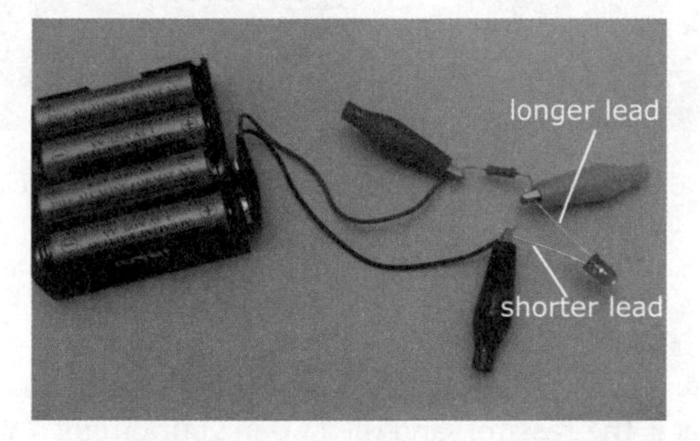

Figure 3-7: Alligator clips connect components in this simple LED circuit.

Building the circuit with a solderless breadboard

Figures 3-8 and 3-9 show the circuit set up on a solderless breadboard. You find out in Chapters 2 and 15 that a solderless breadboard makes connections between holes so that all you have to do is insert components in the right places. On the left and right sides of the breadboard, all the holes in each column are connected to each other. In each of the two center sections of the breadboard, the five holes in each row are all connected to each other.

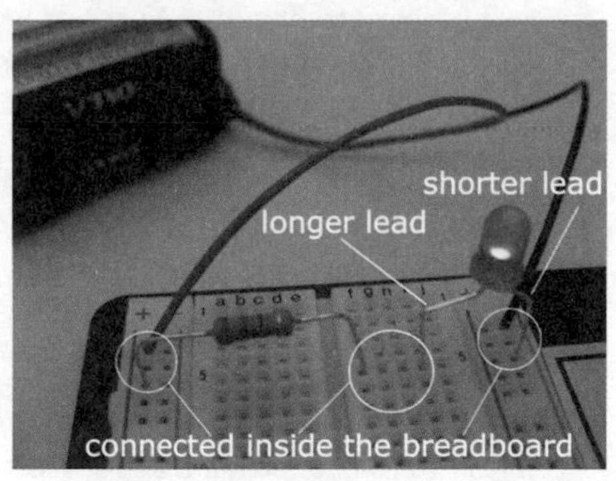

Figure 3-8: The LED circuit is easy to set up on a solderless breadboard.

As you set up the circuit on the breadboard, remember that it doesn't matter how you orient the resistor, but be sure to orient the LED so that the shorter lead is connected to the negative side of the battery pack. If you clip the leads to make your circuit neater (as shown in Figure 3-9), remember to keep track of which lead was shorter to begin with. You use a short jumper wire in your trimmed circuit to connect the resistor to the LED. (In Chapter 9, you find out another way to identify which lead is which on an LED.)

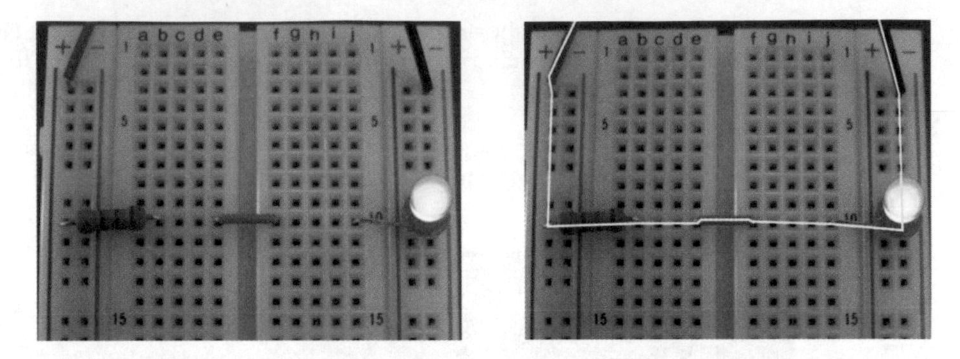

Figure 3-9: A neater way to build your circuit. The yellow line shows the path through which current flows to and from the battery pack.

Examining voltages

In this section, I explain how to use your multimeter to measure the voltage across the battery pack, the resistor, and the LED in your circuit. (You find detailed information about how to use a multimeter in Chapter 16.)

Note that the connection points between components are the same whether you built the circuit using a breadboard or alligator clips. The red lead of your multimeter should be at a higher voltage than the black lead, so take care to orient the probes as described. Set your multimeter to measure DC voltage and get ready to take some measurements!

What's your voltage?

If you ever see references to the voltage *at a single point* in a circuit, it is always with respect to the voltage at another point in the circuit — usually the *reference ground,* or *common ground* (often simply called *ground*), the point in the circuit that is (arbitrarily) said to be at 0 volts. Often, the negative terminal of a battery is used as the reference ground, and all voltages throughout the circuit are measured with respect to that reference point.

An analogy that may help you understand that voltage measurement is distance measurement. If someone were to ask you, "What's your distance?" you'd probably say, "Distance from what?" Similarly, if you're asked, "What's the voltage at the point in the circuit where the current enters the LED?" you should ask, "With respect to what point in the circuit?" On the other hand, you may say, "I'm five miles from home," and you've stated your distance from a reference point (home). So if you say, "The voltage where the current enters the bulb is 1.7 volts with respect to ground," that makes perfect sense.

First, measure the voltage supplied to the circuit by the battery pack. Connect the positive (red) multimeter lead to the point where the positive (red lead) side of the battery pack connects to the resistor, and the negative (black) multimeter lead to the point where the negative (black lead) side of the battery pack connects to the LED. See Figure 3-10. Do you get a voltage reading that is close to the nominal supply voltage of 6 V? (Fresh batteries may supply more than 6 V; old batteries usually supply less than 6 V.)

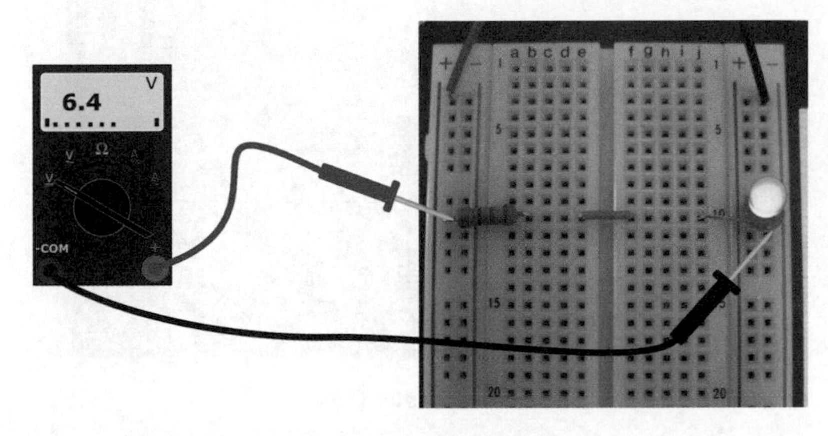

Figure 3-10: Measure the voltage supplied by the battery pack.

Next, measure the voltage across the resistor. Connect the positive (red) multimeter lead to the point where the resistor connects with the positive side of the battery pack, and the negative (black) multimeter lead to the other side of the resistor. See Figure 3-11. Your voltage reading should be close to the one that appears on the multimeter in the figure.

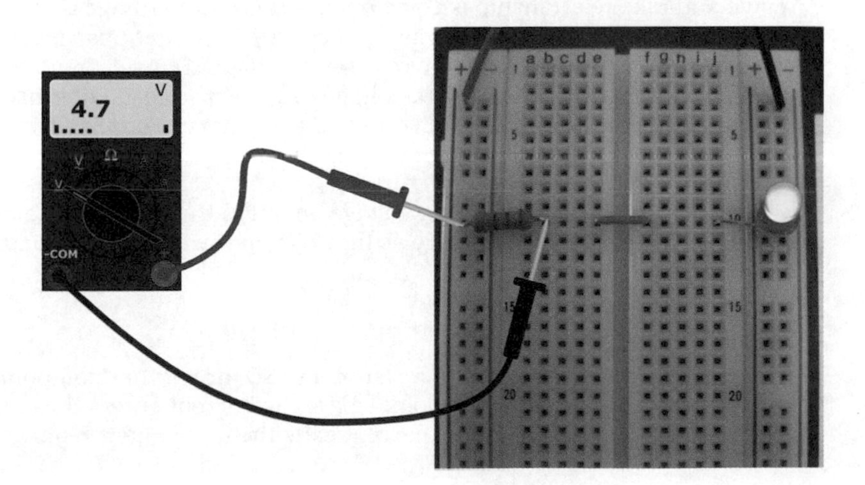

Figure 3-11: Measure the voltage across the resistor.

Finally, measure the voltage across the LED. Place the red multimeter lead to the point where the LED connects with the resistor, and the black multimeter lead to the point where the LED connects to the negative side of the battery pack. See Figure 3-12. Was your voltage reading close to the one in the figure?

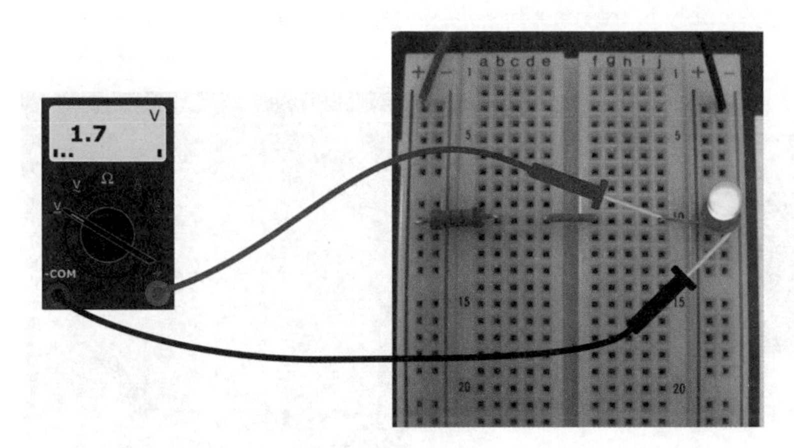

Figure 3-12: Measure the voltage across the LED.

My measurements show that in my circuit, the battery pack is supplying 6.4 volts, and that 4.7 volts are dropped across the resistor and 1.7 volts are dropped across the LED. It's not a coincidence that the sum of the voltage drops across the resistor and the LED is equal to the voltage supplied by the battery pack:

$$4.7\text{ V} + 1.7\text{ V} = 6.4\text{ V}$$

A give-and-take relationship is going on in this circuit: Voltage is the push the battery gives to get current moving, and energy from that push is absorbed when current moves through the resistor and the LED. As current flows through the resistor and the LED, voltage drops across each of those components. The resistor and the LED are using up energy supplied by the force (voltage) that pushes the current through them.

You can rearrange the preceding voltage equation to show that the resistor and the LED are dropping voltage as they use up the energy supplied by the battery:

$$6.4\text{ V} - 4.7\text{ V} - 1.7\text{ V} = 0$$

When you *drop voltage* across a resistor, an LED, or another component, the voltage is more positive at the point where the current enters the component than it is at the point where the current exits the component. Voltage is a relative measurement because it's the force that results from a difference in charge from one point to another. The voltage supplied by a battery represents the difference in charge from the positive terminal to the negative terminal, and that difference in charge has the potential to move current through a circuit; the circuit, in turn, absorbs the energy generated by that force as the current flows, which drops the voltage. No wonder voltage is sometimes called *voltage drop, potential difference,* or *potential drop.*

The important thing to note here is that as you travel around a DC circuit, you gain voltage going from the negative terminal of the battery to the positive terminal (that's known as a *voltage rise*), and you lose, or drop, voltage as you continue in the same direction across circuit components. (See Figure 3-13.) By the time you get back to the negative terminal of the battery, all the battery voltage has been dropped and you're back to 0 volts.

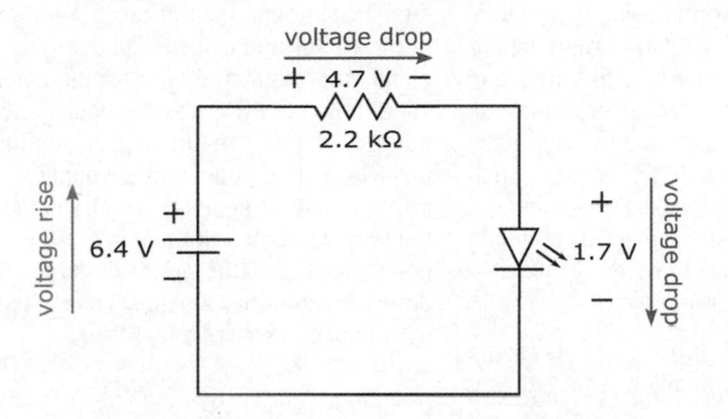

Figure 3-13: The voltage supplied by the battery is dropped across the resistor and the LED.

With all circuits (whether AC or DC), if you start at *any* point in the circuit, and add the voltage rises and drops going around the circuit, you get zero volts. In other words, the net sum of the voltage rises and drops in voltage around a circuit is zero. (This rule is known as *Kirchhoff's Voltage Law.* Kirchhoff is pronounced "keer-cough.")

Keep in mind that these voltage drops have a physical meaning. The electrical energy supplied by the battery is absorbed by the resistor and the LED. The battery will keep supplying electrical energy, and the resistor and LED will keep absorbing that energy, until the battery dies (runs out of energy). That happens when all the chemicals inside the battery have been consumed in the chemical reactions that produced the positive and negative charges. In effect, all the chemical energy supplied by the battery has been converted into electrical energy — and absorbed by the circuit.

Standing your ground

If you hear the word "ground" used in an electronics context, be aware that *ground* can refer to either earth ground or common ground.

Earth ground means pretty much what it says: It's a direct connection to the ground — real ground, the stuff of the planet. The screw in the center of a standard two-prong AC outlet, as well as the third prong in a three-prong outlet, is connected to earth ground. Behind each wall socket is a wire that runs through your house or office and eventually connects to a metal post that makes good contact with the ground. This arrangement provides extra protection for circuits that use large amounts of current; in the event of a short circuit or other hazardous condition, shipping dangerous current directly into the earth gives it a safe place to go. Such was the case when Ben Franklin's lightning rod provided a direct path for dangerous lightning to travel along on its way into the ground — instead of via a house or a person.

In circuits that handle large currents, some point in the circuit is usually connected to a pipe or other metal object that's connected to earth ground. If this connection is missing, the ground is said to be *floating* (or a *floating ground*) and the circuit may be dangerous. You'd be wise to stay away from such a circuit until it is safely grounded (or "earthed," as folks say in the UK)!

Common ground, or simply *common,* isn't a physical ground; rather, it's just a reference point in a circuit for voltage measurements. Certain types of circuits, particularly the circuits commonly used in computers, label the negative terminal of a DC power supply the common ground, and connect the positive terminal of another DC power supply to the same point. That way, the circuit is said to have both positive and negative power supplies. The two physical power supplies may be identical, but the way you connect them in a circuit and the point you choose for the zero voltage reference determine whether a supply voltage is positive or negative. It's all relative!

One of the fundamental laws of physics is that energy cannot be created or destroyed; it can only change form. You witness this law in action with the simple battery-driven LED circuit: Chemical energy is converted to electrical energy, which is converted to heat and light energy, which — well, you get the idea.

Measuring current

To measure the current running through your LED circuit, you must pass the current through your multimeter. The only way to do this is to interrupt the circuit between two components and insert your multimeter, as if it's a circuit component, to complete the circuit.

Switch the multimeter selector to measure DC current in milliamps (mA). Then break the connection between the resistor and LED. (If you are using alligator clips, simply remove the clip that connects the resistor and the

LED. If you are using a breadboard, remove the jumper wire.) The LED should turn off.

Next, touch the positive (red) multimeter lead to the unconnected resistor lead and the negative (black) multimeter lead to the unconnected LED lead, as shown in Figure 3-14. The LED should turn on, because the multimeter has completed the circuit, allowing current to run through it. The current reading I got was 2.14 mA.

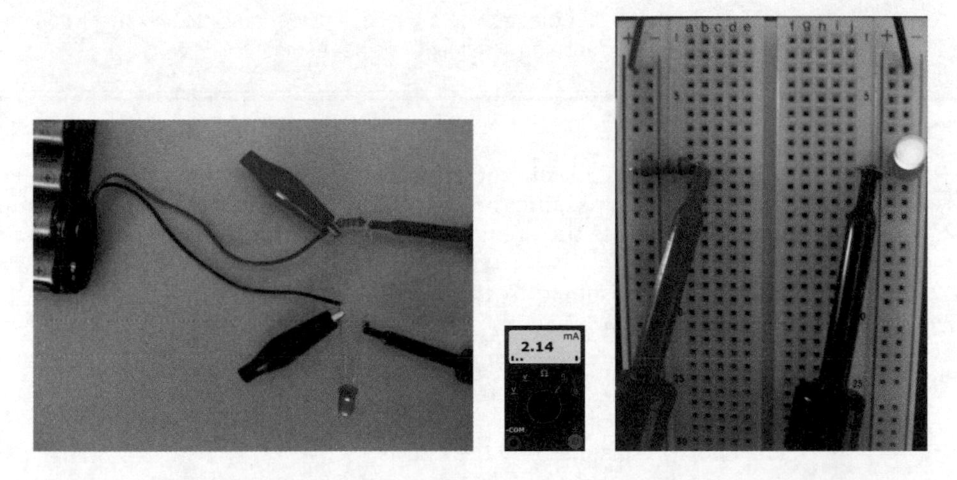

Figure 3-14: To measure current, insert your multimeter into the path through which current flows.

Now, insert your multimeter at another connection point in the circuit (for instance, between the positive battery lead and the resistor), taking care to open the circuit at the point of measurement and to orient the multimeter leads with the positive lead at a more positive voltage point than the negative lead. Do you get the same current reading as before? You should, because this simple circuit provides only one path for current to flow.

Calculating power

The amount of energy consumed by an electronic component is known as *power* (abbreviated *P*), measured in watts (abbreviated *W*). Chapter 1 introduces you to this equation for calculating power:

$$P = V \times I$$

where *V* represents voltage and *I* represents current. When you know the voltage dropped across a component and the current passing through the component, you can use the power equation to calculate the amount of energy consumed by each component.

Empowering you to make the right choices

LEDs, incandescent light bulbs, resistors, and other electronic components have maximum power ratings for a good reason. Send too much current through them, and they overheat and burn or melt. It is good practice to estimate the power requirement for each component in a circuit you design. Using the fact that power is the product of voltage and current, you consider the worst-case scenario — the maximum combination of voltage times current that your component will face during circuit operation — to determine how many watts that component should be able to handle. Then, give yourself some breathing room by selecting a component with a *power rating* that exceeds your maximum power estimate.

For the resistor-LED circuit, you know the voltage drops (refer to Figure 3-13) and the current passing through the circuit (2.14 mA). Using this information, you can calculate the energy supplied or consumed by each component.

The energy consumed by the resistor is

$$4.7\ \text{V} \times 2.14\ \text{mA} = 10.1\ \text{mW}$$

where mW means milliwatts, or thousandths of a watt.

The energy consumed by the LED is

$$1.7\ \text{V} \times 2.14\ \text{mA} = 3.6\ \text{mW}$$

The energy supplied by the battery is

$$6.4\ \text{V} \times 2.14\ \text{mA} = 13.7\ \text{mW}$$

Note that the sum of the power consumed by the resistor and the LED ($10.1\ \text{mW} + 3.6\ \text{mW}$) is equal to the power supplied by the battery (13.7 mW). That's because the battery is providing the electrical energy that the resistor and the LED are using. (Actually, the resistor is converting electrical energy to heat energy, and the LED is converting electrical energy into light energy.)

Say you substitute a 9-volt battery for the 6-volt battery pack. Now you supply more voltage to the circuit, so you can expect to push more current through it and deliver more energy to the resistor and the LED. Because the LED receives more electrical energy to convert into light energy, it will shine more brightly. (There are limits to how much voltage and current you can supply to an LED before it breaks down and no longer works. You find out more about this subject in Chapter 9.)

4

Making Connections

In This Chapter

- Sending current this way and that way
- Examining series and parallel circuits
- Controlling connections with switches
- Seeing the light when the power is on

If you've ever been stuck in a traffic jam and decided to take the less crowded side roads, you know that there's often more than one way to get to a destination. But if your commute requires you to travel, say, over a bridge to cross a major river, you know that sometimes there's just one way for you — and all the other commuters — to go.

In many ways, electronic circuits are like road systems: They provide paths (roads) for electrons (cars) to travel along, sometimes offering alternate paths and sometimes forcing all the electrons to travel along the same path.

This chapter explores different ways to connect electronic components so you can direct — and redirect — electric current. First, you look at the two basic types of circuit structures — series and parallel — and discover that parallel connections are like alternate traffic routes, whereas series connections are like bridge crossings. Then you find out how switches act like traffic cops — allowing, preventing, or redirecting current flow. Finally, you put it all together by building a circuit that imitates a manually controlled three-stage traffic signal.

Creating Series and Parallel Circuits

Just as you can build structures of all shapes and sizes by connecting LEGOs or K'NEX pieces in various ways, so you can build many different kinds of circuits by connecting electronic components in various ways. Exactly how you connect components dictates how current flows through your circuit — and how voltage is dropped throughout the circuit.

In this section, you explore two kinds of connections. If you would like to build the circuits described in this section, you need the following parts:

- Four 1.5-volt AA batteries
- One four-battery holder (for AA batteries)
- One battery clip
- One 2.2 kΩ resistor (identified by a red-red-red stripe pattern and then a gold or silver stripe)
- Two red LEDs (any size)
- Three insulated alligator clips *or* one solderless breadboard

Note that if you've already built the basic LED circuit discussed in Chapter 3, you may alter it to build the circuits shown in this section. You need only one additional red LED and, if you choose to use alligator clips to make the connections, one additional alligator clip.

Series connections

In a *series circuit,* components are arranged along a single path between the positive and negative terminals of a power source. Take a look at Figure 4-1, which shows a series circuit containing a resistor and two LEDs. Current flows from the positive battery terminal through the resistor, through LED1, through LED2, and then back to the negative terminal of the battery. A series circuit has only one path for electrical charges to travel along, so all current passes through each component sequentially.

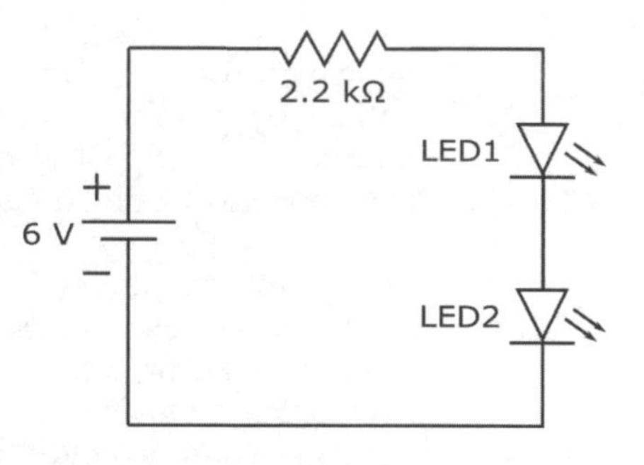

Figure 4-1: In a series circuit, current flows through each component sequentially.

You should remember two important facts about series circuits:

- Each component carries the same current.
- The voltage supplied by the source is divided (though not necessarily evenly) among the components. If you add the voltage drops across each component, you get the total supply voltage.

Using Figure 4-2 as a guide, set up the two-LED series circuit. Use your multimeter set on DC volts to measure the voltage across the battery pack, the resistor, and each of the LEDs. When I did this, I got the following readings:

- Voltage across battery: 6.4 volts
- Voltage across resistor: 3.0 volts
- Voltage across LED1: 1.7 volts
- Voltage across LED2: 1.7 volts

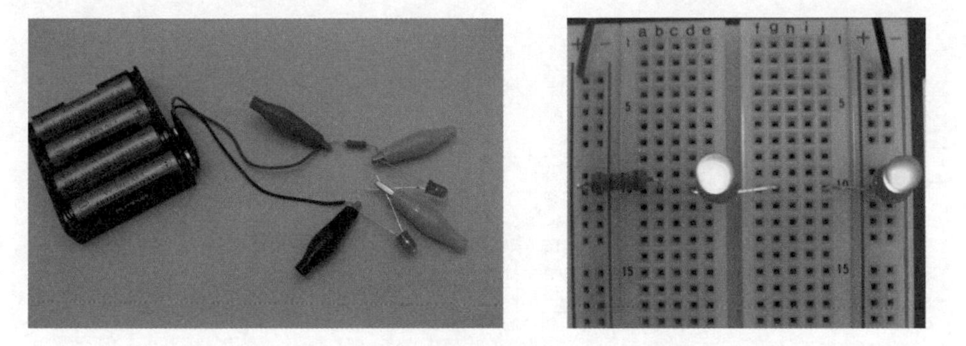

Figure 4-2: Two ways to set up the circuit with two LEDs in series.

By summing the voltage drops across the resistor and the LEDs, you get the total voltage supplied by the battery pack:

$$3.0\ \text{V} + 1.7\ \text{V} + 1.7\ \text{V} = 6.4\ \text{V}$$

Next, set your multimeter to the DC current setting. Interrupt the circuit at any of the connection points and insert your multimeter, remembering to keep the positive lead at a higher voltage than the negative lead. When I did this, I got a reading of 1.4 mA. That amount of current flows through each component in this series circuit because there is only one path for current to flow.

Because the current in a series circuit has only one path to flow through, you may run into a potential problem with this type of circuit. If one component fails, it creates an open circuit, stopping the flow of current to every other component in the circuit. So if your expensive new restaurant sign sports 200 LEDs wired in series to say "BEST FOOD IN TOWN," and a home-run ball knocks out one LED, every one of the LEDs goes dark.

Parallel connections

There's a way to fix the problem of all components in a series circuit blacking out when one component fails. You can wire the components using parallel connections — such as those in the circuit shown in Figure 4-3. With a parallel circuit, current can flow in multiple paths, so even if several baseballs take out bulbs in your sign, the rest of it stays lit. (Of course, you may be left with a glowing sign reading, "BEST FOO I OWN." There are pros and cons to everything.)

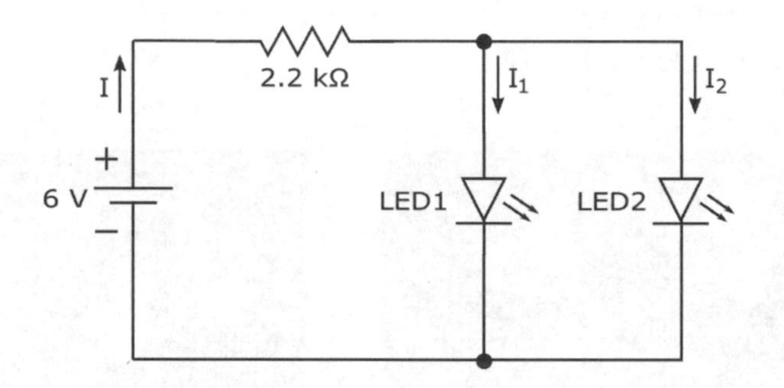

Figure 4-3: Light bulbs are often arranged in a parallel circuit so if one burns out, the rest stay lit.

Here's how the parallel circuit in Figure 4-3 works: Current flows from the positive battery terminal, and then splits at the junction that leads to the parallel branches of the circuit, so each LED gets a share of the supply current. The current flowing through LED1 doesn't flow through LED2. So if your restaurant sign has 200 LEDs wired in parallel and one burns out, light still shines from the other 199 LEDs.

Two important things you need to remember about parallel circuits are

- The voltage across each parallel branch is the same.
- The current supplied by the source is divided among the branches, and the branch currents sum to the total supply current.

Set up the circuit using the example in Figure 4-4 as your guide. With your multimeter set to DC volts, measure the voltage across the battery pack, the resistor, and each of the LEDs. When I did this, I got the following readings:

- Voltage across battery: 6.4 volts
- Voltage across resistor: 4.7 volts
- Voltage across LED1: 1.7 volts
- Voltage across LED2: 1.7 volts

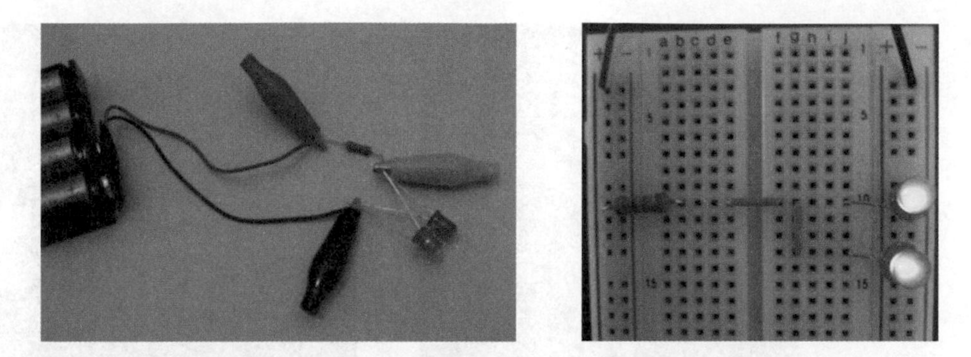

Figure 4-4: Two ways to set up the circuit with two LEDs in parallel.

Now, use your multimeter to measure the current flowing through each of the three circuit components, as follows, remembering to set your multimeter to DC current:

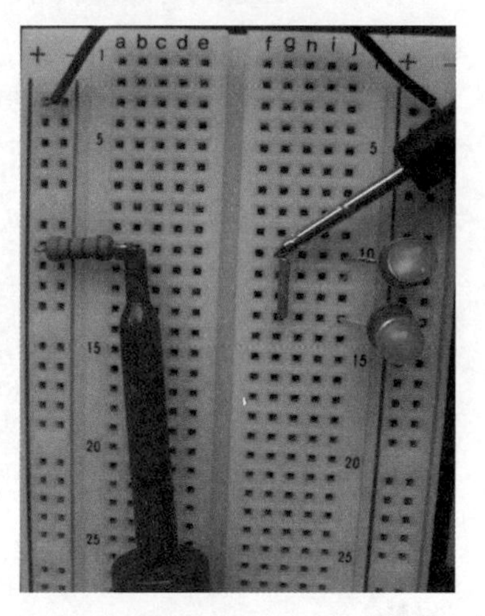

Figure 4-5: Measure the current flowing through the resistor.

- **Resistor current (*I*):** Break the circuit between the resistor and the two LEDs, and insert your multimeter to reconnect the circuit, as shown in Figure 4-5. Because you're placing the multimeter in series with the resistor, you're measuring the current flowing through the resistor. This current is labeled *I* (refer to Figure 4-3).
- **LED1 current (I_1):** Remove the multimeter and reconnect the resistor to the LEDs. Then disconnect the positive lead of LED1 from the resistor. Insert the multimeter leads into the circuit, as shown in the photo on the left in Figure 4-6. Because the multimeter is in series with LED1, you're measuring the current that flows through LED1. This current is labeled I_1 (refer to Figure 4-3).
- **LED2 current (I_2):** Remove the multimeter and reconnect LED1. Then disconnect the positive lead of LED2 from the resistor. Insert the multimeter into the circuit in series with LED2, as shown in the photo on the right in Figure 4-6. (Note that I removed the lower orange jumper to disconnect LED2 from the resistor, and then inserted my multimeter at that opening in the circuit.) You're measuring the current that flows through the LED2, or I_2 (refer to Figure 4-3).

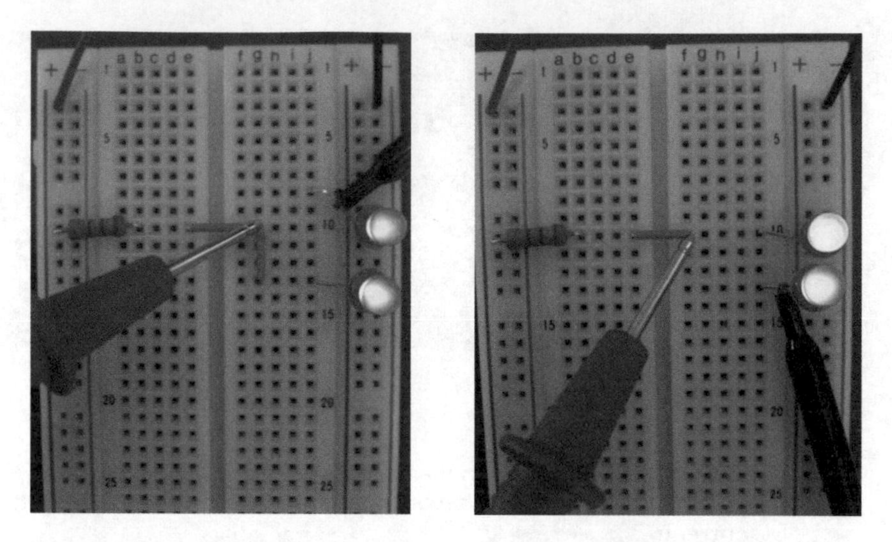

Figure 4-6: Measure the current flowing through LED1 (left) and LED2 (right).

These are the readings I got:

- Resistor current, I: 2.2 mA
- LED1 current, I_1: 1.1 mA
- LED2 current, I_2: 1.1 mA

If you add the two branch currents, I_1 and I_2, you find that their total is equal to the current flowing through the resistor, which is the supply current coming from the battery pack:

$$1.1 \text{ mA} + 1.1 \text{ mA} = 2.2 \text{ mA}$$

Note that the supply current for the parallel circuit, 2.2 mA, is higher than the supply current for the series circuit, 1.4 mA, even though the same components are used in both circuits. Connecting circuit components in parallel draws more current from your source than connecting them in series.

If your circuit is powered by a battery, you need to be aware of just how long your battery can supply the necessary current to your circuit. As I discuss in Chapter 12, batteries have ratings of *amp-hours*. A battery with a rating of one amp-hour (for example) will last for just one hour in a circuit that draws one amp of current — theoretically, anyway. (In practice, even new batteries don't always deliver on their amp-hour promises.) Therefore, when you're deciding what power source to use for a circuit, you must take into account both the current that a circuit draws and how long you want to run the circuit.

Switching Electric Current On and Off

Switching is far and away the most important function in electronics. Think about your TV set: You turn it on and off, select a signal source from several different input choices (such as your DVD player, cable box, or gaming system), and change TV channels. Your TV screen consists of millions of tiny *pixels* (picture elements), each of which is, in essence, a red, blue, or green light that is either on or off. All those TV control and display functions involve switching, whether it's simply on/off switching or what I like to think of as multiple-choice switching — that is, directing one of several input signals to your TV screen. Likewise, your smartphone, computing device, and even your microwave rely on on/off states (for instance, key pressed or not, or transmit sound now or not) for their control and operation.

So what exactly is switching?

Switching is the making or breaking of one or more electrical connections such that the flow of current is either interrupted or redirected from one path to another. Switching is performed by components called (you guessed it!) switches. When a switch is in the *open position,* the electrical connection is broken and you have an open circuit with no current flowing. When a switch is in the *closed position,* an electrical connection is made and current flows.

Tiny semiconductor transistors (which I discuss in Chapter 10) are at the heart of most of the switching that goes on in electronic systems today. The way in which a transistor works is a bit complicated, but the basic idea behind transistor switching is this: You use a small electric current to control the switching action of a transistor, and that switching action controls the flow of a much larger current.

Aside from transistor switches, lots of different kinds of mechanical and electrically operated switches can be used in electronics projects. These switches are categorized by how they are controlled, the type and number of connections they make, and how much voltage and current they can handle.

Controlling the action of a switch

Switches are referred to by names that indicate how the switching action is controlled. You see some of the many different types of switches in Figure 4-7.

Chances are, you encounter one or more of the following types of switches as you go about your daily routine:

- **Slide switch:** You slide a knob back and forth to open and close this type of switch, which you find on many flashlights.

- **Toggle switch:** You flip a lever one way to close the switch and the other way to open the switch. You may see labels on these switches: *on* for the closed position, and *off* for the open position.
- **Rocker switch:** You press one side of the switch down to open the switch, and the other side of the switch down to close the switch. You find rocker switches on many power strips.
- **Leaf switch:** You press a lever or button to temporarily close this type of switch, which is commonly used in doorbells.

Figure 4-7: From top to bottom: two toggle switches, a rocker switch, and a leaf switch.

- **Pushbutton switch:** You push a button to change the state of the switch, but how it changes depends on the type of pushbutton switch you have:
  - **Push on/push off buttons:** Each press of the button reverses the position of the switch.
  - **Normally open (NO):** This momentary switch is normally open (off), but if you hold down the button, the switch is closed (on). When you release the button, the switch becomes open again. This is also known as a *push-to-make switch*.
  - **Normally closed (NC):** This momentary switch is normally closed (on), but if you hold down the button, the switch is open (off). When you release the button, the switch becomes closed again. This is also known as a *push-to-break switch*.
- **Relay:** A relay is an electrically controlled switch. If you apply a certain voltage to a relay, an electromagnet within pulls the switch lever (known as the *armature*) closed. You may hear talk of closing or opening the *contacts* of a relay's coil. That's just the term used to describe a relay's switch.

Making the right contacts

Switches are also categorized by how many connections they make when you "flip the switch" and exactly how those connections are made.

A switch can have one or more *poles,* or sets of input contacts: A *single-pole switch* has one input contact, whereas a *double-pole switch* has two input contacts.

A switch can also have one or more conducting positions, or *throws.* With a *single-throw switch,* you either make or break the connection between each input contact and its designated output contact; a *double-throw switch* allows you to alter the connection of each input contact between each of its two designated output contacts.

Sound confusing? To help clear things up, take a look at the circuit symbols (see Figure 4-8) and descriptions of some common switch varieties:

- **Single-pole, single-throw (SPST):** This is your basic on/off switch, with one input contact and one output contact, for a total of two terminals that connect to your circuit. You either make the connection (switch on) or break the connection (switch off).

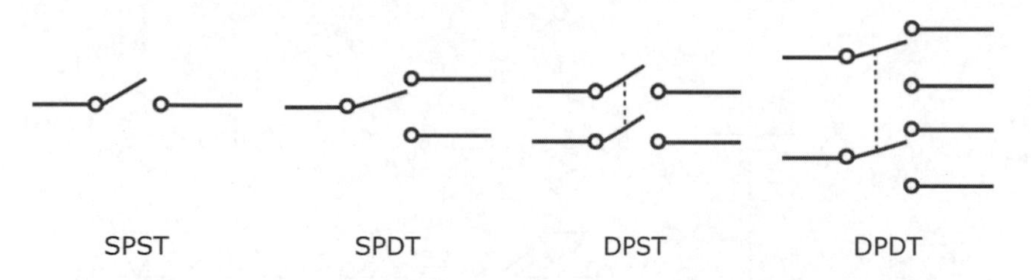

Figure 4-8: Circuit symbols for single-pole, single-throw (SPST), single-pole, double-throw (SPDT), double-pole, single-throw (DPST), and double-pole, double-throw (DPDT) switches.

- **Single-pole, double-throw (SPDT):** This on/on switch contains one input contact and two output contacts (so it has three terminals). It switches the input between two choices of outputs. You use an SPDT switch, or *changeover switch,* when you want to have a circuit turn one device or another on (for example, a green light to let people know they can enter a room, or a red light to tell them to stay out).
- **Double-pole, single-throw (DPST):** This dual on/off switch contains four terminals — two input contacts and two output contacts — and behaves like two separate "make-or-break" SPST switches operating in sync. In the off position, both switches are open and no connections are made. In the on position, both switches are closed and connections are made between each input contact and its corresponding output contact.
- **Double-pole, double-throw (DPDT):** This dual on/on switch contains two input contacts and four output contacts (for a total of six terminals), and behaves like two SPDT (changeover) switches operating in sync. In one position, the two input contacts are connected to one set of output contacts. In the other position, the two input contacts are connected to the other set of output contacts. Some DPDT switches have a third position, which disconnects (or breaks) all contacts. You can use a DPDT switch as a *reversing switch* for a motor, connecting the motor to positive voltage to turn one way, negative voltage to turn the other way, and, if there is a third switch position, zero voltage to stop turning.

In the next section, titled "Creating a Combination Circuit," you see how to use an SPDT switch as an on/off (or SPST) switch.

Creating a Combination Circuit

Most circuits are combinations of series and parallel connections. How you arrange components in a circuit depends on what you're trying to do.

Take a look at the series-parallel circuit in Figure 4-9. Note the three parallel branches, each containing a switch in series with a resistor and an LED. The switches are represented by the symbols at the top of each branch.

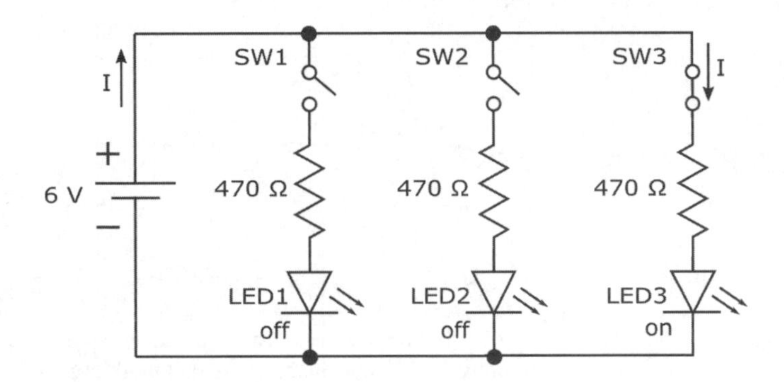

Figure 4-9: By opening and closing switches in this series-parallel circuit, you can direct the supply current through different paths.

If only one switch is in a closed position, as shown in Figure 4-10 (and in the diagram in Figure 4-9), all the supply current flows through just one LED, which lights, and the other LEDs are off.

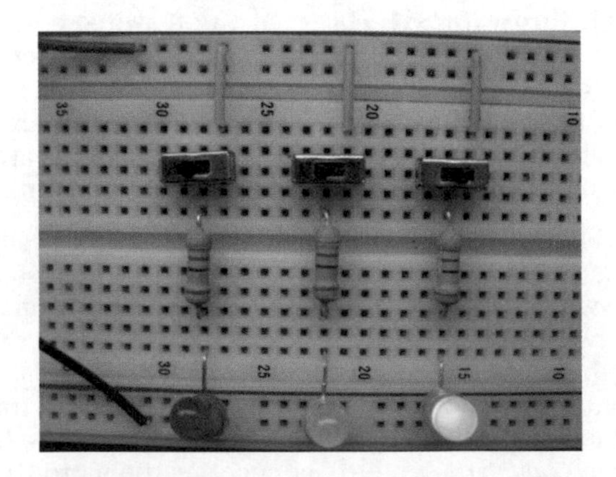

Figure 4-10: By turning just the rightmost switch on, only the green LED receives current.

If all three switches are closed, the supply current travels through the resistor and then splits three different ways — with some current passing through each of the three LEDs. If all three switches are open, the current does not have a complete path to follow, so no current flows out of the battery, as shown in Figure 4-11.

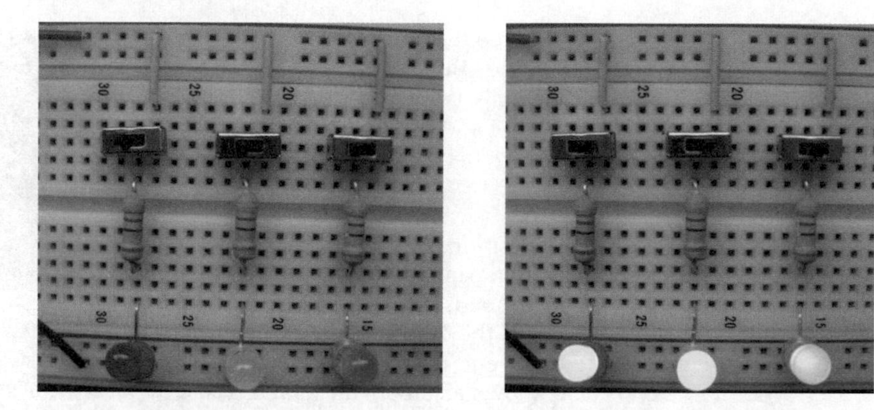

Figure 4-11: With all three switches off, none of the LEDs receives current (left). With all three switches on, all three LEDs receive current and light (right).

By alternating which switch is open at any time, you can control which LED is lit. You can imagine such a circuit controlling the operation of a three-stage traffic light (with additional parts to control the timing and sequencing of the switching action).

To analyze combination circuits, you apply voltage and current rules one step at a time, using series rules for components in series and parallel rules for components in parallel. At this point, you don't quite have enough information to calculate all the currents and voltages in the LED circuits shown here. You need to know about a rule called Ohm's Law, which I explain in Chapter 6, and about how voltage is dropped across diodes, which I cover in Chapter 9. Then you'll have everything you need to analyze simple circuits.

To build the three-LED circuit described in this section, you need the following parts:

- Four 1.5-volt AA batteries
- One four-battery holder (for AA batteries)
- One battery clip
- Three 470 Ω resistors (yellow-violet-brown stripes and then a gold or silver stripe)

- Three LEDs (any size, any color; I used one red, one yellow, and one green)
- Three single-pole, double-throw (SPDT) slide switches designed for solderless breadboard use
- One solderless breadboard and assorted jumper wires

Each SPDT switch has three terminals for making connections, but for the three-LED circuit, you need to use only two of the terminals. (See Figure 4-12.) The slider button controls which end terminal gets connected to the center terminal.

Figure 4-12: An SPDT switch can be used as an on/off switch by connecting just two of its three terminals in your circuit.

With the slider in one position, the center terminal is connected to the terminal at the end where the slider is positioned. Move the slider to the other position, and the center terminal is connected to the other end terminal. This type of switch is also known as an *on/on switch* because it can switch between two circuits, closing one while opening the other.

For the three-LED circuit, you need the switch to function as an on/off switch, so you connect two of the three SPDT terminals in your circuit. Leave the unused end terminal in a breadboard hole but not connected to anything in your circuit, as shown in Figure 4-13. With the slider positioned towards the unused terminal, the switch is off. With the slider positioned towards the other end of the switch, the switch is on.

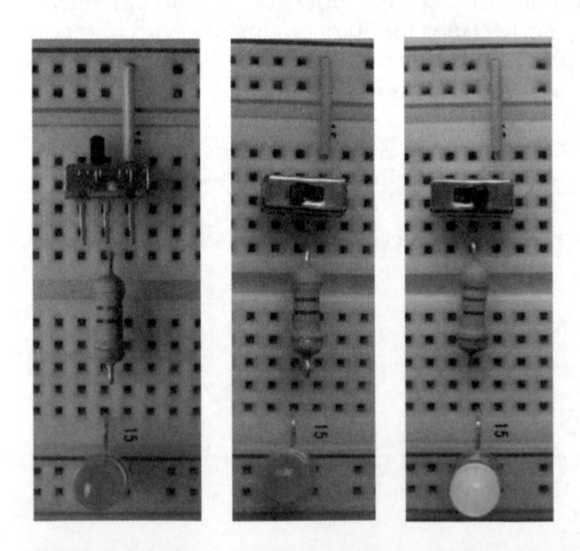

Figure 4-13: Using an SPDT switch as an on/off switch.

The simplest type of on/off switch is a two-terminal, single-pole, single-throw (SPST) switch, which simply connects or disconnects the two terminals when you move the slider. But it's hard to find such a switch with terminals designed to fit into solderless breadboards.

Switching On the Power

You can set up a simple circuit to connect and disconnect your battery pack from any circuits you build on a solderless breadboard without having to physically remove the battery pack from the breadboard.

In Figure 4-14, the positive terminal of the battery is connected to the top terminal of an SPDT switch. The center terminal of the switch is connected to the leftmost column of the breadboard, which is also known as a *power rail*. By moving the slider on the switch, you make or break a connection between the positive terminal of the battery and the positive power rail. As long as you use the power rails to supply power to your circuits, your switch functions as an on/off switch for powering your breadboard circuits.

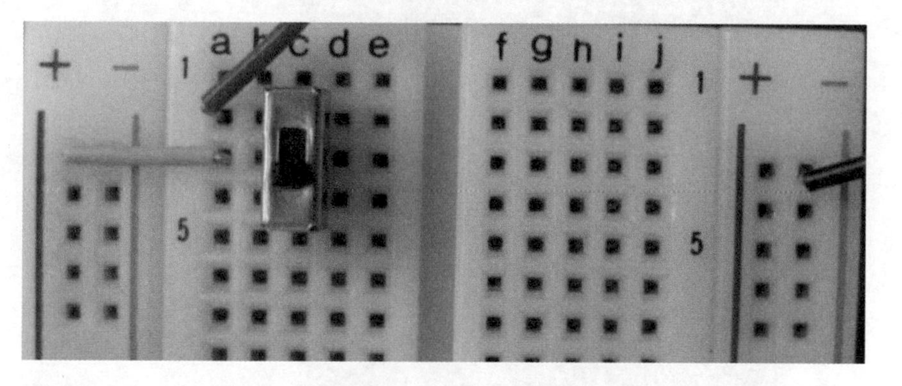

Figure 4-14: A switch connects and disconnects a battery from the power rails of a solderless breadboard.

By adding a 470 Ω resistor and an LED between the center terminal of the switch and the negative power rail, you create an indicator light for your on/off switch. (See Figure 4-15.) If the switch is in the off position, the battery is not connected to the LED, so the LED is off. If the switch is in the on position, the battery is connected to the LED, so the LED is on.

Note that even without this resistor and LED, the switch still functions as an on/off power switch for your breadboard. But it's nice to have a visible indicator that the power is on, as shown in Figure 4-16.

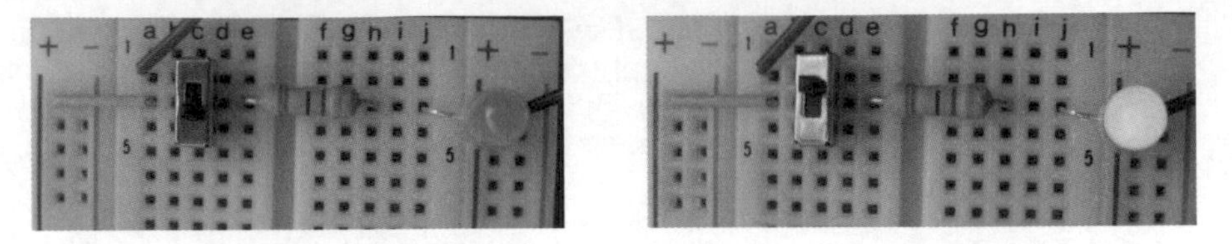

Figure 4-15: A green LED indicates whether the breadboard is powered up or not.

Figure 4-16: The green LED in the top right signals that voltage is applied to the power rails and the three-LED circuit is drawing power.

What Do Circuits Look Like?

Circuits usually don't look as neat and geometric as you might expect. The shape of a circuit is usually not important for its operation. What matters about any circuit — and what you should concern yourself with when building one — is how the components are connected, because the connections show you the path the current takes through the circuit.

The shape of a circuit *does* matter for circuits involving high-frequency signals, such as radio-frequency (RF) and microwave circuits. The *layout,* or placement of circuit components, must be designed with care to reduce noise and other undesired AC signals. Additionally, the proximity of bypass capacitors (which you find out about in Chapter 7) to other circuit components can make a difference in the performance of many circuits.

Figure 4-17 is a photograph of a 1980's style dimmer switch circuit. This simple electronic device uses just a few components to control current flow to a built-in light fixture in my house. But most electronic systems are a lot more complicated than this; they connect lots of individual components in one or more circuits to achieve their ultimate goal.

Figure 4-17: A dimmer switch is a simple electronic circuit with just a few components.

Figure 4-18 gives you an inside view of the circuitry of a computer hard drive. The circuit consists of the following, all attached to a specialized surface known as a *printed circuit board, or PCB*:

- Many *discrete components* (individual parts, such as resistors and capacitors)
- An assortment of *integrated circuits, or ICs* (which look like electronic centipedes)
- *Connectors* (which, not surprisingly, connect the hard drive electronics to the rest of the computer).

Figure 4-18: Computer hard drive electronics.

ICs, which I discuss in Chapter 11, are nothing more than a bunch of tiny circuits that work together to perform a function so commonly desired that it's worthwhile to mass-produce the circuit and package it in a protective case with *leads* (the centipede legs) that enable access to the circuit inside.

After you discover how different types of components control current flow in circuits and can apply voltage and current laws, you can begin to design and construct useful electronic circuits.

Part II
Controlling Current with Components

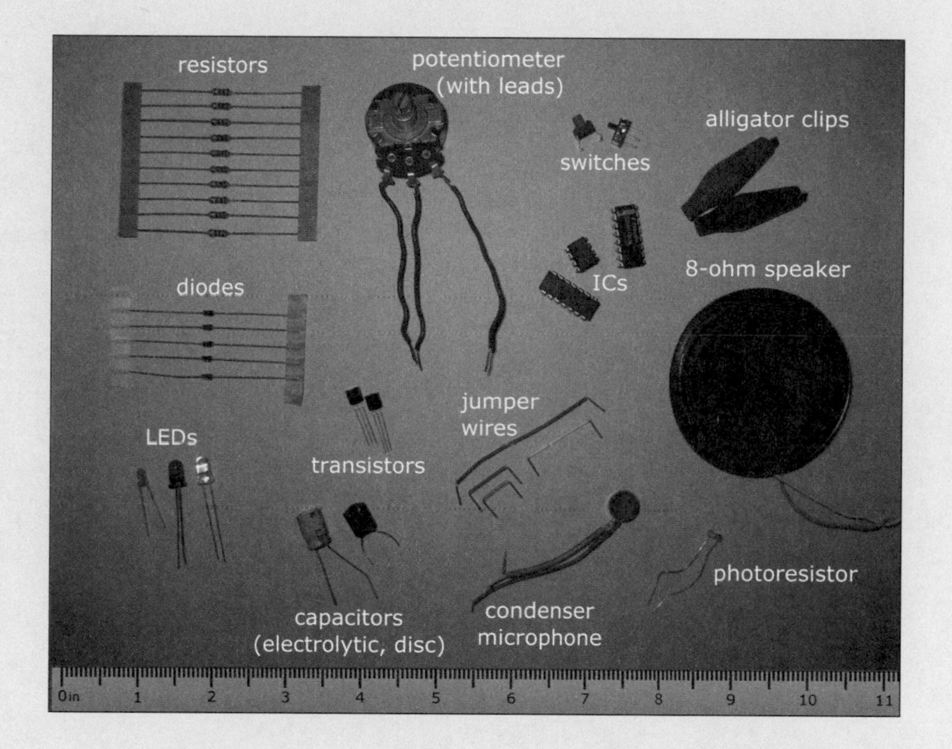

Check out `www.dummies.com/extras/electronics` for an inside look at how semiconductors conduct current.

In this part . . .

- Putting the brakes on current with resistors
- Storing electrical energy in capacitors and inductors
- Allowing current to flow in just one direction with diodes
- Amplifying and switching current with transistors
- Using integrated circuits as amplifiers, counters, oscillators, and more
- Interacting with your environment with sensors and other transducers

5

Meeting Up with Resistance

In This Chapter

- Using resistance to your advantage
- Varying the amount of resistance
- Creating just the right amount of resistance
- Realizing why LEDs need resistors

If you toss a marble into a sandbox, the marble won't go very far. But if you toss a marble onto the surface of a large frozen lake, the marble will enjoy a nice little ride before it eventually comes to a stop. A mechanical force called friction stops that marble on either surface — it's just that the sand provides more friction than the ice.

Resistance in electronics is a lot like friction in mechanical systems: It puts the brakes on electrons (those microscopic moving particles that make up electric current) as they move through materials.

In this chapter, you take a look at exactly what resistance is, where you can find resistance (everywhere), and how you can use it to your advantage by selecting *resistors* (components that provide controlled amounts of resistance) for your electronic circuits. Then you discover how to combine resistors to control current in your circuits. Next, you build and explore a few circuits using resistors and light-emitting diodes (LEDs). Finally, you discover just how important resistors are — and what happens when a critical resistor goes AWOL.

Resisting the Flow of Current

Resistance is a measure of an object's opposition to the flow of electrons. This may sound like a bad thing, but it's actually useful. Resistance is what makes it possible to generate heat and light, restrict the flow of electric current when necessary, and ensure that the correct voltage is supplied to a device. For instance, as electrons travel through the filament of an incandescent light bulb, they meet so much resistance that they slow down a lot. As they fight their way through the filament, the atoms of the filament bump into each other furiously, generating heat — which produces the glow that you see from your light bulb.

Everything — even the best conductors — exhibits a certain amount of resistance to the flow of electrons. (Well, actually, certain materials, called *superconductors,* can conduct current with zero electrical resistance — but only if you cool them down to extremely low temperatures. You won't encounter them in conventional electronics.) The higher the resistance, the more restricted the flow of current.

So what determines how much resistance an object has? Resistance depends on several factors:

- **Material:** Some materials allow their electrons to roam freely, whereas others hold on tight to their electrons. Just how strongly a specific material opposes the flow of electrons determines its resistivity. *Resistivity* is a property of a material that reflects its chemical structure. Conductors have relatively low resistivity, whereas insulators have relatively high resistivity.
- **Cross-sectional area:** Resistance varies inversely with cross-sectional area; the larger the diameter, the easier it is for electrons to move — that is, the lower the resistance to their movement. Think of water flowing through a pipe: The wider the pipe, the easier the water flows. Along the same lines, a copper wire with a large diameter has lower resistance than a copper wire with a small diameter.
- **Length:** The longer the material, the more resistance it has because electrons have more opportunities to bump into other particles along the way. In other words, resistance varies directly with length.
- **Temperature:** For most materials, the higher the temperature, the higher the resistance. Higher temperatures mean that the particles inside have more energy, so they bump into each other a lot more, slowing down the flow of electrons. One notable exception to this is a type of resistor called a *thermistor:* Increase the temperature of a thermistor, and it lowers its resistance in a predictable way. (You can imagine how useful that characteristic is in temperature-sensing circuits.) You can read about thermistors in Chapter 12.

You use the symbol R to represent resistance in an electronic circuit. Sometimes you'll see a subscript next to a resistance, for instance, R_{bulb}. That just means that R_{bulb} represents the resistance of the light bulb (or whatever part of the circuit the subscript refers to). Resistance is measured in units called *ohms* (pronounced "omes"), abbreviated with the Greek letter omega (Ω). The higher the ohm value, the higher the resistance.

A single ohm is so small a unit of resistance that you're likely to see resistance measured in larger quantities, such as *kilohms* (kilo + ohm), which is thousands of ohms and is abbreviated *k*Ω, or *megohms* (mega + ohm), which is millions of ohms and is abbreviated *M*Ω. So 1 kΩ = 1,000 Ω and 1 MΩ = 1,000,000 Ω.

Resistors: Passive Yet Powerful

Resistors are passive electronic components that are specially designed to provide controlled amounts of resistance (for instance, 470 Ω or 1 kΩ). (Refer to Figure 5-1.)

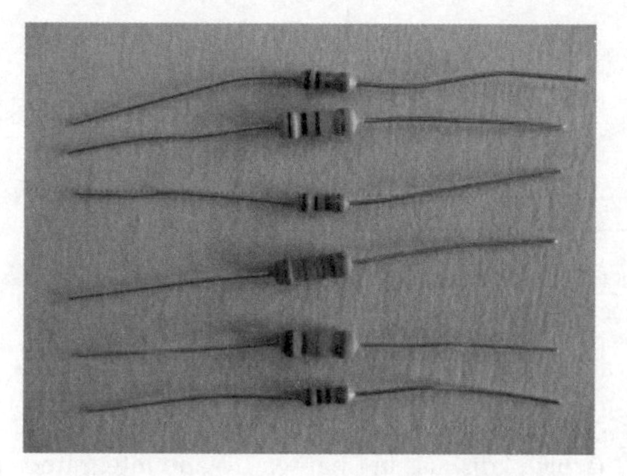

Figure 5-1: Resistors come in a variety of sizes and resistance values.

Although a resistor won't boost current or control its direction (because it's passive), you'll find it to be a powerful little device because it enables you to put the brakes on current flow in a controlled way. By carefully choosing and arranging resistors in different parts of your circuit, you can control just how much — or how little — current each part of your circuit gets.

What are resistors used for?

Resistors are among the most popular electronic components around because they're simple yet versatile. One of the most common uses of a resistor is to limit the amount of current in part of a circuit. However, resistors can also be used to control the amount of voltage provided to part of a circuit and to help create timing circuits.

Limiting current

The circuit in Figure 5-2 shows a 6-volt battery supplying current to a light-emitting diode (LED) through a resistor (shown as a zigzag). LEDs (like many other electronic parts) eat up current like a kid eats candy: They try to gobble up as much as you give them. But LEDs run into a problem — they burn themselves out if they draw too much current. The resistor in the circuit serves the useful function of limiting the amount of current sent to the LED (the way a good parent restricts the intake of candy).

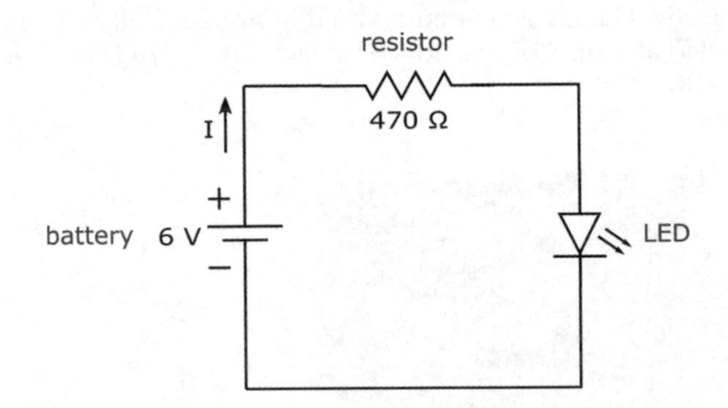

Figure 5-2: The resistor limits the amount of current, *I*, flowing into sensitive components, such as the light-emitting diode (LED) in this circuit.

Too much current can destroy many sensitive electronic components — such as transistors (which I discuss in Chapter 10) and integrated circuits (which I discuss in Chapter 11). By putting a resistor at the input to a sensitive part, you limit the current that reaches the part. (But if you use too high a resistance, say 1 MΩ, which is 1,000,000 ohms, you'll limit the current so much you won't see the light, although it's there!) This simple technique can save you a lot of time and money that you would otherwise lose fixing accidental blow-ups of your circuits.

You can observe how resistors limit current by setting up the circuit shown in Figure 5-2 and trying out resistors of different values. In the "Reading into fixed resistors" section, later in the chapter, I tell you how to decode the stripes on a resistor to determine its value. For now, I tell you what the ones you need look like.

Here's what you use to build the LED-resistor circuit:

- Four 1.5-volt AA batteries
- One four-battery holder (for AA batteries)
- One battery clip

- One 470 Ω resistor (identified by yellow-violet-brown stripes and then a fourth stripe which may be gold, silver, black, brown, or red)
- One 4.7 kΩ resistor (yellow-violet-red and any color for the fourth stripe)
- One 10 kΩ resistor (brown-black-orange and any color for the fourth stripe)
- One 47 kΩ resistor (yellow-violet-orange and any color for the fourth stripe)
- One LED (any size, any color)
- Three insulated alligator clips *or* one solderless breadboard

Use alligator clips or a solderless breadboard to set up the circuit (see Figure 5-3), starting with the 470 Ω resistor. Remember to orient the LED correctly, connecting the shorter lead of the LED to the negative battery terminal. Don't worry about the orientation of the resistor; either way is fine. Note how brightly the LED shines. Then remove the resistor and replace it with the other resistors, one at a time, increasing the amount of resistance each time. Did you notice that the LED shines less brightly each time? That's because higher resistances restrict current more, and the less current an LED receives, the less brightly it shines.

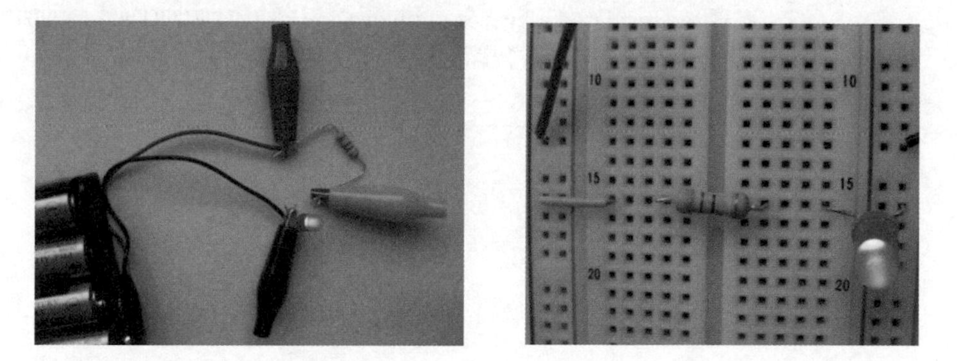

Figure 5-3: Two ways to set up the resistor-LED circuit.

Figure 5-4 shows a parallel circuit (described in Chapter 4) in which each branch contains a different value of resistance. For higher values of resistance, the current that passes through that branch is restricted more, so the LED in that branch emits less light.

Reducing voltage

Resistors can be used also to reduce the voltage supplied to different parts of a circuit. Say, for instance, you have a 9-volt power supply but you need to provide 5 volts to power a particular integrated circuit you're using. You can set up a circuit, such as the one shown in Figure 5-5, to divide the voltage in a way that provides 5 V at the output. Then — voilà — you can use the output voltage, V_{out}, of this *voltage divider* as the supply voltage for your integrated circuit. (You find details of how this works in Chapter 6.)

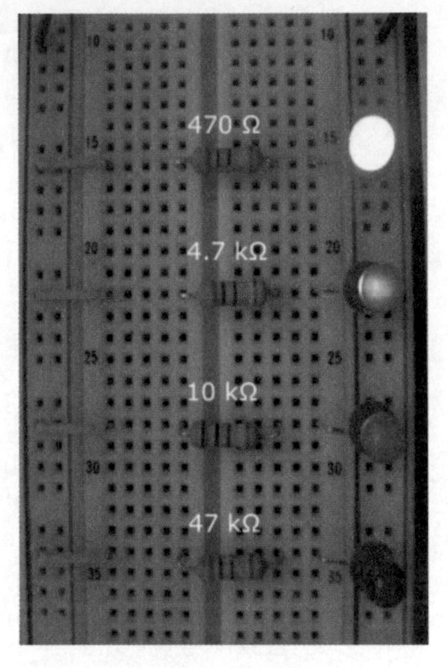

Figure 5-4: Higher values of resistance restrict current more, resulting in less light emitted from the LEDs.

To observe the voltage divider in action, set up the circuit shown in Figure 5-6 by using the following parts:

- One 9-volt battery
- One battery clip
- One 12 kΩ resistor (brown-red-orange and any color for the fourth stripe)
- One 15 kΩ resistor (brown-green-orange and any color for the fourth stripe)
- Three insulated alligator clips *or* one solderless breadboard

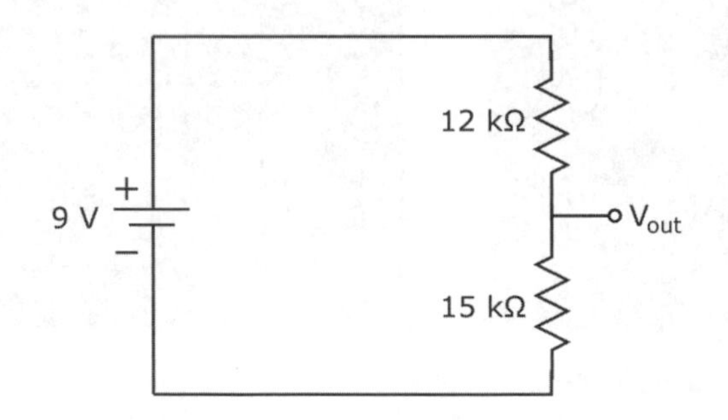

Figure 5-5: Use two resistors to create a voltage divider, a common technique for producing different voltages for different parts of a circuit.

Next, use your multimeter set to volts DC to measure the voltage across the battery and across the 15 kΩ resistor, as shown in Figure 5-7. My measurements show that the actual battery voltage is 9.24 V and V_{out} (the voltage across the 15 kΩ resistor) is 5.15 V.

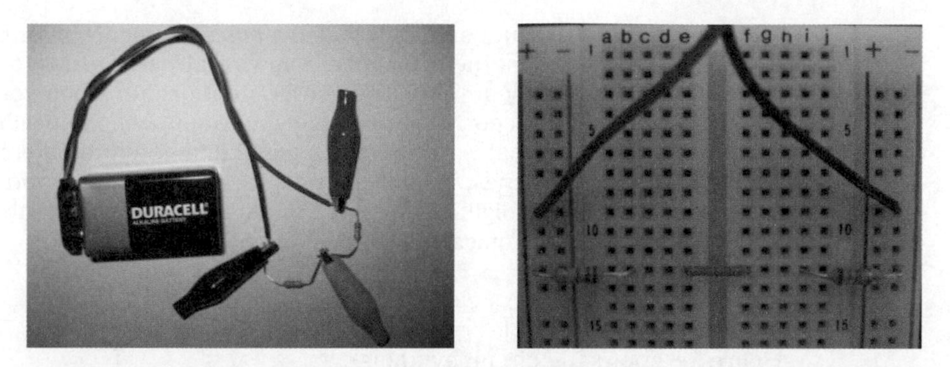

Figure 5-6: Two ways to build the voltage divider circuit.

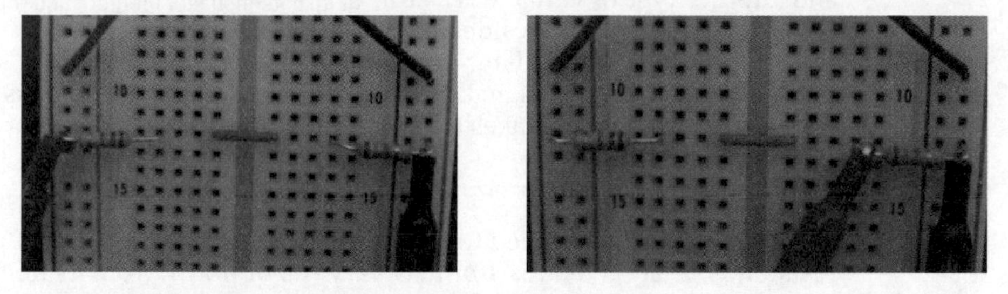

Figure 5-7: Measure the total voltage supplied by the battery (left) and the voltage across the 15 kΩ resistor (right).

Controlling timing cycles

You can also put a resistor to work with another popular component — a capacitor, which I discuss in Chapter 7 — to create predictable up and down voltage swings. You'll find the resistor-capacitor combo helps you create a kind of hourglass timer, which comes in handy for circuits that have time dependencies (for instance, a three-way traffic light). I show how the dynamic duo of resistor and capacitor operates in Chapter 7.

Choosing a type of resistor: Fixed or variable

Resistors come in two basic flavors: fixed and variable. Both types are commonly used in electronic circuits. Here's the lowdown on each type and why you would choose one or the other:

- A ***fixed resistor*** supplies a constant, factory-determined resistance. You use it when you want to restrict current to within a certain range or divide voltage in a particular way. Circuits with LEDs use fixed resistors to limit the current, thus protecting the LED from damage.

- A ***variable resistor,*** commonly called a *potentiometer* (*pot* for short), allows you to adjust the resistance from virtually zero ohms to a factory-determined maximum value. You use a potentiometer when you want to vary the amount of current or voltage you're supplying to part of your circuit. A few examples of where you might find potentiometers are light-dimmer switches, volume controls for audio systems, and position sensors, although digital controls have largely replaced potentiometers in consumer electronics.

In this section, you take a closer look at fixed and variable resistors. Figure 5-8 shows the circuit symbols that are commonly used to represent fixed resistors, potentiometers, and another type of variable resistor called a *rheostat.* (See the sidebar "Recognizing rheostats," later in this chapter.) The zigzag pattern should remind you that resistors make it more difficult for current to pass through, just as a kink in a hose makes it more difficult for water to pass through.

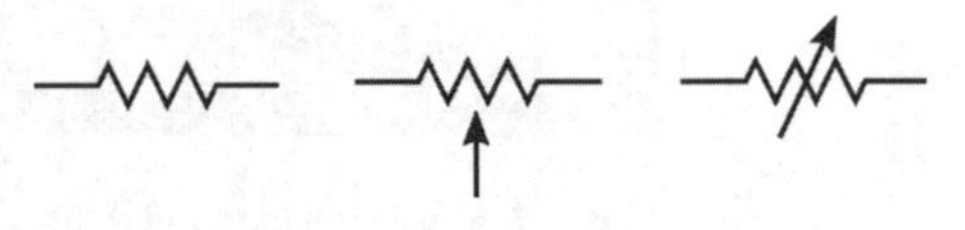

Figure 5-8: Circuit symbols for a fixed resistor (left), potentiometer (center), and rheostat (right).

Fixed resistors

Fixed resistors are designed to have a specific resistance, but the actual resistance of any given resistor may vary (up or down) from its nominal value by some percentage, known as the resistor's *tolerance.*

Say you choose a 1,000 Ω resistor that has a 5% tolerance. The actual resistance it provides could be anywhere from 950 Ω to 1,050 Ω (because 5% of 1,000 is 50). You might say that the resistance is 1,000 Ω, give or take 5%.

I used my multimeter set to ohms to measure the actual resistance of five 1 k Ω, 5% tolerance resistors. These were the readings: 985 Ω, 980 Ω, 984 Ω, 981 Ω, and 988 Ω.

There are two categories of fixed resistors:

- ***Standard-precision resistors*** can vary anywhere from 2% to (gulp) 20% of their nominal values. Markings on the resistor package will tell you just how far off the actual resistance may be (for instance, ±2%, ±5%, ±10%, or ±20%). You use standard-precision resistors in most hobby projects because (more often than not) you're using resistors to limit current or divide voltages to within an acceptable range. Resistors with 5% or 10% tolerance are commonly used in electronic circuits.
- ***High-precision resistors*** come within just 1% of their nominal value. You use these in circuits where you need extreme accuracy, as in a precision timing or voltage reference circuit.

Fixed resistors often come in a cylindrical package with two leads sticking out (refer to Figure 5-1) so you can connect them to other circuit elements. (see the "Identifying resistors on circuit boards" sidebar for exceptions.) Feel free to insert fixed resistors either way in your circuits — there's no left or right, up or down, or to or from when it comes to these little two-terminal devices.

Most fixed resistors are color-coded with their nominal value and tolerance (see the "Reading into fixed resistors" section), but some resistors have their values stamped right onto the tiny package, along with a bunch of other letters and numbers guaranteed to cause confusion. If you aren't sure of the value of a specific resistor, pull out your multimeter, set it to measure resistance in ohms, and place its probes across the resistor (either way), as shown in Figure 5-9. Make sure your resistor is not wired into a circuit when you measure its resistance; otherwise, you won't get an accurate reading.

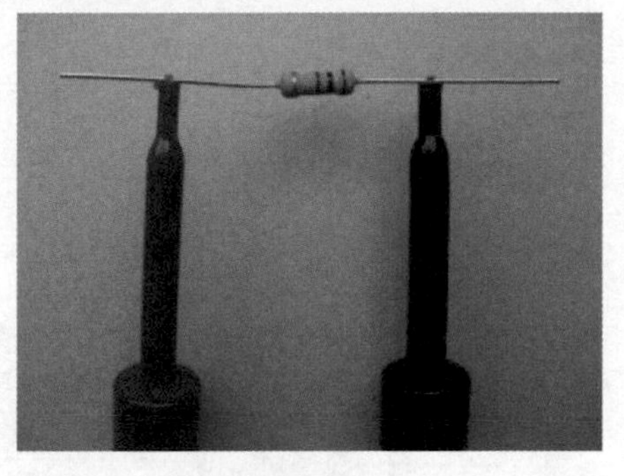

Figure 5-9: Use your multimeter set on ohms to measure the actual resistance of a fixed resistor.

Circuit designs usually tell you the safe resistor tolerance to use, whether for each individual resistor or for all the resistors in the circuit. Look for a notation in the parts list or as a footnote in the circuit diagram. If the schematic doesn't state a tolerance, you can assume it's okay to use standard-tolerance resistors (±5% or ±10%).

Reading into fixed resistors

The attractive rainbow colors adorning most fixed resistors serve a purpose beyond catching your eye. Color-coding identifies the *nominal value* and *tolerance* of most resistors; the others are drab and boring, and have their values stamped on them. The color code starts near the edge of one side of the resistor and consists of several stripes, or *bands,* of color. Each color represents a number, and the position of the band indicates how you use that number.

Standard-precision resistors use four color bands: The first three bands indicate the nominal value of the resistor, and the fourth indicates the tolerance. Using Table 5-1, you can decipher the nominal value and tolerance of a standard-precision resistor as follows:

- The **first band** gives you the first digit.
- The **second band** gives you the second digit.

- The **third band** gives you the multiplier as the number of zeros to tack on to the end of the first two digits — except if the band is gold or silver.
  - If the third band is **gold,** take the first two digits and divide by 10.
  - If the third band is **silver,** take the first two digits and divide by 100.
- The **fourth band** tells you the tolerance, as shown in the fourth column of Table 5-1. If there is no fourth band, you can assume the tolerance is ±20%.

You get the nominal value of the resistance in ohms by putting the first two digits together (side by side) and applying the multiplier.

Table 5-1 Resistor Color Coding

Color	*Band 1 (First Digit)*	*Band 2 (Second Digit)*	*Band 3 (Multiplier, or Number of Zeros)*	*Band 4 (Tolerance)*
Black	0	0	$10^0 = 1$ (no zeros)	±20%
Brown	1	1	$10^1 = 10$ (1 zero)	±1%
Red	2	2	$10^2 = 100$ (2 zeros)	±2%
Orange	3	3	$10^3 = 1,000$ (3 zeros)	±3%
Yellow	4	4	$10^4 = 10,000$ (4 zeros)	±4%
Green	5	5	$10^5 = 100,000$ (5 zeros)	–
Blue	6	6	$10^6 = 1,000,000$ (6 zeros)	–
Violet	7	7	$10^7 = 10,000,000$ (7 zeros)	–
Gray	8	8	$10^8 = 100,000,000$ (8 zeros)	–
White	9	9	$10^9 = 1,000,000,000$ (9 zeros)	–
Gold	–		0.1 (divide by 10)	±5%
Silver	–		0.01 (divide by 100)	±10%

Take a look at a couple of examples:

- **Red-red-yellow-gold:** A resistor with red (2), red (2), yellow (4 zeros), and gold (±5%) bands (see Figure 5-10, top) provides a nominal resistance of 220,000 Ω, or 220 kΩ, which could vary up or down by as much as 5% of that value. So it could have a resistance of anywhere between 209 kΩ and 231 kΩ.
- **Brown-black-gold-silver:** A resistor with brown (1), black (0), gold (0.1), and silver (±10%) bands (see Figure 5-10, bottom) provides a resistance of 10 × 0.1, or 1 Ω, which could vary by up to 10% of that value. So the actual resistance could be anywhere from 0.9 Ω to 1.1 Ω.

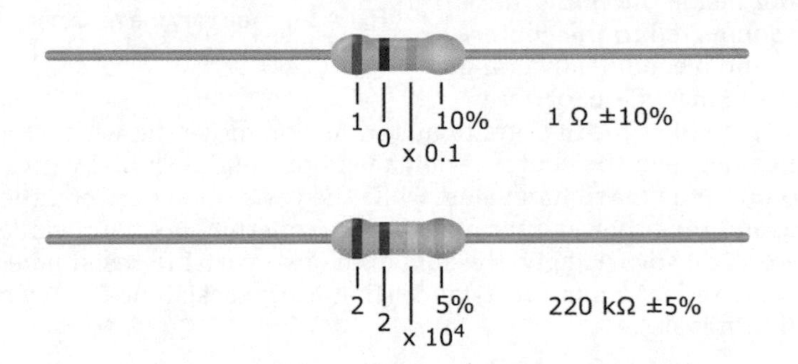

Figure 5-10: Decode the resistor's stripe pattern to determine the resistance.

A resistor with five bands of color is a high-precision resistor. The first three bands of color give you the first three digits, the fourth band gives you the multiplier, and the fifth band represents the tolerance (typically ±1%).

Colors vary greatly on resistor packaging, and some resistors don't use the color code at all, so you'd be wise to verify the actual resistance by using a multimeter set to ohms.

Variable resistors (potentiometers)

Potentiometers, or pots, allow you to adjust resistance continuously. Pots are three-terminal devices, meaning that they provide three places to connect to the outside world. (See Figure 5-11.) Between the two outermost terminals is a fixed resistance — the maximum value of the pot. Between the center terminal and either end terminal is an amount of resistance that varies depending on the position of a rotatable shaft or other control mechanism on the outside of the pot.

Inside a potentiometer is a resistance track with connections at both ends and a wiper that moves along the track. (See Figure 5-12.) Each end of the resistance track is electrically connected to one of the two end terminals on the outside of the pot, which is why the resistance between the two end terminals is fixed and equal to the maximum value of the pot.

Figure 5-11: You vary the resistance of these 10 kΩ dial pots by rotating a shaft.

The wiper inside the pot is electrically connected to the center terminal and mechanically connected to a shaft, slide, or screw, depending on the type of potentiometer. As you move the wiper, the resistance between the center terminal and one end terminal varies from 0 (zero) up to the maximum value, while the resistance between the center terminal and the other end terminal varies from the maximum value down to 0 (zero). Not surprisingly, the sum of the two variable resistances always equals the fixed maximum resistance (that is, the resistance between the two end terminals).

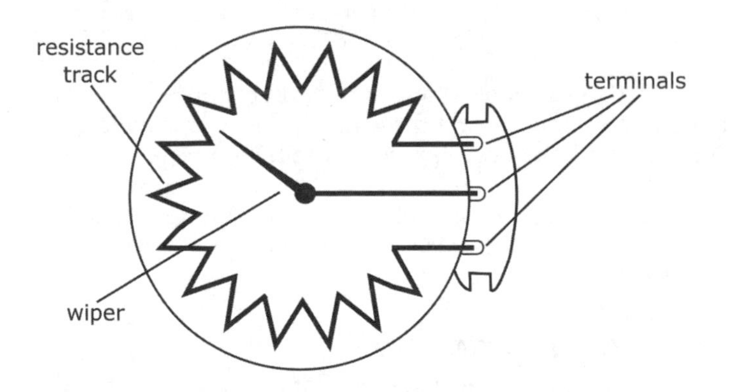

Figure 5-12: A potentiometer has a wiper that moves along a resistance track.

Most often, potentiometers are marked with their maximum value — 10 kΩ, 50 kΩ, 100 kΩ, 1 MΩ, and so forth — and they don't always include the little ohm symbol (Ω). For example, with a 50 k pot, you can dial up any resistance from 0 to 50,000 Ω.

Potentiometers are available in various packages known as dial pots, slide pots, and trim pots:

- ***Dial pots*** contain rotary resistance tracks and are controlled by turning a shaft or knob. Commonly used in electronics projects, dial pots are designed to be mounted through a hole cut in a case that houses a circuit, with the knob accessible from outside the case. Dial pots are popular for adjusting volume in sound circuits.
- ***Slide pots*** contain a linear resistance track and are controlled by moving a slide along the track. You see them on stereo equipment (for instance, faders) and some dimmer switches.
- ***Trim pots*** (also known as *preset pots*) are smaller, are designed to be mounted on a circuit board, and provide a screw for adjusting resistance. They are typically used to fine-tune a circuit — for instance, to set the sensitivity of a light-sensitive circuit — rather than to allow for variations (such as volume adjustments) during the operation of a circuit.

If you use a potentiometer in a circuit, bear in mind that if the wiper is dialed down all the way, you have zero resistance, and you aren't limiting current with this device. It's common practice to insert a fixed resistor in series with a potentiometer as a safety net to limit current. You just choose a value for the fixed resistor so it works with your variable resistor to produce the range of resistance you need. (Look for details about figuring out the total resistance of multiple resistors in series in the "Combining Resistors" section, later in this chapter.)

Recognizing rheostats

The word *potentiometer* is often used to categorize all variable resistors, but another type of variable resistor, known as a *rheostat*, is different from a true potentiometer. Rheostats are two-terminal devices, with one lead connected to the wiper and the other lead connected to one end of the resistance track. Although a potentiometer is a three-terminal device — its leads connect to the wiper and to *both* ends of the resistance track — you can use a potentiometer as a rheostat (as is quite common) by connecting only two of its leads, or you can connect all three leads in your circuit — and get both a fixed and variable resistor for the price of one!

Rheostats typically handle higher levels of voltage and current than potentiometers. This makes them ideal for industrial applications, such as controlling the speed of electric motors in large machines. However, rheostats have largely been replaced by circuits that use semiconductor devices (see Chapter 9), which consume much less power.

The circuit symbol used to represent a rheostat is shown in Figure 5-8, right.

Note that the range on the potentiometer is approximate. If the potentiometer lacks markings, use a multimeter (set to ohms) to figure out the component's value. You can also use a multimeter to measure the variable resistance between the center terminal and either end terminal. (Chapter 16 explains in detail how to test resistances using a multimeter.)

The circuit symbol commonly used to represent a potentiometer (refer to Figure 5-8, center) consists of a zigzag pattern representing the resistance and an arrow representing the wiper.

Rating resistors according to power

Quiz time! What do you get when you let too many electrons pass through a resistor at the same time? If you answered "a charred mess and no money-back guarantee," you're right! Whenever electrons flow through something with resistance, they generate heat — and the more electrons, the higher the heat.

Electronic components (such as resistors) can stand only so much heat (just how much depends on the size and type of component) before they have a meltdown. Because heat is a form of energy, and power is a measure of the energy consumed over a period of time, you can use the *power rating* of an electronic component to tell you how many watts (what? *Watts,* abbreviated W, are units of electric power) a component can safely handle.

Identifying resistors on printed circuit boards

As you learn more about electronics, you may get curious enough to take a look inside some of the electronics in your house. Warning: Be careful! Follow the safety guidelines given in Chapter 13. You might (for example) open up the remote control for your TV and see some components wired up between a touchpad and an LED. On *printed circuit boards (PCBs)* — which serve as platforms for building the mass-produced circuits commonly found in computers and other electronic systems — you may have trouble recognizing the individual circuit components. That's because manufacturers use fancy techniques to populate PCBs with components, aiming to eke out efficiencies and save space (known as *real estate*) on the boards.

One such technique, *surface-mount technology (SMT),* allows components to be mounted directly to the surface of a board (think of them as "hitting the deck"). *Surface-mount devices,* such as the SMT resistors shown in the photograph, look a bit different from the components you would use to build a circuit in your garage because they don't require long leads to connect them within a circuit. Such components use their own coding system to label the value of the part.

All resistors (including potentiometers) come with power ratings. Standard, run-of-the-mill fixed resistors can handle 1/8 W or 1/4 W, but you can easily find 1/2 W and 1 W resistors — and some are even flameproof. (Does that make you nervous about building circuits?) Of course, you won't see the power rating indicated on the resistor itself (that would make it too easy), so you have to figure it out by the size of the resistor (the bigger the resistor, the more power it can handle) or get it from the manufacturer or your parts supplier.

So how do you use the power rating to choose a particular resistor for your circuit? You estimate the peak power that your resistor will be expected to handle, and choose a power rating that meets or exceeds it. Power is calculated as follows:

$$P = V \times I$$

V represents the voltage (in volts, abbreviated V) measured across the resistor and *I* represents the current (in amps, abbreviated A) flowing through the resistor. For example, suppose the voltage is 5 V and you want to pass 25 mA (milliamps) of current through the resistor. To calculate the power, first you convert 25 mA to 0.025 A (remember, *milli*amps are thousandths of an amp). Then you multiply 5 by 0.025, and you get 0.125 W, or 1/8 W. So you know that a 1/8 W resistor may be okay, but you can be sure that a 1/4 W resistor *will* take the heat just fine in your circuit.

For most hobby electronics projects, 1/4 W or 1/8 W resistors are fine. You need high-wattage resistors for *high-load* applications, where loads, such as a motor or a lamp control, require higher-than-hobby-level currents to operate. High-wattage resistors take many forms, but you can bet they are bigger and bulkier than your average resistor. Resistors with power ratings over 5 W are wrapped in epoxy (or another waterproof and flameproof coating) and have a rectangular, rather than cylindrical, shape. A high-wattage resistor may even include its own metal *heat sink,* with fins that conduct heat away from the resistor.

Combining Resistors

When you start shopping for resistors, you'll find that you can't always get exactly what you want. It would be impractical for manufacturers to make resistors with every possible value of resistance. Instead, they make resistors with a limited set of resistance values, and you work around it (as you're about to see). For instance, you can search far and wide for a 25 kΩ resistor, but you may never find it; however, 22 kΩ resistors are as common as the day is long! The trick is to figure out how to get the resistance you need using standard available parts.

As it turns out, you can combine resistors in various ways to create an *equivalent resistance* value that will come pretty darn close to whatever resistance you need. And because standard precision resistors are accurate to 5%–10% of their nominal value anyway, combining resistors works out just fine.

There are certain "rules" for combining resistances, which I cover in this section. Use these rules not only to help you choose off-the-shelf resistors for your own circuits, but also as a key part of your effort to analyze other people's electronic circuits. For instance, if you know that a light bulb has a certain amount of resistance, and you place a resistor in series with the bulb to limit the current, you'll need to know what the total resistance of the two components is before you can calculate the current passing through them.

Resistors in series

When you combine two or more resistors (or resistances) in series, you connect them end to end (as shown in Figure 5-13), so that the same current passes sequentially through each resistor. By doing this, you restrict the current somewhat with the first resistor, you restrict it even more with the next resistor, and so forth. So the effect of the series combination is an *increase* in the overall resistance.

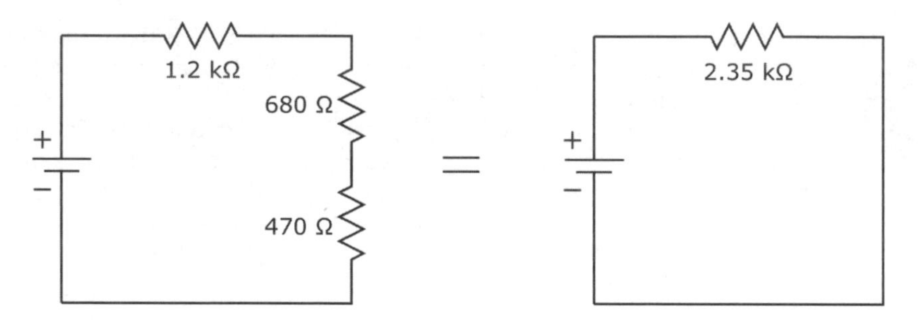

Figure 5-13: The combined resistance of two or more resistors in series is the sum of the individual resistances.

To calculate the combined (equivalent) resistance of multiple resistors in series, you simply add the values of the individual resistances. You can extend this rule to any number of resistances in series:

$$R_{series} = R1 + R2 + R3 + \ldots$$

$R1$, $R2$, $R3$, and so forth represent the values of the resistors, and R_{series} represents the total equivalent resistance. Remember that the same current flows through all resistors connected in series and that all the resistors contribute to the overall restriction of the current.

You can apply this concept of equivalent resistance to help you select resistors for a specific circuit need. For example, suppose you need a 25 kΩ resistor, but cannot find a standard resistor with that value. You can combine two standards resistors — a 22 kΩ resistor and a 3.3 kΩ resistor — in series to get 25.3 kΩ of resistance. That's less than 2% different from the 25 kΩ you seek — well within typical resistor tolerance levels (which are 5%–10%).

Be careful with your units of measurement when you add resistance values. For example, suppose you connect the following resistors in series: 1.2 kΩ, 680 Ω, and 470 Ω. (Refer to Figure 5-13.) Before you add the resistances, you need to convert the values to the same units, for instance, ohms. In this case, the total resistance, R_{series}, is calculated as follows:

$$\begin{aligned} R_{series} &= 1{,}200\ \Omega + 680\ \Omega + 470\ \Omega \\ &= 2{,}350\ \Omega \\ &= 2.35\ \text{k}\Omega \end{aligned}$$

The combined resistance will *always* be greater than any individual resistances. This fact comes in handy when you're designing circuits. For example, if you want to limit current going into a light bulb but don't know the resistance of the bulb, you can place a resistor in series with the bulb and be secure in the knowledge that the total resistance to current flow is *at least* as much as the value of the resistor you added. For circuits that use variable resistors (such as a light-dimmer circuit), putting a fixed resistor in series with the variable resistor guarantees that the current will be limited even if the pot is dialed down to zero ohms. (The lowdown on how to calculate just what the current will be for a given voltage/resistance combo appears later in this chapter.)

See for yourself how a small series resistor can save an LED. Set up the circuit shown in Figure 5-14, left, using the following parts:

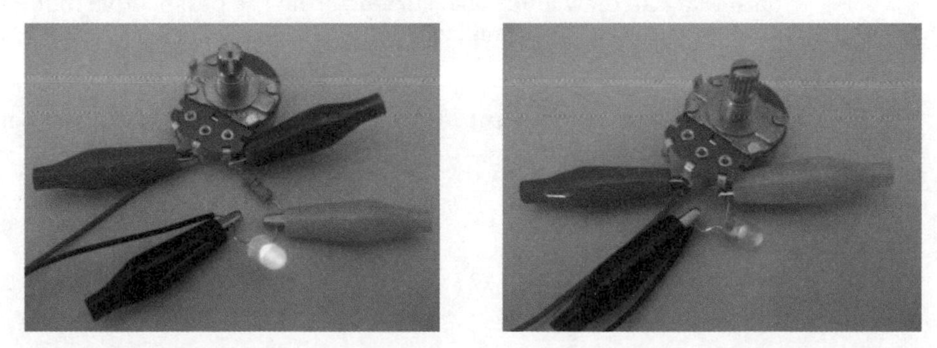

Figure 5-14: A resistor in series with a potentiometer ensures that current to the LED is restricted even if the pot is dialed down to zero ohms. Without that resistor, the LED is toast.

- One 9-volt battery
- One battery clip
- One 470 Ω resistor (yellow-violet-brown)
- One 10 kΩ potentiometer
- Four alligator clips
- One LED, any size, any color

Be sure to connect the center and one end terminal of the potentiometer into the circuit, leaving the other terminal unconnected, and to orient the LED correctly, so that its shorter lead is connected to the negative battery terminal.

Rotate the shaft of the potentiometer and observe what happens to the LED. You should see the light vary in intensity from fairly bright to very dim (or vice versa) as you adjust the resistance of the pot.

With the potentiometer shaft positioned somewhere in the middle of its range, remove the 470 Ω resistor and connect the LED directly to the potentiometer, as shown in Figure 5-14, right. Now rotate the shaft slowly in the direction that makes the light grow brighter. Rotate it all the way to the end and observe what happens to the LED. As you dial the pot down to 0 Ω, you should see the LED get brighter and brighter until finally it stops shining completely. With no resistance to limit current, the LED could burn out. (If it does, you should throw it away because it won't work anymore.)

Resistors in parallel

When you combine two resistors in parallel, you connect both sets of ends together (see Figure 5-15) so each resistor has the same voltage. By doing so, you provide two different paths for current to flow. Even though each resistor is restricting current flow through one circuit path, there's still another path that can draw additional current. From the perspective of the source voltage, the effect of arranging resistors in parallel is a *decrease* in the overall resistance.

To calculate the equivalent resistance, $R_{parallel}$, of two resistors in parallel, you use the following formula:

$$R_{parallel} = \frac{R1 \times R2}{R1 + R2}$$

where $R1$ and $R2$ are the values of the individual resistors.

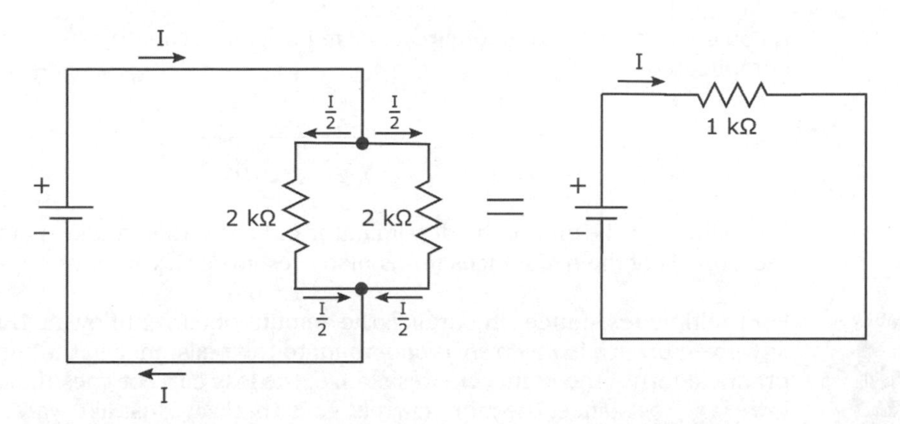

Figure 5-15: The combined resistance of two or more resistors in parallel is always lower than any of the individual resistances.

You may remember that the line separating the numerator and the denominator in a fraction represents division, so another way to think of this formula is like this:

$$R_{parallel} = (R1 \times R2) \div (R1 + R2)$$

In the example in Figure 5-14, two 2 kΩ resistors are placed in parallel. The equivalent resistance is as follows:

$$\begin{aligned} R_{parallel} &= \frac{2{,}000\ \Omega \times 2{,}000\ \Omega}{2{,}000\ \Omega + 2{,}000\ \Omega} \\ &= \frac{4{,}000{,}000\ \Omega^2}{4{,}000\ \Omega} \\ &= 1{,}000\ \Omega \\ &= 1\ \text{k}\Omega \end{aligned}$$

In this example, because the two resistors have equal resistance, connecting them in parallel results in an equivalent resistance of *half the value of either one.* The result is that each resistor draws half the supply current. If two resistors of unequal value are placed in parallel, *more* current will flow through the path with the *lower* resistance than the path with the higher resistance.

If your circuit calls for a resistor with a somewhat higher power rating, say 1 W, but you have only 1/2 W resistors on hand, you can combine two 1/2 W resistors in parallel instead. Just select resistor values that combine to create the resistance you need. Because each one draws half the current that a single resistor would draw, it dissipates half the power (remember that power = voltage × current).

If you combine more than one resistor in parallel, the math gets a little more complicated:

$$R_{parallel} = \frac{1}{\frac{1}{R1} + \frac{1}{R2} + \frac{1}{R3} + \ldots}$$

The ellipsis at the end of the denominator indicates that you keep adding the reciprocals of the resistances for as many resistors as you have in parallel.

For multiple resistances in parallel, the amount of current flowing through any given branch is *inversely proportional* to the resistance in that branch. In practical terms, the higher the resistance, the less current goes that way; the lower the resistance, the more current goes that way. Just like water, electric current favors the path of least resistance.

Measuring combined resistances

Using your multimeter set to measure resistance in ohms, you can verify the equivalent resistance of resistors in series and resistors in parallel.

The following photos show how to measure the equivalent resistance of three resistors in series (left), two resistors in parallel (center), and a combination of one resistor in series with two parallel resistors (right). Select any three resistors and try this out for yourself!

The resistors in the photos have nominal values of 220 kΩ, 33 kΩ, and 1 kΩ. In both the center and right photos, the parallel resistors are the 220 kΩ and 33 kΩ resistors. The lone resistor in series in the photo on the right is the 1 kΩ resistor.

For the series resistors (left photo), the calculated equivalent resistance in kΩ was $(220+33+1)=254$, and the actual resistance I measured was 255.4 kΩ.

For the parallel resistors (center photo), the calculated equivalent resistance in kΩ was $(220\times 33)\div(220+33)=28.7$, and the actual resistance I measured was 28.5 kΩ.

For the series-parallel combination (right photo), the calculated resistance in kΩ was $(28.7+1)=29.7$, and the actual resistance I measured was 29.4 kΩ.

Remember, the values of most resistors vary somewhat from their nominal values; That's why the actual measured resistance for each resistor combination shown here is a bit different (in these cases, < 2%) from its calculated equivalent resistance.

As a shorthand in electronic equations, you may see the symbol || used to represent the formula for resistors in parallel. For example:

$$R_{parallel} = R1 \,||\, R2 = \frac{R1 \times R2}{R1 + R2}$$

or

$$R_{parallel} = R1 \,||\, R2 \,||\, R3 = \frac{1}{\frac{1}{R1} + \frac{1}{R2} + \frac{1}{R3}}$$

Combining series and parallel resistors

Many circuits combine series resistors and parallel resistors in various ways to restrict current in some parts of the circuit while splitting current in other parts of the circuit. In some cases, you can calculate equivalent resistance by combining the equations for resistors in series and resistors in parallel.

For instance, in Figure 5-16, resistor *R2* (3.3 kΩ) is in parallel with resistor *R3* (3.3 kΩ), and that parallel combination is in series with resistor *R1* (1 kΩ). You can calculate the total resistance (in kΩ) as follows:

$$\begin{aligned} R_{equivalent} &= R1 + \left(R2 \,||\, R3\right) \\ &= R1 + \frac{R2 \times R3}{R2 + R3} \\ &= 1\text{ k}\Omega + \frac{3.3\text{ k}\Omega \times 3.3\text{ k}\Omega}{3.3\text{ k}\Omega + 3.3\text{ k}\Omega} \\ &= 1\text{ k}\Omega + 1.65\text{ k}\Omega \\ &= 2.65\text{ k}\Omega \end{aligned}$$

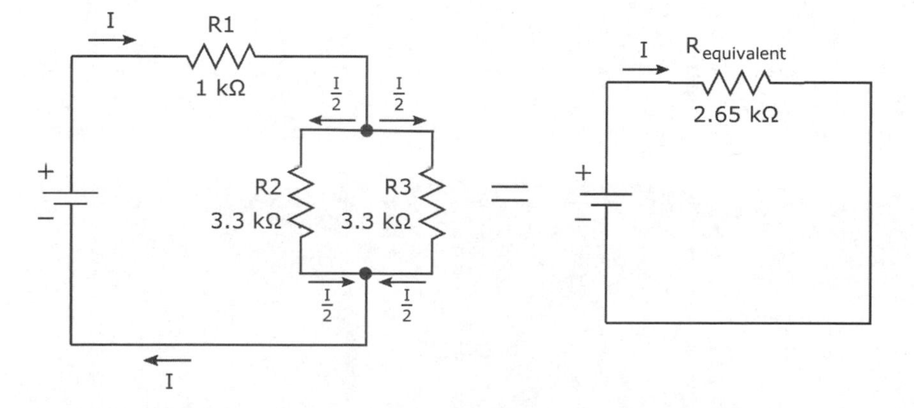

Figure 5-16: Many circuits include a combination of parallel and series resistances.

In this circuit, the current supplied by the battery is limited by the *total* resistance of the circuit, which is 2.65 kΩ. Supply current flows from the positive battery terminal through resistor *R1*, splits — with half flowing through resistor *R2* and half flowing through resistor *R3* (because those resistances are the same) — and then combines again to flow into the negative battery terminal.

Circuits often have more complex combinations of resistances than simple series or parallel relationships, and figuring out equivalent resistances isn't always easy. You have to use matrix mathematics to analyze them, and because this book isn't a math book, I'm not going to detour to explore the complexities of matrix math.

6

Obeying Ohm's Law

In This Chapter

- Understanding how current, voltage, and resistance are governed by Ohm's Law
- Practicing Ohm's Law by analyzing circuits
- Using power as your guide in choosing circuit components

An intimate relationship exists between voltage (the electrical force that pushes electrons) and current in components that have resistance. That relationship is summed up nicely in a simple equation with an authoritative name: Ohm's Law. In this chapter, you put Ohm's Law to work to find out what's going on in some basic circuits. Then you get a look at the role of Ohm's Law and related power calculations in the design of electronic circuits.

Defining Ohm's Law

One of the most important concepts to understand in electronics is the relationship between voltage, current, and resistance in a circuit, summarized in a simple equation known as Ohm's Law. When you understand this equation thoroughly, you'll be well on your way to analyzing circuits that other people have designed, as well as successfully designing your own circuits. Before diving into Ohm's Law, it may help to take a quick look at the ebbs and flows of current.

Driving current through a resistance

If you place a voltage source across an electronic component that has measurable resistance (such as a light bulb or a resistor), the force of the voltage will push electrons through the component. The movement of gobs of electrons is what constitutes electric current. By applying a greater voltage, you exert a stronger force on the electrons, which creates a stronger flow of electrons — a larger current — through the resistance. The stronger the force (voltage V), the stronger the flow of electrons (current I).

This is analogous to water flowing through a pipe of a certain diameter. If you exert a certain water pressure on the water in the pipe, the current will flow at a certain rate. If you increase the water pressure, the current will flow faster through that same pipe, and if you decrease the water pressure, the current will flow slower through the pipe.

It's constantly proportional!

The relationship between voltage (V) and current (I) in a component with resistance (R) was discovered in the early 1800s by Georg Ohm (does his name sound familiar?). He figured out that for components with a fixed resistance, voltage and current vary in the same way: Double the voltage, and the current is doubled; halve the voltage, and the current is halved. He summed up this relationship quite nicely in the simple mathematical equation that bears his name: Ohm's Law.

Ohm's Law states that voltage equals current multiplied by resistance, or

$$V = I \times R$$

What this really means is that the voltage (V), measured across a component with a fixed resistance, is equal to the current (I) flowing through the component multiplied by the value of the resistance (R).

For example, in the simple circuit in Figure 6-1, a 9-volt battery applied across a 1 kΩ resistor produces a current of 9 mA (which is 0.009 A) through the circuit:

$$9 \text{ V} = 0.009 \text{ A} \times 1{,}000 \text{ } \Omega$$

Ohm's Law is so important in electronics that you'd be wise to repeat it over and over again, like a mantra, until you've mastered it! To help you remember, think of Ohm's Law as a **V**ery **I**mportant **R**ule.

When using Ohm's Law, watch your units of measurement carefully. Make sure that you convert any *kilos* and *millis* before you get out your calculator. If you think of Ohm's Law as volts = amps × ohms, you'll be okay. And if you're brave, you can also use volts = milliamps × kilohms, which works just as well (because the *millis* cancel out the *kilos*).

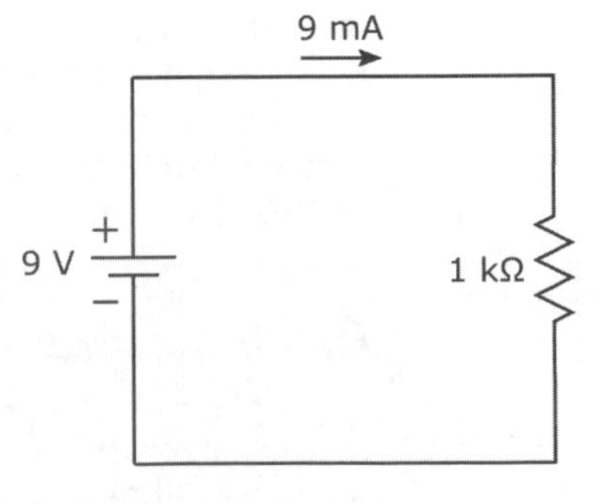

Figure 6-1: A voltage of 9 V applied to a resistor of 1 kΩ produces a current of 9 mA.

But if you aren't careful and you mix units, you may be in for a shock! For instance, a lamp with a resistance of 100 Ω passes a current of 50 mA. If you forget to convert milliamps to amps, you'll multiply 100 by

50 to get 5,000 V as the voltage across the lamp! Ouch! The correct way to perform the calculation is to convert 50 mA to 0.05 A, and *then* multiply by 100 Ω, to get 5 V. Much better!

Ohm's Law is so important (did I say that already?) that I've created the following list to help you remember how to use it:

$$voltage = current \times resistance$$

$$V = I \times R$$

$$\text{volts} = \text{amps} \times \text{ohms}$$

$$\text{V} = \text{A} \times \Omega$$

$$\text{volts} = \text{milliamps} \times \text{kilohms}$$

$$\text{V} = \text{mA} \times \text{k}\Omega$$

There's a reason Georg Ohm has his name associated with resistance values as well as the law. The definition of an ohm, or unit of resistance, came from Georg Ohm's work. The *ohm* is defined as the resistance between two points on a conductor when one volt, applied across those points, produces one amp of current through the conductor. I just thought you might like to know that. (Good that Georg's last name wasn't Wojciehowicz!)

One law, three equations

Remember your high-school algebra? Remember how you can rearrange the terms of an equation containing variables (such as the familiar *x* and *y*) to solve for one variable, as long as you know the values of the other variables? Well, the same rules apply to Ohm's Law. You can rearrange its terms to create two more equations, for a total of three equations from that one law!

$$V = I \times R \qquad I = \frac{V}{R} \qquad R = \frac{V}{I}$$

These three equations all say the same thing but in different ways. You can use them to calculate one quantity when you know the other two. Which one you use at any given time depends on what you're trying to do. For example:

- **To calculate an unknown voltage,** multiply the current times the resistance ($V = I \times R$). For instance, if you have a 2 mA current running through a 2 kΩ resistor, the voltage across the resistor is 2 mA × 2 kΩ (or 0.002 A × 2,000 Ω) = 4 V.
- **To calculate an unknown current,** take the voltage and divide it by the resistance ($I = V \div R$). For example, if 9 V is applied across a 1 kΩ resistor, the current is 9 V ÷ 1,000 Ω = 0.009 A or 9 mA.

- **To calculate an unknown resistance,** take the voltage and divide it by the current ($R = V \div I$). For instance, if you have 3.5 V across an unknown resistor with 10 mA of current running through it, the resistance is $3.5\text{ V} \div 0.01\text{ A} = 350\ \Omega$.

Using Ohm's Law to Analyze Circuits

When you have a good handle on Ohm's Law, you'll be ready to put it into practice. Ohm's Law is like a master key, unlocking the secrets to electronic circuits. Use it to understand circuit behavior and to track down problems within a circuit (for instance, why the light isn't shining, the buzzer isn't buzzing, or the resistor isn't resisting because it melted). You can also use it to design circuits and choose the right parts for use in your circuits. I get to these topics in a later section of this chapter. In this section, I discuss how to apply Ohm's Law to analyze circuits.

Calculating current through a component

In the simple circuit you saw in Figure 6-1, a 9-volt battery is applied across a 1 kΩ resistor. You calculate the current through the resistor as follows:

$$I = \frac{9\text{ V}}{1{,}000\ \Omega} = 0.009\text{ A} = 9\text{ mA}$$

If you add a 220 Ω resistor in series with the 1 kΩ, as shown in Figure 6-2, you're restricting the current even more.

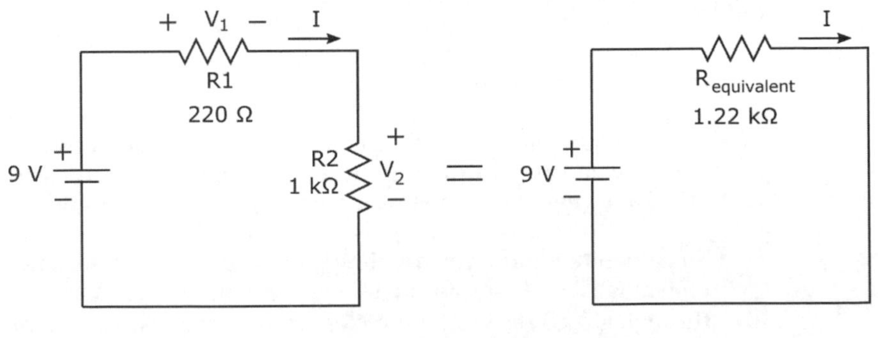

Figure 6-2: To calculate the current through this circuit, determine the equivalent resistance and apply Ohm's Law.

To calculate the current flowing through the circuit, you need to determine the total resistance that the 9-volt battery is facing in the circuit. Because the resistors are in series, the resistances add up, for a total equivalent resistance of 1.22 kΩ. You use this equivalent resistance to calculate the new current, as follows:

$$I = \frac{9\text{ V}}{1{,}220\ \Omega} \approx 0.0074\text{ A} \approx 7.4\text{ mA}$$

By adding the extra resistor, you've reduced the current in your circuit from 9 mA to 7.4 mA.

TIP

That double squiggle symbol (≈) in the equation just given means "is approximately equal to," and I used that because I rounded off the current to the nearest tenth of a milliamp. It's usually okay to round off the tinier parts of values in electronics — unless you're working on the electronics that control a subatomic particle smasher or other high-precision industrial device.

Calculating voltage across a component

In the circuit that was shown in Figure 6-1, the voltage across the resistor is simply the voltage supplied by the battery: 9 V. That's because the resistor is the only circuit element other than the battery. Adding a second resistor in series (refer to Figure 6-2) changes the voltage picture. Now *some* of the battery voltage is dropped across the 220 Ω resistor (*R1*), and the *rest* of the battery voltage is dropped across the 1 kΩ resistor (*R2*). I labeled these voltages V_1 and V_2, respectively.

To figure out how much voltage is dropped across each resistor, you use Ohm's Law for each individual resistor. You know the value of each resistor, and *you now know the current flowing through each resistor.* Remember that current (*I*) is the battery voltage (9 V) divided by the total resistance (*R1* + *R2*, or 1.22 kΩ), or approximately 7.4 mA. Now you can apply Ohm's Law to each resistor to calculate its voltage drop:

$$\begin{aligned} V_1 &= I \times R1 \\ &= 0.0074\text{ A} \times 220\ \Omega \\ &= 1.628\text{ V} \\ &\approx 1.6\text{ V} \end{aligned}$$

$$\begin{aligned} V_2 &= I \times R2 \\ &= 0.0074\text{ A} \times 1{,}000\ \Omega \\ &= 7.4\text{ V} \end{aligned}$$

Note that if you add the voltage drops across the two resistors, you get 9 volts, which is the total voltage supplied by the battery. That isn't a coincidence; the battery is supplying voltage to the two resistors in the circuit, and the supply voltage is divided between the resistors proportionally, according to the values of the resistors. This type of circuit is known as a *voltage divider.*

There's a quicker way to calculate either of the "divided voltages" (V_1 or V_2) in Figure 6-2. You know that the current passing through the circuit can be expressed as

$$I = \frac{V_{battery}}{R1 + R2}$$

You also know that:

$$V_1 = I \times R_1$$

and

$$V_2 = I \times R_2$$

To calculate V_1, for example, you can substitute the expression for *I* shown above, and you get

$$V_1 = \frac{V_{battery}}{R1 + R2} \times R1$$

You can rearrange the terms, without changing the equation, to get

$$V_1 = \frac{R1}{R1 + R2} \times V_{battery}$$

Similarly, the equation for V_2 is

$$V_2 = \frac{R2}{R1 + R2} \times V_{battery}$$

By plugging in the values of *R1*, *R2*, and $V_{battery}$, you get $V_1 = 1.628$ V and $V_2 = 7.4$ V, just as calculated earlier.

The following general equation is commonly used for the voltage across a resistor (*R1*) in a voltage divider circuit:

$$V_1 = \frac{R1}{R1 + R2} \times V_{battery}$$

Many electronic systems use voltage dividers to bring down a supply voltage to a lower level, after which they feed that reduced voltage into the input of another part of the overall system that requires that lower voltage.

In Chapter 5, you see an example of a voltage divider that reduces a 9-volt supply to 5 volts using 15 kΩ and 12 kΩ resistors. You can use the voltage divider equation to calculate the output voltage, V_{out}, of that voltage divider circuit, which is shown in Figure 6-3, as follows:

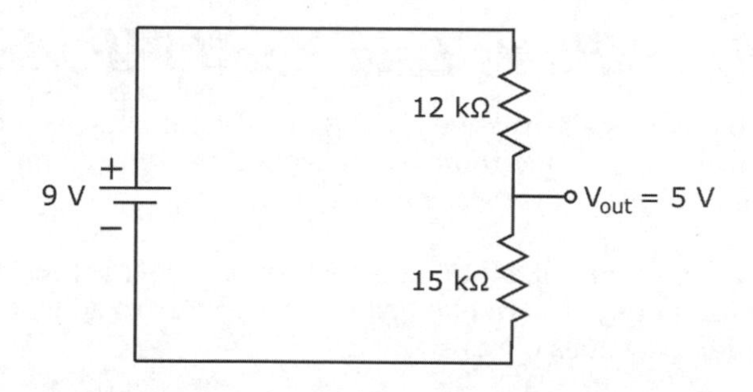

Figure 6-3: This voltage divider circuit reduces the 9-volt supply to 5 volts at V_{out}.

$$
\begin{aligned}
V_{out} &= \frac{15{,}000\ \Omega}{(12{,}000+15{,}000)\ \Omega} \times 9\ \text{V} \\
&= \frac{15{,}000}{27{,}000} \times 9\ \text{V} \\
&= 5\ \text{V}
\end{aligned}
$$

The circuit in Figure 6-3 divides a 9 V supply down to 5 V.

Calculating an unknown resistance

Say you have a large flashlight that you're running off a 12-volt battery, and you measure a current of 1.3 A through the circuit. (I discuss how to measure current in Chapter 16.) You can calculate the resistance of the incandescent bulb by taking the voltage across the bulb (12 V) and dividing it by the current through the bulb (1.3 A). It's a pretty quick number-crunch:

$$R_{bulb} = \frac{12\ \text{V}}{1.3\ \text{A}} = 9\ \Omega$$

Using Ohm's Law

Ohm's Law is useful in analyzing voltage and current for resistors and other components that behave like resistors, such as light bulbs. But you have to be careful about applying Ohm's Law to other electronic components, such as capacitors (detailed in Chapter 7) and inductors (covered in Chapter 8), which don't exhibit a *constant* resistance under all circumstances. For such components, the opposition to current — known as *impedance* — will vary depending on what's going on in the circuit. So you can't just use a multimeter to measure the "resistance" of a capacitor, for instance, and then try to apply Ohm's Law willy-nilly.

Seeing Is Believing: Ohm's Law Really Works!

Ohm's Law, which governs all resistive electronic components, is one of the most important principles in electronics. In this section, you can put Ohm's Law to the test and take your first steps in circuit analysis.

Figure 6-4 shows a series circuit containing a 9 V battery, a 1 kΩ resistor (*R1*), and a 10 kΩ potentiometer, or variable resistor (*R2*). You are going to test Ohm's Law for different values of resistance.

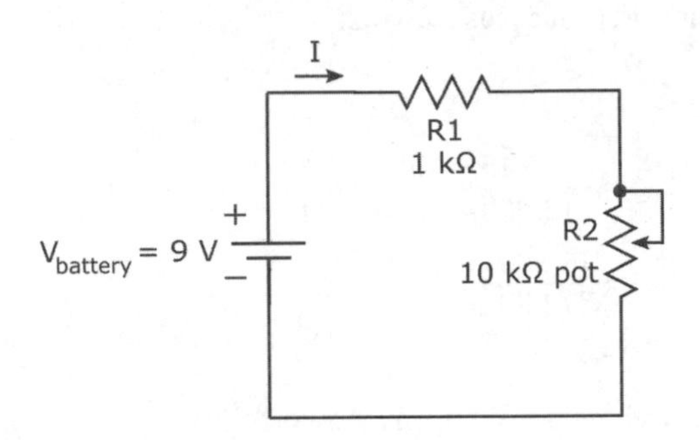

Figure 6-4: With a simple series circuit, you can witness Ohm's Law in action.

To build this circuit, you need the following parts:

- One 9-volt battery
- One battery clip
- One 1 kΩ 1/4 W (minimum) resistor (brown-black-red stripes)
- One 10 kΩ potentiometer
- One solderless breadboard

Refer to Chapter 2 for information on buying these parts and Chapter 5 for details on resistors and potentiometers. You'll need to attach wires to your potentiometer terminals so that you can connect the pot into your circuit. Refer to Chapter 15 for tips on attaching wires to pots and using a solderless breadboard. Because you will be measuring current, you may want to refer to Chapter 16 for details on how to use a multimeter.

Here are the steps to follow to build the circuit and test Ohm's Law:

1. **Connect the middle (wiper) lead and one outer (fixed) lead of the potentiometer.**

   When using a potentiometer (pot) as a two-terminal variable resistor, it's common practice to connect the wiper and one fixed lead together.

This way, *R2* is the resistance from the combined leads to the other outer (fixed) lead. By dialing the pot, you can vary the resistance of *R2* between 0 (zero) Ω and 10 kΩ. For now, simply twist together the ends of the wires coming off the pot terminals.

2. **Zero the pot.**

   With your multimeter set on ohms, measure the resistance of the pot from the wiper lead to the fixed lead that is not connected to the wiper. Then, dial the pot all the way in one direction or the other until your multimeter reads 0 Ω. This value is the pot resistance you will start with in your circuit.

3. **Build the circuit using Figure 6-5 as your guide.**

   Note that it doesn't matter which way you orient the 1 kΩ resistor or the pot (as long as you keep the same two pot leads together as you did when you measured the pot resistance). You can untwist the pot leads and insert them into neighboring holes on the breadboard.

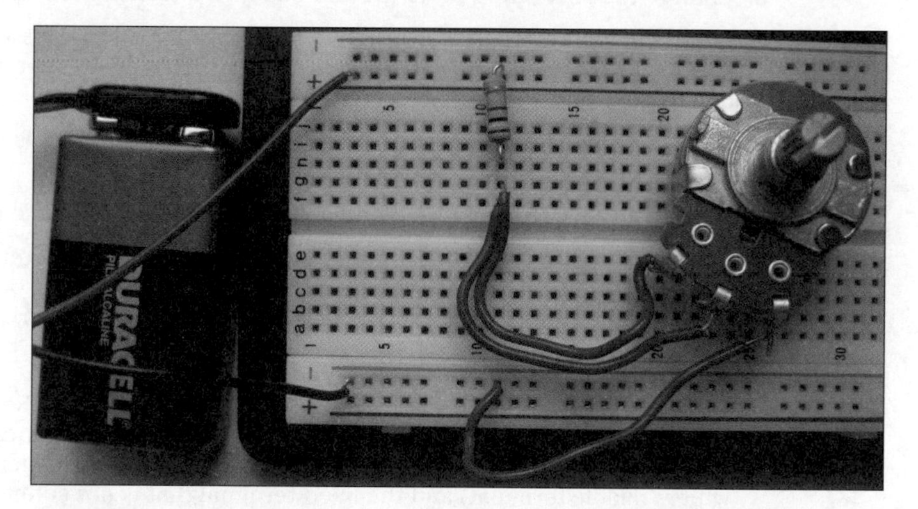

Figure 6-5: Your solderless breadboard makes the connections between components in this simple series circuit.

4. **Measure the current flowing through the circuit.**

   To measure current, you need to break the circuit and insert your multimeter in series with whatever you're measuring current through. In a series circuit, the current flowing through each component is the same, so you can measure the current anywhere you want in the circuit. For this example, I tell you how to insert your multimeter between the resistor and the pot.

Before you insert your multimeter into the circuit, switch your multimeter to measure DC current in milliamps (a range of 20 mA is fine). Then, move the resistor lead that is connected to the pot to another column in your breadboard (or just leave the disconnected lead hanging). Now you've broken the circuit.

Connect the positive multimeter lead to the open side of the 1 kΩ resistor and the negative multimeter lead to the open side of the pot. Note the current reading.

Is this the reading you expect to get by applying Ohm's Law to your circuit? Remember that because the pot is dialed all the way down to 0 Ω, the total resistance in your circuit — let's call that R_{total} — is roughly 1 kΩ.

You should expect a current of roughly 9 mA because $V_{battery} / R_{total} = 9\text{ V} / 1\text{ k}\Omega = 9\text{ mA}$. Any discrepancies are due to variations in the supply voltage, the tolerance of the resistor, and the slight resistance of the multimeter.

5. **Change the pot setting to 10 kΩ and observe the change in current.**

   With your multimeter still inserted in your circuit, dial the pot all the way to the other end so that its resistance is 10 kΩ. What current reading do you get? Is the actual current what you expected?

   You should expect a current of roughly 0.82 mA because R_{total} is now 11 kΩ and $V_{battery} / R_{total} = 9\text{ V} / 11\text{ k}\Omega = 0.82\text{ mA}$.

6. **Change the pot to some intermediate setting and observe the current.**

   With your multimeter still inserted in your circuit, dial the potentiometer to some position in the middle of its range. Don't worry about the exact value. What current do you measure now? Write it down.

7. **Measure the resistance of the pot.**

   Remove the pot from the circuit *without repositioning its dial*. Remove your multimeter from the circuit, switch it to measure resistance in ohms, and measure the resistance of the potentiometer between the wiper (middle terminal) and the fixed terminal that is not connected to the wiper. If Ohm's Law really works (which it should), you should find that the following equation is true:

$$I = \frac{V_{battery}}{R_{total}} = \frac{9\text{ V}}{1\text{ k}\Omega + R_{pot}}$$

   where I is the current you measured in Step 6 and R_{pot} is the resistance you measured across the potentiometer.

You can experiment as much as you want, varying the pot and measuring the current and the pot resistance to verify that Ohm's Law really does work.

What Is Ohm's Law Really Good For?

Ohm's Law also comes in handy when you're analyzing all kinds of circuits, whether simple or complex. You'll use it in designing and altering electronic circuits, to make sure you get the right current and voltage to the right places in your circuit. You'll use Ohm's Law so much, it will become second nature to you.

Analyzing complex circuits

Ohm's Law comes in handy when analyzing more complex circuits than the simple light bulb circuit discussed earlier. You often need to incorporate your knowledge of equivalent resistances to apply Ohm's Law and figure out exactly where current is flowing and how voltages are being dropped throughout your circuit.

Look at the series-parallel circuit shown in Figure 6-6. Say you need to know exactly how much current is flowing through each resistor in the circuit.

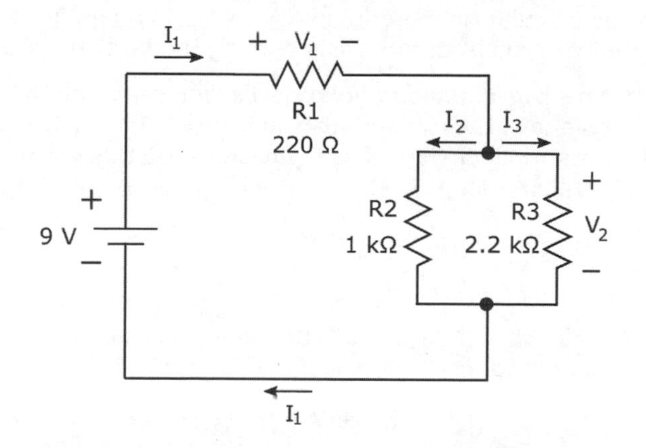

Figure 6-6: Analyze complex circuits by applying Ohm's Law and calculating equivalent resistances.

You can calculate the current running through each resistor, step by step, as follows:

1. **Calculate the equivalent resistance of the circuit.**

   You can find this value by applying the rules for resistors in parallel and resistors in series (refer to Chapter 5 for details), like this:

$$\begin{aligned} R_{equivalent} &= R1 + R2 \parallel R3 \\ &= R1 + \frac{R2 \times R3}{R2 + R3} \\ &= 220\ \Omega + \frac{1{,}000\ \Omega \times 2{,}200\ \Omega}{1{,}000\ \Omega + 2{,}200\ \Omega} \\ &\approx 220\ \Omega + 688\ \Omega \\ &\approx 908\ \Omega \end{aligned}$$

2. **Calculate the total current supplied by the battery.**

   Apply Ohm's Law, using the battery voltage and the equivalent resistance of the circuit:

$$I_1 = \frac{9\ \text{V}}{908\ \Omega} \approx 0.0099\ \text{A or } 9.9\ \text{mA}$$

3. **Calculate the voltage dropped across the parallel resistors.**

   You can do this calculation either of two ways, and you get pretty much the same result (the slight difference is due to rounding error):

   - *Apply Ohm's Law to the parallel resistors.* You calculate the equivalent resistance of the two resistors in parallel, and then multiply that by the supply current. The equivalent resistance is 688 Ω, as shown in the first step. So the voltage is

$$\begin{aligned} V_2 &= 0.0099\ \text{A} \times 688\ \Omega \\ &\approx 6.81\ \text{V} \end{aligned}$$

   - *Apply Ohm's Law to R1 (the 220 Ω resistor), and subtract its voltage from the supply voltage.* The voltage, V_1, across *R1* is

$$\begin{aligned} V_1 &= 0.0099\ \text{A} \times 220\ \Omega \\ &\approx 2.18\ \text{V} \end{aligned}$$

     So the voltage, V_2, across the parallel resistors is

$$\begin{aligned} V_2 &= V_{battery} - V_1 \\ &\approx 9\ \text{V} - 2.18\ \text{V} \\ &\approx 6.82\ \text{V} \end{aligned}$$

4. **Finally, calculate the current through each parallel resistor.**

   To get that result, you apply Ohm's Law to each resistor, using the voltage you just calculated (V_2). Here's what it looks like:

$$I_2 = \frac{6.82\ \text{V}}{1{,}000\ \Omega} = 0.0682\ \text{A} \approx 6.8\ \text{mA}$$

$$I_3 = \frac{6.82\ \text{V}}{2{,}200\ \Omega} = 0.0031\ \text{A} = 3.1\ \text{mA}$$

   Note that the two branch currents, I_2 and I_3, add up to the total supply current, I_1: 6.8 mA + 3.1 mA = 9.9 mA. That's a good thing (and a good way to check that you've performed your calculations correctly).

Designing and altering circuits

You can use Ohm's Law to determine what components to use in a circuit design. For instance, you may have a series circuit consisting of a 9 V power supply, a resistor, and an LED, as shown in Figure 6-7.

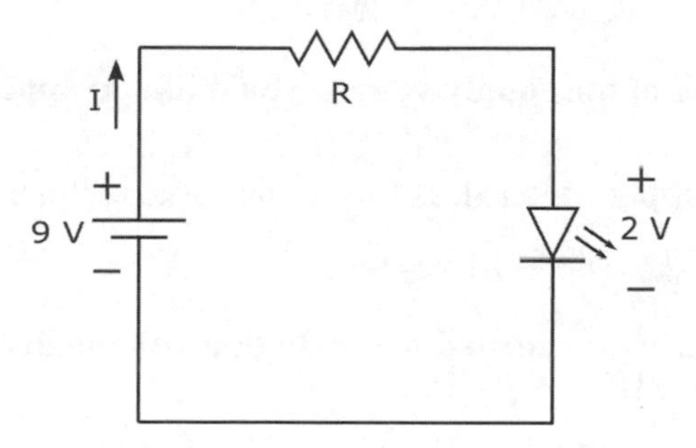

Figure 6-7: You can use Ohm's Law to determine the minimum resistance you need to protect an LED.

As you see in Chapter 9, the voltage drop across an LED remains constant for a certain range of current passing through it, but if you try to pass too much current through the LED, it will burn out. For example, suppose your LED voltage is 2 V and the maximum current it can handle is 20 mA. What resistance should you put in series with the LED to limit the current so it never exceeds 20 mA?

To figure this out, first you have to calculate the voltage drop across the resistor when the LED is on. You already know that the supply voltage is 9 V and the LED eats up 2 V. The only other component in the circuit is the resistor, so you know that it will eat up the remaining supply voltage — all 7 V of it. If you want to limit the current to be no more than 20 mA, you need a resistor that is *at least* 7 V ÷ 0.020 A = 350 Ω. Because it's hard to find a 350 Ω resistor, suppose you choose a commonly available resistor that is close to, but higher in value than, 350 Ω, such as a 390 Ω resistor. The current will be $7\text{ V} \div 390\ \Omega = 0.0179\text{ A}$, or about 18 mA. The LED may burn a little less brightly, but that's okay.

Ohm's Law also comes in handy when tweaking an existing circuit. Say your spouse is trying to sleep but you want to read, so you get out your big flashlight. The bulb in your flashlight has a resistance of 9 Ω and is powered by a 6 V battery, so you know that the current in the flashlight circuit is $6\text{ V} \div 9\ \Omega \approx 0.67\text{ A}$. Your spouse thinks the light is too bright, so to reduce the brightness (and save your marriage), you want to restrict the current flowing through the bulb a bit. You think that bringing it down to 0.45 A will do the trick, and you know that inserting a resistor in series between the battery and the bulb will restrict the current.

But what value of resistance do you need? You can use Ohm's Law to figure out the resistance value as follows:

1. **Using the desired new current, calculate the desired voltage drop across the bulb:**

   $$V_{bulb} = 0.45\text{ A} \times 9\ \Omega \approx 4.1\text{ V}$$

2. **Calculate the portion of the supply voltage you'd like to apply across the new resistor.**

   This voltage is the supply voltage less the voltage across the bulb:

   $$V_{resistor} = 6\text{ V} - 4.1\text{ V} = 1.9\text{ V}$$

3. **Calculate the resistor value needed to create that voltage drop given the desired new current:**

   $$R = \frac{1.9\text{ V}}{0.45\text{ A}} \approx 4.2\ \Omega$$

4. **Choose a resistor value that is close to the calculated value, and make sure it can handle the power dissipation:**

   As you see in the next section, to calculate the power dissipated in an electronic component, you multiply the voltage dropped across the component by the current passing through it. So the power that your 4.2 Ω resistor needs to be able to handle is

   $$P_{resistor} = 1.9\text{ V} \times 0.45\text{ A} \approx 0.9\text{ W}$$

   Result: Because you won't find a 4.2 Ω resistor, you can use a 4.7 Ω 1 W resistor to reduce the brightness of the light. Your spouse will sleep soundly; let's hope the snoring won't interfere with your reading!

The Power of Joule's Law

Another scientist hard at work in the early 1800s was the energetic James Prescott Joule. Joule is responsible for coming up with the equation that gives you power values; it's known as *Joule's Law:*

$$P = V \times I$$

This equation states that the power (in watts) equals the voltage (in volts) across a component times the current (in amps) passing through that component. The nice thing about this equation is that it applies to every electronic component, whether it's a resistor, a light bulb, a capacitor, or something else. It tells you the rate at which electrical energy is consumed by the component — what that power is.

Using Joule's Law to choose components

You've already seen how to use Joule's Law to ensure that a resistor is big enough to resist a meltdown in a circuit, but you should know that this equation comes in handy also when you're selecting other electronic parts.

Lamps, diodes (discussed in Chapter 9), and other components also come with maximum power ratings. If you expect them to perform at power levels higher than their ratings, you're going to be disappointed when too much power makes them pop and fizzle. When you select the part, you should consider the *maximum possible* power the part will need to handle in the circuit. You do this by determining the maximum current you'll be passing through the part and the voltage across the part, and then multiplying those quantities together. Then you choose a part with a power rating that exceeds that estimated maximum power.

Joule and Ohm: perfect together

You can get creative and combine Joule's Law and Ohm's Law to derive more useful equations to help you calculate power for resistive components in circuits. For instance, if you substitute $I \times R$ for V in Joule's Law, you get:

$$P = (I \times R) \times I = I^2 R$$

That gives you a way to calculate power if you know the current and resistance but not the voltage. Similarly, you can substitute V/R for I in Joule's Law to get

$$P = V \times \frac{V}{R} = \frac{V^2}{R}$$

Using that formula, you can calculate power if you know the voltage and the resistance but not the current.

Joule's Law and Ohm's Law are used in combination so often that Georg Ohm sometimes gets the credit for both laws!

7

Getting Charged Up about Capacitors

In This Chapter

- Storing electrical energy in capacitors
- Charging and discharging capacitors
- Saying "no" to DC and "yes" to AC
- Creating the dynamic duo: capacitors and resistors
- Blocking, filtering, smoothing, and delaying signals

If resistors are the most popular electronic component, capacitors are a close second. Skilled at storing electrical energy, capacitors are important contributors to all sorts of electronic circuits — and your life would be a lot duller without them.

Capacitors make it possible to change the *shape* (the pattern over time) of electrical signals carried by current — a task resistors alone cannot perform. Although they're not as straightforward to understand as resistors, capacitors are essential ingredients in many of the electronic and industrial systems you enjoy today, such as radio receivers, computer memory devices, and automobile airbag-deployment systems, so it's well worth investing your time and brain power in understanding how capacitors operate.

This chapter looks at what capacitors are made of, how they store electrical energy, and how circuits use that energy. You get to watch a capacitor charge, store energy, and later release its energy. Then you find out how capacitors work closely with resistors to perform useful functions. Finally, I showcase the various uses of capacitors in electronic circuits — proving beyond a shadow of a doubt that capacitors are worth getting charged up about.

Capacitors: Reservoirs for Electrical Energy

When you're thirsty for water, you can generally get a drink in two ways: by catching water that originated from a source and flows out of pipes when you turn on the tap, or by getting water from a storage container, such as a water cooler. You can think about electrical energy in a similar vein: You can get electrical energy directly from a source (such as a battery or a generator), or you can get it from a device that stores electrical energy — a capacitor.

Just as you fill a water cooler by connecting it to a water source, you fill a capacitor with electrical energy by connecting it to a source of electrical energy. And just as water stored in a water cooler remains even after the source is removed, so too the electrical energy stored in a capacitor remains even after the source is removed. In each case, the stuff (water or electrical energy) stored in the device stays there until something comes along and taps into it — whether it's a thirsty consumer or an electronic component in need of electrical energy.

A *capacitor* is a two-terminal electronic device that stores electrical energy transferred from a voltage source. (See Figure 7-1.) If you remove the voltage source and electrically isolate the capacitor (so that it isn't connected in a complete circuit), it holds onto the stored electrical energy. If you connect it to other components in a complete circuit, it will release some or all of the stored energy. A capacitor is made from two metal plates separated by an insulator, which is known as a *dielectric*.

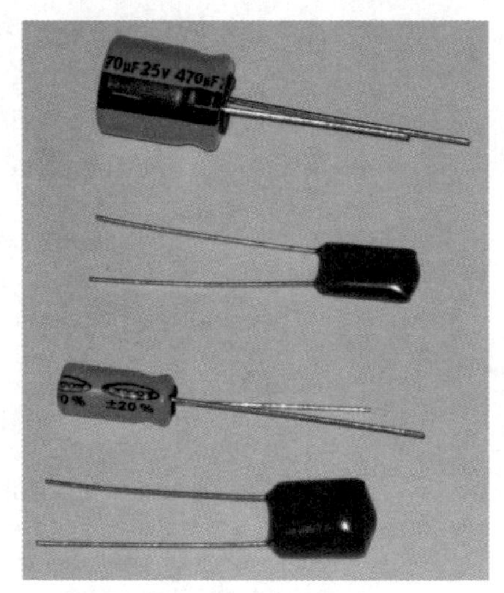

Figure 7-1: Capacitors come in a variety of shapes and sizes.

Capacitors and batteries: What's the difference?

Capacitors and batteries both store electrical energy, but in different ways. A battery uses electrochemical reactions to produce charged particles, which build up on its two metal terminals, creating a voltage. A capacitor doesn't produce charged particles, but it does allow charged particles to build up on its plates, creating a voltage across the plates (see the section "Charging and Discharging Capacitors"). A battery's electrical energy is the result of an energy conversion process originating from the chemicals stored inside the battery, whereas a capacitor's electrical energy is supplied by a source outside the capacitor.

Charging and Discharging Capacitors

If you supply a DC voltage to a circuit containing a capacitor in series with a light bulb (as shown in Figure 7-2), current flow cannot be sustained because there's no complete conductive path across the plates. In other words, capacitors block DC current. However, electrons do move around this little circuit — temporarily — in an interesting way.

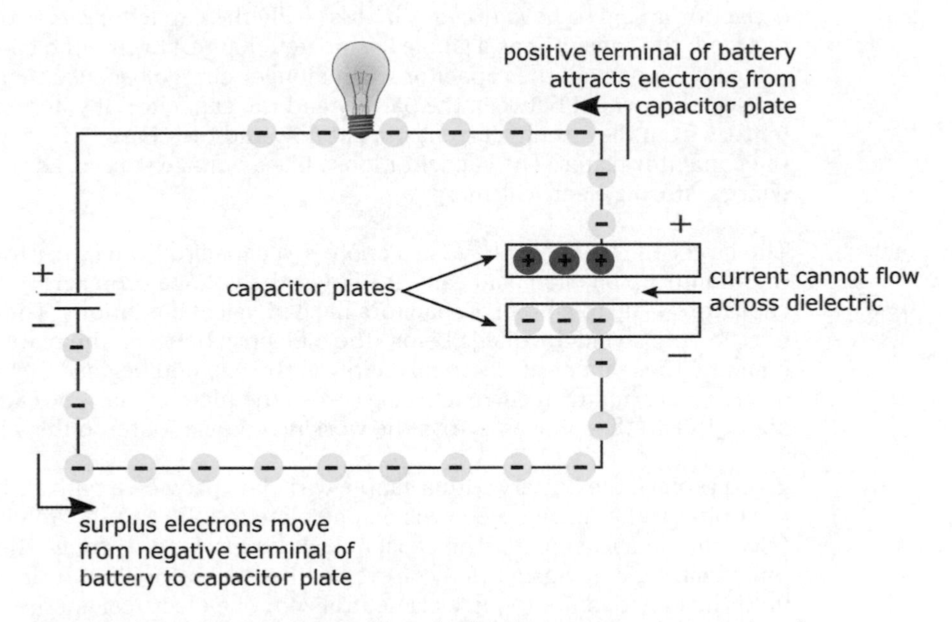

Figure 7-2: When a battery is placed in a circuit with a capacitor, the capacitor charges. A charged capacitor stores electrical energy, much like a battery.

Remember that the negative terminal of a battery has a surplus of electrons. So in the circuit shown in Figure 7-2, the surplus electrons begin to move away from the battery toward one side of the capacitor. After they reach the capacitor, they're stopped in their tracks, with no conductive path to follow across the capacitor. The result is an excess of electrons on that plate.

At the same time, the positive terminal of the battery attracts electrons from the other capacitor plate, so *they* begin to move. As they pass through the light bulb, they light it (but only for a split second, which I explain in the next paragraph). A net positive charge (due to a deficiency of electrons) is produced on that plate. With a net negative charge on one plate, and a net positive charge on the other plate, the result is a voltage difference across the two plates. This voltage difference represents the electrical energy stored in the capacitor.

The battery keeps pushing electrons onto one plate (and pulling electrons off the other plate) until the voltage drop across the capacitor plates is equal to the battery voltage. At this equilibrium point, there is no voltage differential between the battery and the capacitor, so there's no push for electrons to flow from the battery to the capacitor. The capacitor stops charging, and electrons stop moving through the circuit — and the light bulb goes out.

When the voltage drop across the plates is equal to the battery voltage, the capacitor is said to be *fully charged.* (It's really the capacitor *plates* that are charged; the capacitor as a whole has no net charge.) Even if the battery remains connected, the capacitor will no longer charge because there is no voltage differential between the battery and the capacitor. If you remove the battery from the circuit, current will not flow and the charge will remain on the capacitor plates. The capacitor looks like a voltage source, as it holds the charge, storing electrical energy.

The larger the battery voltage you apply to a capacitor, the larger the charge that builds up on each plate, and the larger the voltage drop across the capacitor — up to a point. Capacitors have physical limitations: They can handle only so much voltage before the dielectric between the plates is overcome by the amount of electrical energy in the cap and begins to give up electrons, resulting in current arcing across the plates. You can read more about this in "Keeping an eye on the working voltage," later in this chapter.

If you replace the battery with a simple wire, you provide a path through the bulb for the surplus electrons on one plate to follow to the other (electron-deficient) plate. The capacitor plates *discharge* through the light bulb, lighting it up again briefly — even without a battery in the circuit — until the charge on both plates is neutralized. The electrical energy that had been stored in the capacitor is consumed by the light bulb. When the capacitor is discharged (again, it's really the *plates* that discharge), no more current will flow.

A capacitor can store electrical energy for hours on end. You'd be wise to make sure a capacitor is discharged before handling it, lest it discharges through you. To discharge a capacitor, carefully place a small incandescent bulb across its terminals, using insulated alligator clips (refer to Chapter 2) to make the connection. If the bulb lights up, you know the capacitor was charged, and the light should dim and go out in a few seconds as the capacitor discharges. If you don't have a bulb handy, place a 1 MΩ 1 W resistor across the terminals and wait at least 30 seconds. (Refer to Chapter 16 for details.)

Watching a capacitor charge

The circuit shown in Figure 7-3 enables you to observe a capacitor charging and discharging right before your very eyes. The capacitor is symbolized by a straight line facing a curved line.

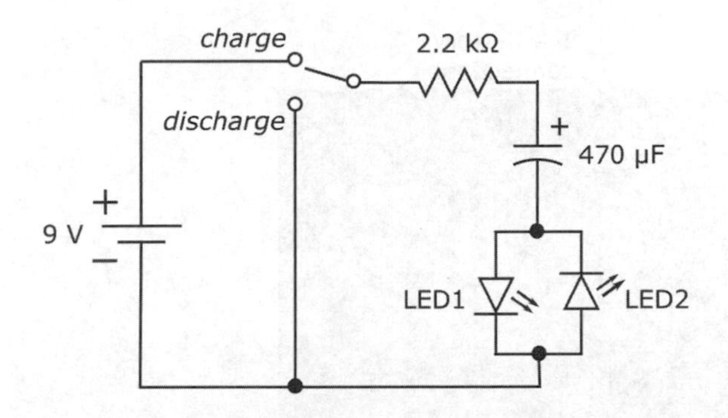

Figure 7-3: This circuit enables you to watch a capacitor charging and discharging.

You need these parts to build the circuit:

- One 9-volt battery with battery clip
- One 470 µF electrolytic capacitor
- One 2.2 kΩ 1/4 W resistor (red-red-red)
- Two light-emitting diodes (LEDs), any size, any colors you have on hand
- One single-pole, double-throw (SPDT) switch
- Two or three jumper wires
- One solderless breadboard

Chapter 2 tells you where to find the necessary parts. I walk you through the circuit setup in this section, but if you want to know more before you begin, refer to Chapter 9 for the lowdown on LEDs and Chapter 15 for detailed information about constructing circuits using a solderless breadboard.

Using Figure 7-4 as your guide, set up the circuit by following these steps:

1. **Insert the SPDT switch vertically across any three rows in an inner column of the left-of-center section of your breadboard.**

   In one position (labeled *charge* in Figure 7-3), this changeover switch will connect the 9 V battery to the circuit, allowing the capacitor to charge up. In the other position (labeled *discharge* in Figure 7-3), the switch will disconnect the battery and replace it with a wire, so that the capacitor can discharge.

2. **Insert a small jumper wire between the positive power rail and the lower end terminal of the SPDT switch.**

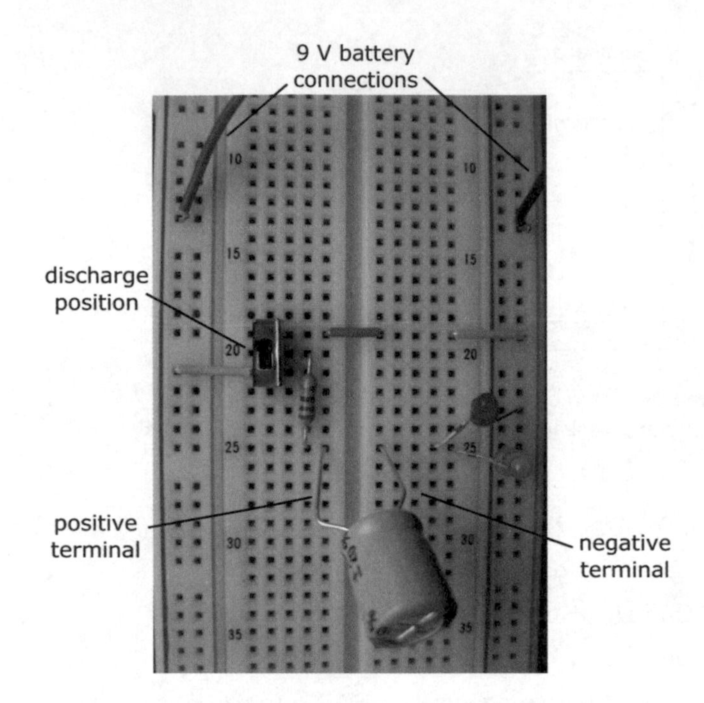

Figure 7-4: Setting up the capacitor charging/discharging circuit.

3. **Insert a jumper wire between the upper end terminal of the switch and the negative power rail.**

   I used two small jumper wires and internal breadboard connections to connect the switch to the negative power rail (refer to Figure 7-4), but you can use one longer wire to connect the switch terminal directly to the negative power rail.

4. **Position the switch slider towards the upper end terminal.**

   This discharge position is shown in Figure 7-4.

5. **Insert the 2.2 kΩ resistor into the breadboard.**

   Connect one side (either one will do) of the resistor to a hole in the same row as the center terminal of the SPDT switch. Connect the other side of the resistor to a hole in an open row (for neatness, make this connection in the same column as the connection for the first side of the resistor).

6. **Insert the 470 μF electrolytic capacitor into the breadboard.**

   A 470 μF electrolytic capacitor is polarized, so it matters which way you insert the capacitor. The capacitor should have a minus sign (–) or arrow indicating the negative terminal. If it doesn't, check the length of the leads; the shorter lead is on the negative side.

Connect the positive side to the side of the resistor not connected to the switch by inserting the cap lead into a hole in the same row as the resistor lead. Connect the negative side of the cap to a hole in the right-of-center section of the breadboard. Be sure to check the orientation of the capacitor because if you insert it the wrong way, you may damage — or even blow up — the capacitor.

7. **Insert the LEDs in the breadboard in parallel with each other but flipped in orientation.**

LEDs are polarized, so it matters which way you orient them. Current flows from the positive side to the negative side but not vice versa (refer to Chapter 9 for details). The negative side of an LED has a shorter lead.

Insert the positive side of LED1 in the same row as the negative side of the capacitor. Connect the negative side of LED1 to the negative power rail. Insert the negative side of LED2 into the same row as the negative cap lead and the positive side of LED2 into the negative power rail.

8. **Connect the 9-volt battery.**

Connect the negative battery terminal to the negative power rail. Connect the positive battery terminal to the positive power rail.

After you have the circuit set up, you are ready to observe the capacitor charging and discharging.

Keeping an eye on the LEDs, move the slider on the switch to the charge position (towards the lower end terminal). Did LED1 light? Did it stay lit? (It should light immediately, and then grow dimmer until it goes out, in roughly 5 seconds.)

Then, move the slider on the switch back to the discharge position (towards the upper end terminal). Did LED2 light, then grow dimmer and go out after about 5 seconds?

With the switch in the charge position, the battery is connected to the circuit and current flows as the capacitor charges. LED1 lights temporarily, as current flows through it, for as long as it takes the capacitor to charge. After the capacitor stops charging, current no longer flows and LED1 goes out. Note that LED2 is oriented "backwards" so current does not pass through it as the capacitor charges.

With the switch in the discharge position, the battery is replaced by a wire (or, rather, two jumper wires and an internal connection in the breadboard, which together are the equivalent of a single wire connection). The capacitor discharges and current flows through the resistor and LED2, which lights as long as the capacitor is discharging. After the capacitor stops discharging, current stops flowing and LED2 goes out.

You can flip the switch back and forth as much as you like to observe the LEDs lighting and dimming as the capacitor charges and discharges.

The resistor protects the LEDs by limiting the current that flows through them (refer to Chapter 5 for details). The resistor also slows down the capacitor's charging/discharging process so that you can observe the LEDs lighting. By carefully selecting the resistor and capacitor values, you can control the charge/discharge timing, as you see in the section titled "Teaming Up with Resistors," later in this chapter.

Opposing voltage change

Because it takes time for charges to build on the capacitor plates when a DC voltage is applied, and it takes time for the charge to leave the plates after the DC voltage is removed, capacitors are said to "oppose voltage change." This just means that if you suddenly change the voltage applied to a capacitor, it can't react right away; the voltage across the capacitor changes more slowly than the voltage you applied.

Think about being in your car, stopped at a red light. When the light turns green, you get your car moving again, building speed until you reach the speed limit. It takes time to get to that speed, just like it takes time for a capacitor to get to a certain voltage level. If you switch on a battery across a resistor, however, the voltage across the resistor changes almost instantaneously.

It takes some time for the capacitor voltage to "catch up" to the source voltage. That isn't a bad thing; many circuits use capacitors because it takes time to charge them. This is the crux of the reason why capacitors can change the shape (pattern) of electrical signals.

Giving alternating current a pass

Although capacitors cannot pass direct current (DC) — except temporarily, as you saw in the preceding section — because of the dielectric providing a barrier to electron flow, they can pass alternating current (AC).

Suppose you apply an AC voltage source across a capacitor. Remember that an AC voltage source varies up and down, rising from 0 volts to its peak voltage, then falling back through 0 volts and down to its negative peak voltage, then rising back through 0 volts to its peak voltage, and so on. Imagine being an atom on one of the capacitor plates and looking at the AC source terminal nearest you. Sometimes you feel a force pulling your electrons in one direction, and other times you feel a force pulling your electrons in the other direction. In each case, the strength of the force will vary over time as the source voltage varies in intensity. You and the other atoms on the capacitor plate will alternate between giving up electrons and receiving electrons as the source voltage swings up and down.

What's really happening is that as the AC source voltage rises from 0 volts to its peak voltage, the capacitor charges, just as it does when you apply a DC voltage. When the supply voltage is at its peak, the capacitor may or may not be fully charged (it depends on a bunch of factors, such as the size of the capacitor plates). Then the source voltage starts to decrease from its peak down to 0 volts. As it does, at some point, the source voltage becomes lower than the capacitor voltage. When this happens, the capacitor starts to discharge through the AC source. Then the source voltage reverses polarity and the capacitor discharges all the way. As the source voltage keeps heading down toward its negative peak voltage, charges start to build up *in reverse* on the capacitor plates: The plate that previously held more negative charges now holds positive charges, and the plate that previously held more positive charges now holds more negative charges. As the source voltage rises from its negative peak, the capacitor again discharges through the AC source, but in the direction opposite to that of its original discharge, and the cycle repeats. This continuous charge/discharge cycle can occur thousands — even millions — of times per second, as the capacitor tries to keep up, so to speak, with the ups and downs of the AC source.

Because the AC source is constantly changing direction, the capacitor goes through a continuous cycle of charging, discharging, and recharging. As a result, electrical charges move back and forth through the circuit, and even though virtually no current flows across the dielectric (except a small *leakage current*), the effect is the same as if current were flowing through the capacitor. These amazing capacitors are said to pass alternating current (AC) even though they block direct current (DC).

If you add a light bulb to your capacitor circuit powered by an AC voltage source, the bulb will light and *will stay lit* as long as the AC source is connected. Current alternates its direction through the bulb, but the bulb doesn't care which way current flows through it. (Not so for an LED, which cares very much which way current flows.) Although no current ever passes *through* the capacitor, the charging/discharging action of the capacitor creates the effect of current flowing back and forth through the circuit.

Discovering Uses for Capacitors

Capacitors are put to good use in most electronic circuits you encounter every day. The key capabilities of capacitors — storing electrical energy, blocking DC current, and varying opposition to current depending on applied frequency — are commonly exploited by circuit designers to set the stage for extremely useful functionality in electronic circuits.

Here are some of the ways capacitors are used in circuits:

- **Storing electrical energy:** Many devices use capacitors to store energy temporarily for later use. Uninterruptible power supplies (UPSs) and alarm clocks keep charged capacitors on hand in case there's a power failure. The energy stored in the capacitor is released the moment the charging circuit is disconnected (which it will be if the power goes out!).

  Cameras use capacitors for temporary storage of the energy used to generate the flash, and many electronic devices use capacitors to supply energy while the batteries are being changed. Car audio systems commonly use capacitors to supply energy when the amplifier needs more than the car's electrical system can give. Without a capacitor in your system, every time you hear a heavy bass note, your lights would dim!
- **Preventing DC current from passing between circuit stages:** When connected in series with a signal source (such as a microphone), capacitors block DC current but pass AC current. This is known as *capacitive coupling* or *AC coupling,* and when used this way, the capacitor is known as a *coupling capacitor.*

  Multistage audio systems commonly use this functionality between stages so that only the AC portion of the audio signal — the part that carries the encoded sound information — from one stage is passed to the next stage. Any DC current used to power components in a previous stage is removed before the audio signal is amplified.
- **Smoothing out voltage:** Power supplies that convert AC to DC, such as your cellphone charger, often take advantage of the fact that capacitors don't react quickly to sudden changes in voltage. These devices use large electrolytic capacitors to smooth out varying DC supplies. These *smoothing capacitors* keep the output voltage at a relatively constant level by discharging through the load when the DC supply falls below a certain level. This is a classic example of using a capacitor to store electrical energy until you need it: When the DC supply can't maintain the voltage, the capacitor gives up some of its stored energy to take up the slack.
- **Creating timers:** Because it takes time to charge and discharge a capacitor, capacitors are often used in timing circuits to create ticks and tocks when the voltage rises above or falls below a certain level. The timing of the ticking and tocking can be controlled through the selection of the capacitor and other circuit components. (For details, see the section "Teaming Up with Resistors," later in the chapter.)

- **Tuning in (or out) frequencies:** Capacitors are often used to help select or reject certain electrical signals — which are time-varying electric currents that carry encoded information — depending on their frequency.

  For instance, a tuning circuit in a radio-receiver system relies on capacitors and other components to allow the signal from just one radio station at a time to pass through to the amplifier stage, while blocking signals from all other radio stations. Each radio station is assigned a specific broadcast frequency, and it's the radio builder's job to design circuits that tune in target frequencies. Because capacitors behave differently for different signal frequencies, they are a key component of these tuning circuits. The net effect is a kind of electronic filtering.

Characterizing Capacitors

Capacitors can be built in lots of ways by using different materials for the plates and dielectric and by varying the size of the plates. The particular makeup of a capacitor determines its characteristics and influences its behavior in a circuit.

Defining capacitance

Capacitance is the capability of a body to store an electric charge. The same term — capacitance — is used to describe just how much charge a capacitor can store on either one of its plates. The higher the capacitance, the more charge the capacitor can store at any one time.

The capacitance of any given capacitor depends on three things: the surface area of the metal plates, the thickness of the dielectric between the plates, and the type of dielectric used (more about dielectrics later in this section).

You don't need to know how to calculate capacitance (and, yes, there is a scary-looking formula), because any capacitor worth its salt will come with a documented capacitance value. It just helps to understand that how much charge a capacitor's plates can hold depends on how the capacitor is made.

Capacitance is measured in units called *farads*. One farad (abbreviated F) is defined as the capacitance needed to get one amp of current to flow when the voltage changes at a rate of one volt per second. Don't worry about the details of the definition; just know that one farad is a humongous amount of capacitance. You're more likely to run across capacitors with much smaller capacitance values — hovering in the microfarad (μF) or picofarad (pF) range. A *microfarad* is a millionth of a farad, or 0.000001 farad, and a *picofarad* is a millionth of a millionth of a farad, or 0.000000000001 farad.

Here are some examples:

- A 10 µF capacitor is 10 millionths of a farad.
- A 1 µF capacitor is 1 millionth of a farad.
- A 100 pF capacitor is 100 millionths of a millionth of a farad, or you could say it is 100 millionths of a microfarad. Whew!

Larger capacitors (1 F or more) are used for system energy storage, and smaller capacitors are used in a variety of applications, as shown in Table 7-1.

Table 7-1 Capacitor Characteristics

Type	***Typical Range***	***Application***
Ceramic	1 pF to 2.2 µF	Filtering, bypass
Mica	1 pF to 1 µF	Timing, oscillator, precision circuits
Metalized foil	0.01 to 100 µF	DC blocking, power supply, filtering
Polyester (Mylar)	0.001 to 100 µF	Coupling, bypass
Polypropylene	100 pF to 50 µF	Switching power supply
Polystyrene	10 pF to 10 µF	Timing, tuning circuits
Tantalum (electrolytic)	0.001 to 1,000 µF	Bypass, coupling, DC blocking
Aluminum electrolytic	10 to 220,000 µF	Filtering, coupling, bypass, smoothing

Most capacitors are inexact beasts. The actual capacitance of the capacitor can vary quite a bit from the nominal capacitance. Manufacturing variations cause this problem; capacitor makers aren't just out to confuse you. Fortunately, the inexactness is seldom an issue in home-brewed circuits. Still, you need to know about these variations so that if a circuit calls for a higher-precision capacitor, you know what to buy. As with resistors, capacitors are rated by their tolerance, which is expressed as a percentage.

Keeping an eye on the working voltage

The *working voltage,* sometimes abbreviated as WV, is the highest voltage that the manufacturer recommends placing across a capacitor safely. If you exceed the working voltage, you may damage the dielectric, which could result in current arcing between the plates, like a lightning strike during a storm. You could short out your capacitor and allow all sorts of unwanted current to flow — and maybe even damage nearby components.

Capacitors designed for DC circuits are typically rated for a WV of no more than 16 V to 35 V. That's plenty for DC circuits that are powered by sources ranging from 3.3 V to 12 V. If you build circuits that use higher voltages, be sure to select a capacitor that has a WV of at least 10% to 15% more than the supply voltage in your circuit, just to be on the safe side.

Choosing the right dielectric for the job

Designers of electronic circuitry specify capacitors for a project by the dielectric material in them. Some materials are better in certain applications and are inappropriate for other applications. For instance, electrolytic capacitors can handle large currents but perform reliably only for signal frequencies of less than 100 kHz, so they are commonly used in audio amplifiers and power supply circuits. Mica capacitors, however, exhibit exceptional frequency characteristics and are often used in radio frequency (RF) transmitter circuits.

The most common dielectric materials are aluminum electrolytic, tantalum electrolytic, ceramic, mica, polypropylene, polyester (or Mylar), and polystyrene. If a circuit diagram calls for a capacitor of a certain type, be sure you get one that matches.

Refer to Table 7-1 for a list of the most common capacitor types, their typical value range, and common applications.

Sizing up capacitor packaging

Capacitors come in a variety of shapes and sizes (refer to Figure 7-1). Aluminum electrolytic and paper capacitors commonly come in a cylindrical shape. Tantalum electrolytic, ceramic, mica, and polystyrene capacitors have a more bulbous shape because they typically get dipped into an epoxy or plastic bath to form their outside skin. However, not all capacitors of any particular type (such as mica or Mylar) get manufactured the same way, so you can't always tell the component by its cover.

Your parts supplier may label capacitors according to the way their leads are arranged: axially or radially. Axial leads extend from each end of a cylindrically shaped capacitor, along its axis; radial leads extend from one end of a capacitor and are parallel to each other (until you bend them for use in a circuit).

If you go searching for capacitors inside your tablet or laptop, you may not recognize some of them. That's because many of the capacitors in your PC don't have any leads! *Surface-mount packages* for capacitors are extremely small and are designed to be soldered directly to printed circuit boards (PCBs). Since the 1980s, high-volume manufacturing processes have been using surface-mount technology (SMT) to connect capacitors and other components directly to the surface of PCBs, saving space and improving circuit performance.

Being positive about capacitor polarity

Some larger-value electrolytic capacitors (1 μF and up) are *polarized* — meaning that the positive terminal must be kept at a higher voltage than the negative terminal, so it matters which way you insert the capacitor into your circuit. Polarized capacitors are designed for use in DC circuits.

Many polarized capacitors sport a minus (–) sign or a large arrow pointing toward the negative terminal. For radial capacitors, the negative lead is often shorter than the positive lead.

If a capacitor is polarized, you *really, really* need to make sure to install it in the circuit with the proper orientation. If you reverse the leads, say, by connecting the + side to the ground rail in your circuit, you may cause the dielectric inside the capacitor to break down, which could effectively short-circuit the capacitor. Reversing the leads may damage other components in your circuit (by sending too much current their way), and your capacitor may even explode.

Reading into capacitor values

Some capacitors have their values printed directly on them, either in farads or portions of a farad. You usually find this to be the case with larger capacitors because there is enough real estate for printing the capacitance and working voltage.

Most smaller capacitors (such as 0.1 μF or 0.01 μF mica disc capacitors) use a three-digit marking system to indicate capacitance. Most folks find the numbering system easy to use. But there's a catch! (There's always a catch.) The system is based on *pico*farads, not microfarads. A number using this marking system, such as 103, means 10, followed by three zeros, as in 10,000, for a total of 10,000 picofarads. Some capacitors are printed with a two-digit number, which is simply its value in picofarads. For instance, a value of 22 means 22 picofarads. No third digit means no zeros to tag on to the end.

For values over 1,000 picofarads, your parts supplier will most likely list the capacitor in microfarads, even if the markings on it indicate picofarads. To convert the picofarad value on the capacitor into microfarads, just move the decimal point six places to the *left*. So a capacitor marked with a 103 (say, the example in the preceding paragraph) has a value of 10,000 pF or 0.01 μF.

Suppose you're building a circuit that calls for a 0.1 μF disc capacitor. You can convert microfarads into picofarads to figure out what marking to look for on the capacitor package. Just move the decimal point six places to the *right,* and you get 100,000 pF. Because the three-digit marking consists of the first two digits of your pF value (10) followed by the additional number of zeros (4), you'll need a mica disc capacitor labeled 104.

You can use Table 7-2 as a reference guide to common capacitor markings that use this numbering system.

Table 7-2 Capacitor Value Reference

Marking	*Value*
nn (a number from 01 to 99)	*nn* pF
101	100 pF
102	0.001 μF
103	0.01 μF
104	0.1 μF
221	220 pF
222	0.0022 μF
223	0.022 μF
224	0.22 μF
331	330 pF
332	0.0033 μF
333	0.033 μF
334	0.33 μF
471	470 pF
472	0.0047 μF
473	0.047 μF
474	0.47 μF

Another, less-often-used numbering system uses both numbers and letters, like this: 4R1. The placement of the letter R tells you the position of the decimal point: 4R1 is really 4.1. This numbering system doesn't indicate the units of measure, however, which can be either microfarads or picofarads.

You can test capacitance with a capacitor meter or a multimeter with a capacitance input (refer to Chapter 16 for details). Most meters require that you plug the capacitor directly into the test instrument because the capacitance can increase with longer leads, making the reading less accurate.

On many capacitors, a single-letter code indicates the tolerance. You may find that letter placed by itself on the body of the capacitor or placed after the three-digit mark, like this: 103Z.

Here the letter Z denotes a tolerance of +80% to –20%. This tolerance means that the capacitor, rated at 0.01 μF, may have an actual value that's as much as 80% higher or 20% lower than the stated value. Table 7-3 lists the meanings of common code letters used to indicate capacitor tolerance. Note that the letters B, C, and D represent tolerances expressed in absolute capacitance values, rather than percentages. These three letters are used on only very small (pF range) capacitors.

Table 7-3 Capacitor Tolerance Markings

Code	*Tolerance*
B	±0.1 pF
C	±0.25 pF
D	±0.5 pF
F	±1%
G	±2%
J	±5%
K	±10%
M	±20%
P	+100%, –0%
Z	+80%, –20%

Varying capacitance

Variable capacitors allow you to adjust capacitance to suit your circuit's needs. The most common type of variable capacitor is the *air dielectric,* which is found frequently in the tuning controls of AM radios. Smaller-variable capacitors are often used in radio receivers and transmitters, and they work in circuits that use quartz crystals to provide an accurate reference signal. The value of such variable capacitors typically falls in the 5 pF to 500 pF range.

Mechanically controlled variable capacitors work by changing the distance between the capacitor plates or by changing the amount of overlap between the surfaces of the two plates. A specially designed *diode* (a semiconductor device, discussed in Chapter 9) can function as an electronically controlled variable capacitor; such devices are known as *varactors* or *varicaps* — and you can change their capacitance by changing the DC voltage you apply to them.

Chances are, you interact with variable capacitors more than you would with a spouse. Variable capacitors are behind many touch-sensitive devices, such as certain smartphone screens, computer keyboard keys, control panels on many appliances and in some elevators, and the buttons on your favorite remote control. One type of microphone uses a variable capacitor to convert sound into electrical signals, with the diaphragm of the mic acting as a movable capacitor plate. Sound fluctuations make the diaphragm vibrate, which varies the capacitance, producing voltage fluctuations. This device is known as a *condenser microphone,* so-named because capacitors used to be called condensers.

Interpreting capacitor symbols

Figure 7-5 shows the circuit symbols for different types of capacitors. There are two commonly used capacitor circuit symbols: one with two parallel lines (representing the capacitor plates) and the other with one line and one curve. The curved side represents the more negative side of a polarized capacitor; however, some folks use the two types of symbols interchangeably. Usually, if a circuit has a polarized capacitor, you'll see a plus sign (+) on one side of the capacitor symbol, showing you how to orient the capacitor in the circuit. An arrow through either type of symbol represents a variable capacitor.

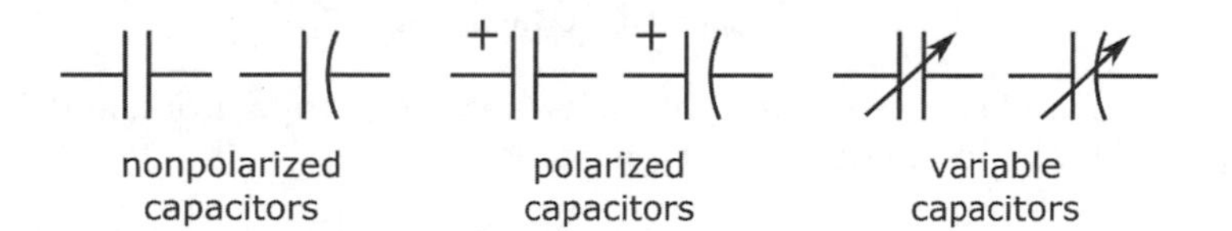

Figure 7-5: Circuit symbols for capacitors.

Combining Capacitors

You can combine capacitors in series, parallel, or a combination of both to get whatever value of capacitance you need. But as you'll see, the rules for combining capacitors are different from the rules for combining resistors.

Capacitors in parallel

Figure 7-6 shows two capacitors in parallel, with the common connection points labeled A and B. Note that point A is connected to one plate of capacitor C1 and one plate of capacitor C2. Electrically speaking, point A is connected to a metal plate that is the size of the two plates combined. Likewise for point B, which is connected to both the other plate of capacitor C1 and the other plate of capacitor C2. The larger the surface area of a capacitor plate, the larger the capacitance.

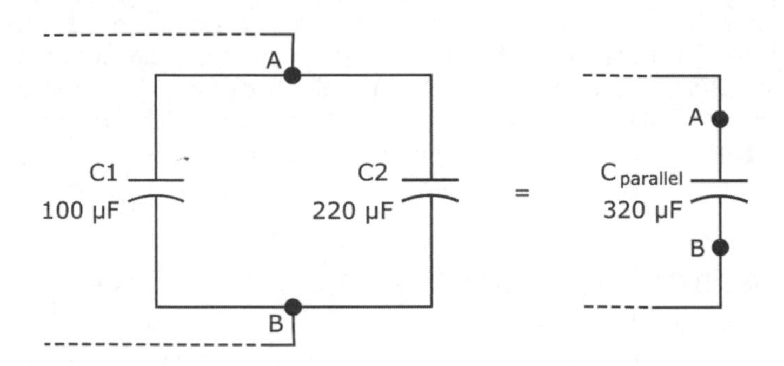

Figure 7-6: Capacitors in parallel add up.

Capacitors in parallel add up: Each metal plate of one capacitor is tied electrically to one metal plate of the parallel capacitor. Each pair of plates behaves as a single larger plate with a higher capacitance (refer to Figure 7-6).

The equivalent capacitance of a set of capacitors in parallel is

$$C_{parallel} = C1 + C2 + C3 + C4 + \ldots$$

C1, *C2*, *C3*, and so forth represent the values of the capacitors, and $C_{parallel}$ represents the total equivalent capacitance.

For the capacitors in Figure 7-6, the total capacitance is

$$\begin{aligned} C_{parallel} &= 100\ \mu F + 220\ \mu F \\ &= 320\ \mu F \end{aligned}$$

If you place the capacitors in Figure 7-6 in a working circuit, the voltage across each capacitor would be the same, and current flowing in to point A would split up to travel through each capacitor and then join together again at point B.

Capacitors in series

Capacitors placed in series work against each other, reducing the effective capacitance in the same way that resistors in parallel reduce the overall resistance. The calculation looks like this:

$$C_{series} = \frac{C1 \times C2}{C1 + C2}$$

C1 and *C2* are the values of the individual capacitors, and C_{series} is the equivalent capacitance. The total capacitance (in µF) of a 100 µF capacitor in series with a 220 µF capacitor, as shown in Figure 7-7, is

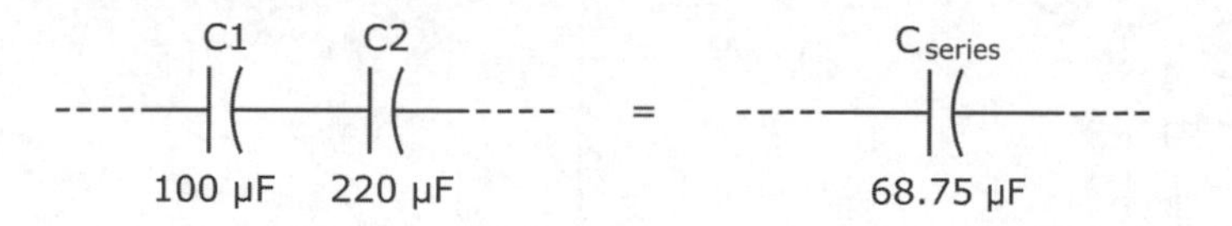

Figure 7-7: Capacitors in series work against each other, reducing overall capacitance.

$$\begin{aligned} C_{series} &= \frac{100 \times 220}{100 + 220} \\ &= \frac{22{,}000}{320} \\ &= 68.75 \\ C_{series} &= 68.75\ \mu\text{F} \end{aligned}$$

TIP

You can temporarily ignore the µ part of µF while you're performing the calculation just shown — as long as all the capacitance values are in µF and you remember that the resulting *total* capacitance is also in µF.

The equivalent capacitance of a set of capacitors in series is

$$C_{series} = \frac{1}{\frac{1}{C1} + \frac{1}{C2} + \frac{1}{C3} + \ldots}$$

As for any components in series, the current running through each capacitor in series is the same, but the voltage dropped across each capacitor may be different.

Teaming Up with Resistors

Capacitors are often found working hand in hand with resistors in electronic circuits, combining their talent for storing electrical energy with a resistor's control of electron flow. Put these two capabilities together and you can control how fast electrons fill (or charge) a capacitor — and how fast those electrons empty out (or discharge) from a capacitor. This dynamic duo is so popular that circuits containing both resistors and capacitors are known by a handy nickname: *RC circuits.*

Timing is everything

Take a look at the RC circuit in Figure 7-8. The battery will charge the capacitor through the resistor when the switch is closed.

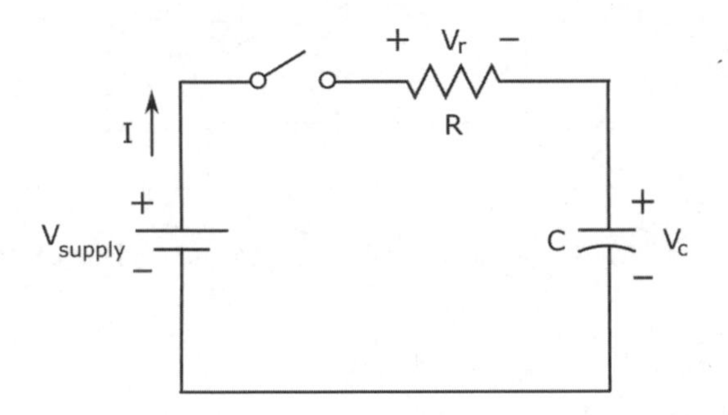

Figure 7-8: The capacitor charges until its voltage equals the supply voltage.

Assuming the capacitor was discharged to begin with, the initial voltage across the capacitor, V_c, is zero. When you close the switch, current starts to flow and charges start to build up on the capacitor plates. Ohm's Law (refer to Chapter 6) tells you that the charging current, *I*, is determined by the voltage across the resistor, V_r, and the value of the resistor, *R*, as follows:

$$I = \frac{V_r}{R}$$

And because the voltage drops equal the voltage rises around the circuit, you know that the resistor voltage is the difference between the supply voltage, V_{supply}, and the capacitor voltage, V_c:

$$V_r = V_{supply} - V_c$$

Using those two facts, you can analyze what is going on in this circuit over time, as follows:

- **Initially:** Because the capacitor voltage is initially zero, the resistor voltage is initially equal to the supply voltage.

- **Charging:** As the capacitor begins to charge, it develops a voltage, so the resistor voltage begins to fall, which in turn reduces the charging current. The capacitor continues to charge, but at a slower rate because the charging current has decreased. As V_c continues to increase, V_r continues to decrease, so the current continues to decrease.
- **Fully charged:** When the capacitor is fully charged, current stops flowing, the voltage drop across the resistor is zero, and the voltage drop across the capacitor is equal to the supply voltage.

If you remove the battery and replace it with a wire so that the circuit consists of just the resistor and the capacitor, the capacitor will discharge through the resistor. This time, the voltage across the resistor is equal to the voltage across the capacitor ($V_r = V_c$), so the current is V_c/R. Here's what happens:

- **Initially:** Because the capacitor is fully charged, its voltage is initially V_{supply}. Because $V_r = V_c$, the resistor voltage is initially V_{supply}, so the current jumps up immediately to V_{supply}/R. This means the capacitor is shuffling charges from one plate to the other pretty quickly.
- **Discharging:** As charges begin to flow from one capacitor plate to the other, the capacitor voltage (and so V_r) starts to drop, resulting in a lower current. The capacitor continues to discharge, but at a slower rate. As V_c (and so V_r) continues to decrease, so does the current.
- **Fully discharged:** When the capacitor is fully discharged, current stops flowing, and no voltage is dropped across either the resistor or the capacitor.

The waveform in Figure 7-9 shows how, when a constant voltage is applied and then removed from the circuit, the capacitor voltage changes over time as the capacitor charges and discharges through the resistor.

How fast the capacitor charges (and discharges) depends on the resistance and capacitance of the RC circuit. The larger the resistance, the smaller the current that flows for the same supply voltage — and the longer it takes the capacitor to charge. A smaller resistance allows more current to flow, charging the capacitor faster.

Likewise, the larger the capacitance, the more charges it takes to fill the capacitor plates, so the longer it takes to charge the capacitor. During the discharge cycle, a larger resistor slows down the electrons more as they move from one plate to the other, increasing the discharge time, and a larger capacitor holds more charge, taking longer to discharge.

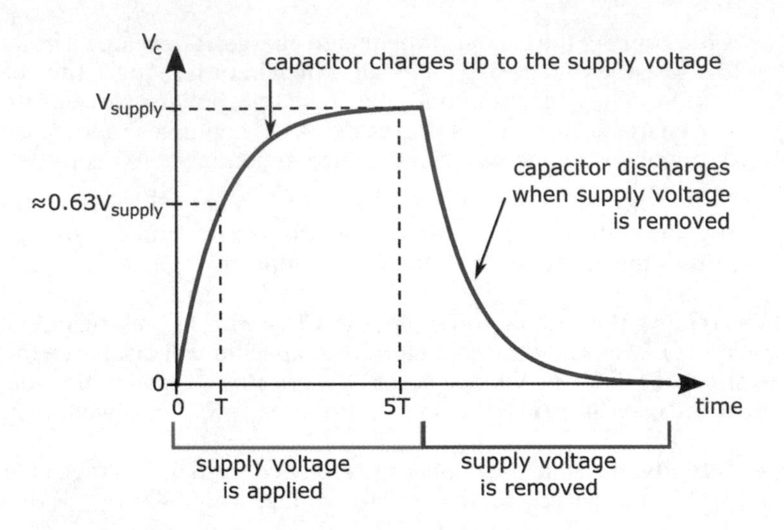

Figure 7-9: The voltage across a capacitor changes over time as the capacitor charges and discharges.

Calculating RC time constants

By choosing the values of the capacitor and the resistor carefully, you can adjust a capacitor's charge and discharge time. As it turns out, your choice of resistance, *R*, and capacitance, *C*, *defines* the time it takes to charge and discharge your chosen capacitor through your chosen resistor. If you multiply *R* (in ohms) by *C* (in farads), you get what is known as the *RC time constant* of your RC circuit, symbolized by T. And that makes another handy formula:

$$\mathrm{T} = R \times C$$

A capacitor charges and discharges almost completely after five times its RC time constant, or 5*RC* (which means 5×*R*×*C*). After time has passed that's equivalent to one time constant, a discharged capacitor will charge to roughly two-thirds its capacity — and a charged capacitor will discharge nearly two-thirds of the way.

For instance, suppose you choose a 2 MΩ resistor and a 15 μF capacitor for the circuit in Figure 7-7. You calculate the RC time constant as follows:

$$\begin{aligned}\text{RC time constant} &= R \times C \\ &= 2{,}000{,}000\ \Omega \times 0.000015\ \text{F} \\ &= 30\ \text{seconds}\end{aligned}$$

Now you know that it will take about 150 seconds (or 2½ minutes) to fully charge or discharge the capacitor. If you'd like a shorter charge/discharge cycle time, you can reduce the value you choose for the resistor or the capacitor (or both). Suppose that you have only a 15 μF capacitor, and you want to charge it in five seconds. You can figure out what resistor you need to make this happen as follows:

- **Find the RC time constant:** You know that it takes five times the RC time constant to fully charge the capacitor, and you want to fully charge your capacitor in five seconds. That means that $5RC = 5$ seconds, so $R \times C = 1$ second.
- **Calculate *R*:** If $R \times C = 1$ second, and C is 15 μF, then you know that $R = 1 \text{ second} \div 0.000015 \text{ F}$, which is approximately 66,667 Ω or 67 kΩ.

Varying the RC time constant

To see that you really can control the time it takes to charge and discharge a capacitor, you can build the circuit shown in Figure 7-10 and use your multimeter to observe the voltage changes across the capacitor. You can also observe the capacitor holding its charge (that is, storing electrical energy).

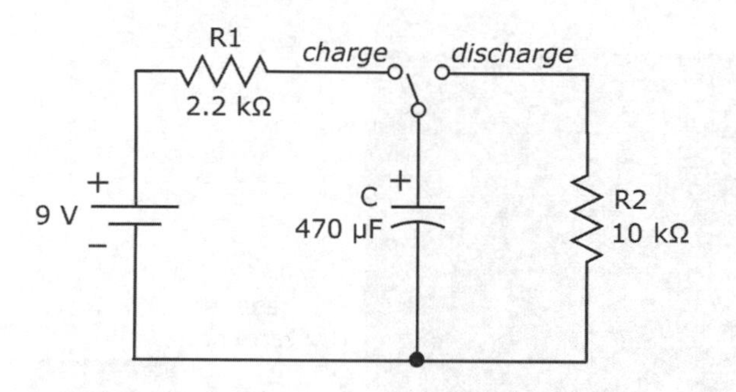

Figure 7-10: By choosing different resistance values, you can alter the charge time and discharge time of a capacitor.

The circuit in Figure 7-10 is really two circuits in one. The changeover switch alternates between positions labeled *charge* and *discharge,* creating two circuit options:

- **Charging circuit:** When the switch is in the charge position, the circuit consists of the battery, resistor *R1*, and the capacitor, *C*. Resistor *R2* is not connected to the circuit.
- **Discharging circuit:** When the switch is in the discharge position, the capacitor is connected to resistor *R2* in a complete circuit. The battery and resistor *R1* are disconnected from the circuit (they are open).

To build this circuit, you need the following parts:

- One 9-volt battery with battery clip
- One 470 μF electrolytic capacitor
- One 2.2 kΩ resistor (red-red-red)
- One 10 kΩ resistor (brown-black-orange)
- One jumper wire (which will play the role of the changeover switch)
- One solderless breadboard

Set up the circuit using Figure 7-11 as your guide. Make sure that you place one end of the jumper wire into your breadboard so that it's electrically connected to the positive side of the capacitor. Then you can use the other end to connect the capacitor to *R1* (to complete the charging circuit) or to *R2* (to complete the discharging circuit). You can also leave the other end of the jumper wire unconnected (as it is in Figure 7-11), which I suggest you do later in this section. You'll see why soon.

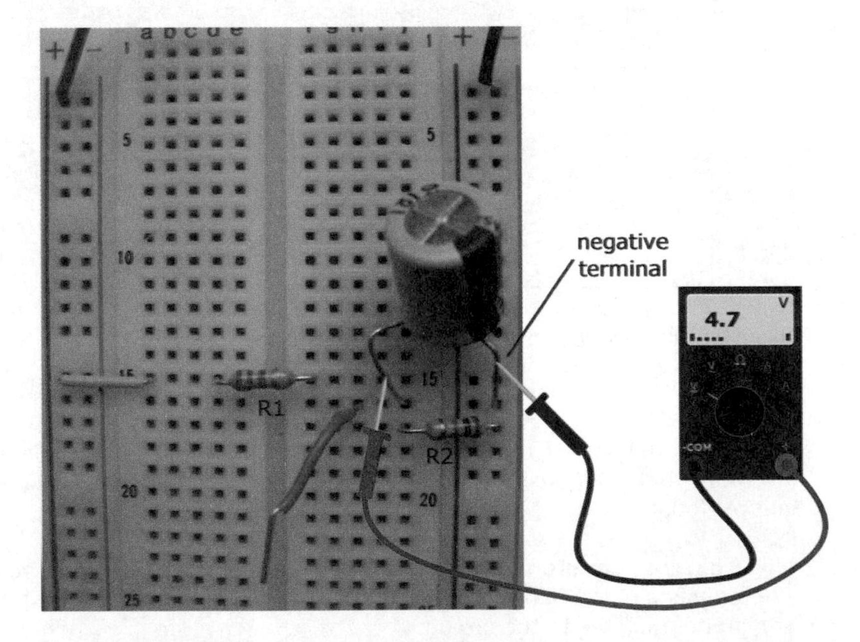

Figure 7-11: Circuit setup for observing different RC time constants for charging and discharging a capacitor.

To watch the capacitor charge, hold its charge, and then discharge, follow these steps:

1. **Set your multimeter to DC volts with a range of 10 V and connect it across the capacitor.**

   Connect the red multimeter lead to the positive side of the cap and the black multimeter lead to the negative side of the cap (see Figure 7-11).

2. **Charge the capacitor.**

   Connect the changeover switch to the charge position by placing the free end of the jumper wire into a hole in the same row as the unconnected side of *R1*. Observe the voltage reading on your meter. You should see the meter reading rise to approximately 9 V as the capacitor charges through resistor *R1*. It should take about 5 seconds for the capacitor to charge completely.

3. **Put the capacitor in a holding pattern.**

   Remove the end of the jumper wire that you connected in Step 2 and just let it hang. Observe the voltage reading on your meter. Your meter should continue to read 9 V or thereabouts. (You may see the reading decrease a tiny bit, as the capacitor discharges slowly through the internal resistance of your meter.) The capacitor is holding its charge (really, storing electrical energy), even without the battery connected.

4. **Let the capacitor discharge.**

   Connect the changeover switch to the discharge position by placing the free end of the jumper wire into a hole in the same row as the unconnected side of *R2*. Observe the voltage reading on your meter. You should see the reading decrease, as the capacitor discharges through resistor *R2* to 0 V. It should take roughly 25 seconds for the capacitor to fully discharge.

In the preceding section ("Calculating RC time constants"), you find out that a capacitor in a simple RC circuit reaches nearly its full charge at approximately five times the RC time constant, T. T is simply the value of the resistance (in ohms) times the value of the capacitance (in farads). So you can calculate the time it takes to charge and discharge the capacitor in your circuit as follows:

$$\begin{aligned}\text{Charge time} &= 5 \times R1 \times C \\ &= 5 \times 2,200\ \Omega \times 0.000470\ \text{F} \\ &= 5.17\ \text{seconds}\end{aligned}$$

$$\begin{aligned}\text{Discharge time} &= 5 \times R2 \times C \\ &= 5 \times 10,000\ \Omega \times 0.000470\ \text{F} \\ &= 23.5\ \text{seconds}\end{aligned}$$

Do those charge/discharge times seem close to what you observed? Repeat the charging and discharging experiment using a timer, and see if your calculations seem about right.

If you want to explore further, try replacing *R1* and *R2* with different resistances or use a different capacitor. Then calculate the charge/discharge times you expect to see and measure the actual times as you charge and discharge the capacitor. You'll find that the RC time constant doesn't disappoint!

8

Identifying with Inductors

In This Chapter

- Inducing currents in coils with a changing magnetic field
- Opposing changes in current with an inductor
- Using inductors in filter circuits
- Resonating with RLC circuits
- Making frequencies crystal-clear
- Coupling magnetic flux to transfer energy between circuits

Many of the best inventions in the world, including penicillin, sticky notes, champagne, and the pacemaker, were the result of pure accidental discovery. One such serendipitous discovery — the interaction between electricity and magnetism — led to the development of two amazingly useful electronic components: the induction coil and the transformer.

The *induction coil,* or *inductor,* stores electrical energy in a magnetic field and shapes electrical signals in a different way than does a capacitor. Whether operating alone, in special pairs known as *transformers,* or as part of a team along with capacitors and resistors, inductors are at the heart of many modern-day conveniences you might not want to live without, including radio systems, television, and the electric power transmission network.

This chapter exposes the relationship between electricity and magnetism and explains how 19th-century scientists purposely exploited that relationship to create inductors and transformers. You get a look at what happens when you try to change the direction of current through an inductor too quickly. Then you explore how inductors are used in circuits and why crystals ring at just one frequency. Finally, you get a handle on how transformers transfer electrical energy from one circuit to another — without any direct contact between the circuits.

Kissing Cousins: Magnetism and Electricity

Magnetism and electricity were once thought to be two separate phenomena, until a 19th-century scientist named Hans Christian Ørsted discovered that a compass needle moved away from magnetic north when current supplied by a nearby battery was switched on and off. Ørsted's keen observation led to lots of research and experimentation, ultimately confirming the fact that electricity and magnetism are closely related. After several years (and many more accidental discoveries), Michael Faraday and other 19th-century scientists figured out how to capitalize on the phenomenon known as *electromagnetism* to create the world's first electromechanical devices. Today's power transformers, electromagnetic generators, and many industrial motors are based on the principles of electromagnetism.

This section looks at how electricity and magnetism interact.

Drawing the (flux) lines with magnets

Just as electricity involves a force (voltage) between two electrical charges, magnetism involves a force between two magnetic poles. If you've ever performed the classic grade-school science experiment in which you place a magnet on a surface and toss a bunch of iron filings near the magnet, you've seen the effects of magnetic force. Remember what happened to the filings? They settled into curved linear paths from the north pole of the magnet to its south pole. Those filings showed you the magnetic lines of force — also known as *flux lines* — within the magnetic field created by the magnet. You may have seen more filings closer to the magnet because that's where the magnetic field is strongest. Figure 8-1 shows the pattern produced by invisible lines of flux around a magnet.

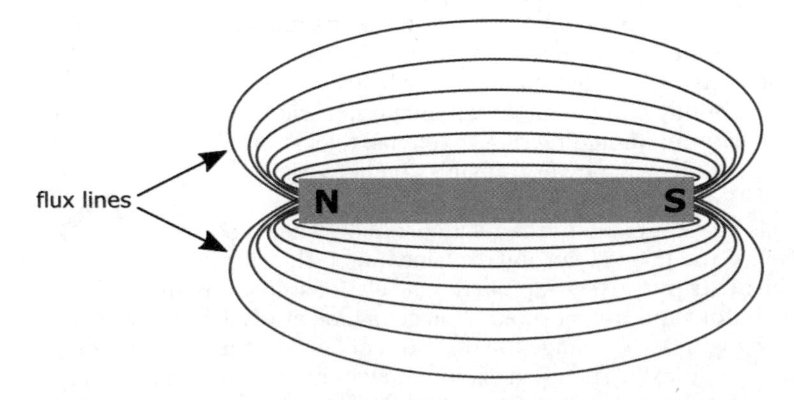

Figure 8-1: Magnetic lines of force exist in parallel flux lines from a magnet's north pole to its south pole.

Magnetic flux is just a way to represent the strength and direction of a magnetic field. To understand magnetic lines of flux, think about the effects of air on a sailboat's sail. The stronger the wind and the larger the sail, the greater the force of air on the sail. But if the sail is oriented parallel to the direction of wind flow, air slips by the sail and even a strong wind will not move the sail. The effect of the wind is greatest when it hits the sail head-on — that is, when the surface of the sail is perpendicular to the direction of wind flow. If you try to represent the strength and direction of the wind and the orientation of the sail in a diagram, you might draw arrows showing the force of the wind extending through the surface of the sail. Likewise, lines of magnetic flux illustrate the strength and orientation of a magnetic field, showing you how the force of the magnetic field will act on an object placed within the field. Objects placed in the magnetic field will be maximally affected by the force of the magnetic field if they are oriented perpendicular to the flux lines.

Producing a magnetic field with electricity

As Ørsted discovered, electrical current running through a wire produces a weak magnetic field surrounding the wire. This is why the compass needle moved when the compass was close to Ørsted's circuit. Stop the current from flowing, and the magnetic field disappears. This temporary magnet is electronically controllable — that is, you can turn the magnet on and off by switching current on and off — and it's known as an *electromagnet*.

With the current on, the lines of force encircle the wire and are spaced evenly along the length of the wire, as shown in Figure 8-2. Picture a roll of paper towels with a wire running through its exact center. If you pass current through the wire, invisible flux lines will wrap around the wire along the surface of the roll, and along similar rings around the wire at various distances from the wire. The strength of the magnetic force decreases as the flux lines get farther away from the wire. If you wind the current-carrying wire into a uniform coil of wire, the flux lines align and reinforce each other: You've strengthened the magnetic field.

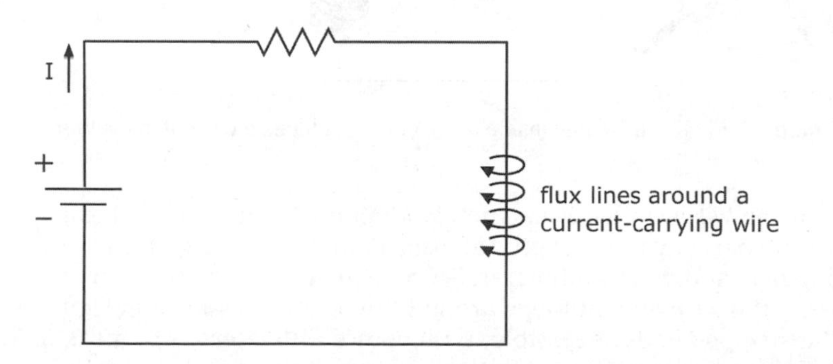

Figure 8-2: Current flowing through a wire produces a weak magnetic field around the wire.

Inducing current with a magnet

Hmmm . . . if electricity running through a wire produces a magnetic field, what happens if you place a closed loop of wire near a permanent magnet? Actually, nothing happens — unless you move the magnet. A moving magnetic field will *induce* a voltage across the ends of the wire, causing current to flow through the wire. *Electromagnetic induction* seems to make current magically appear — without any direct contact with the wire. The strength of the current depends on a lot of things, such as the strength of the magnet, the number of flux lines intercepted by the wire, the angle at which the wire cuts across flux lines, and the speed of the magnet's motion. You can increase your chances of inducing a strong current by wrapping the wire into a coil and placing the magnet through the center *(core)* of the coil. The more turns of wire you wrap, the stronger the current will be.

Suppose you place a strong permanent magnet in the center of a coil of wire connected as in Figure 8-3. (Note that the multimeter and the wire form a complete path.) If you move the magnet up, current is induced in the wire and flows in one direction. If you move the magnet down, current is also induced, but it flows in the other direction. By moving the magnet up and down repeatedly, you can produce an alternating current (AC) in the wire. Alternatively, you can move the *wire* up and down around the magnet, and the same thing will happen. As long as there is relative motion between the wire and the magnet, current will be induced in the wire.

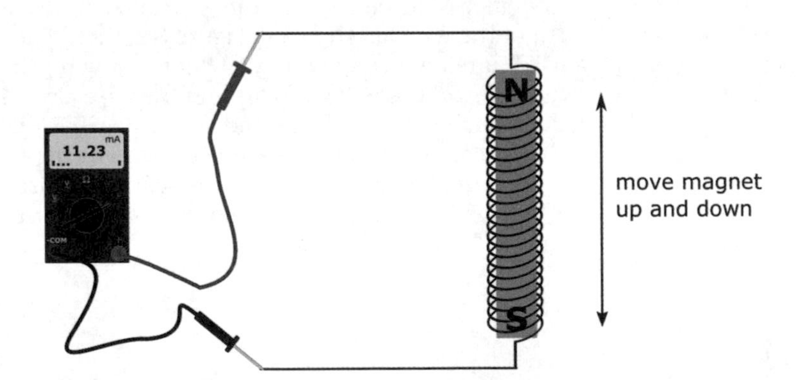

Figure 8-3: Moving a magnet inside a coil of wire induces a current in the wire.

You can witness electromagnetic induction firsthand by using a relatively strong bar magnet, a length (at least 12 inches) of 22 gauge (or thinner) wire, a pencil, and your multimeter. Strip the insulation off the ends of the wire and wrap the wire in tight loops around the pencil, as shown in Figure 8-4, leaving the stripped ends accessible. With your multimeter set to measure milliamps, connect your meter leads to the wire. You've completed a path for current to flow, but because there is no power source, no current flows. Next, place the magnet right next to the wire. Still no current, right? Finally, move the magnet back and forth along the wire and observe the multimeter display. You should see a tiny current reading (mine was a few hundredths of a milliamp) alternating between positive and negative.

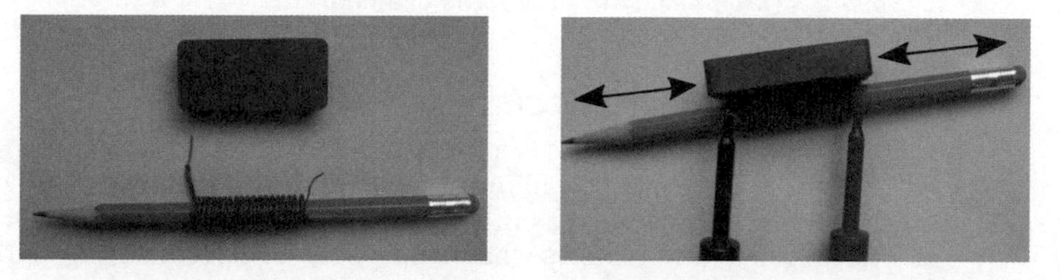

Figure 8-4: Moving a magnet near a coil of wire induces a current in the wire.

Many power plants generate AC by rotating conductive coils inside a strong horseshoe-shaped magnet. The coils are attached to a rotating turbine, which turns as water or steam applies pressure to its fins. As the coils make one full rotation inside the magnet, the magnet pulls electrons first in one direction and then in the other direction, producing alternating current.

Introducing the Inductor: A Coil with a Magnetic Personality

An *inductor* is a passive electronic component made from a coil of wire wrapped around a core — which could be air, iron, or ferrite (a brittle material made from iron). Iron-based core materials increase the strength of the magnetic field induced by current several hundred times. Inductors are sometimes known as *coils, chokes, electromagnets,* and *solenoids,* depending on how they're used in circuits. The circuit symbol for an inductor is shown in Figure 8-5.

Figure 8-5: Circuit symbol for an inductor.

If you pass current through an inductor, you create a magnetic field around the wire. If you *change* the current, increasing it or decreasing it, the magnetic flux around the coil changes, and a voltage is induced across the inductor. That voltage, sometimes called *back voltage,* causes a current flow that opposes the main current. This property of inductors is known as *self-inductance,* or simply *inductance.*

Measuring inductance

Inductance, symbolized by L, is measured in units called *henrys* (named for Joseph Henry, a New Yorker who liked to play with magnets and discovered the property of self-inductance). An inductance of one henry (abbreviated H) will induce one volt when the current changes its rate of flow by one ampere per second. One henry is much too large for everyday electronics, so you're more likely to hear about millihenrys (mH), because inductance measured in *thousandths* of a henry is more commonplace. You'll also run across microhenrys (μH), which are millionths of a henry.

Opposing current changes

In Figure 8-6, a DC voltage is applied to a resistor in series with an inductor. If there were no inductor in the circuit, a current equal to V_{supply}/R would flow instantaneously as soon as the DC voltage was switched on. However, introducing an inductor affects what happens to the current flowing in the circuit.

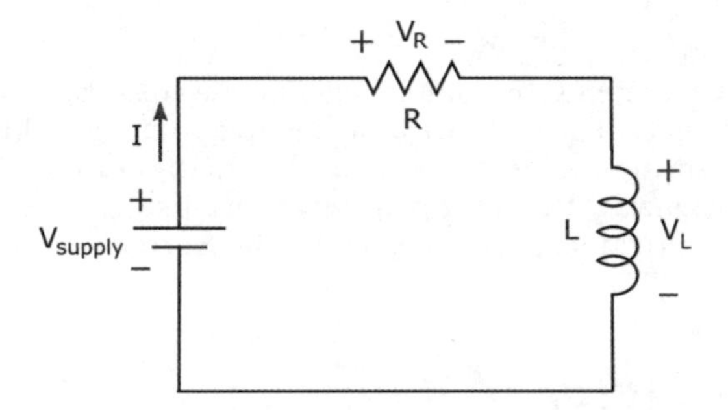

Figure 8-6: An inductor delays changes in current.

When the DC voltage is first switched on, the current that starts to flow induces a magnetic field around the coils of the inductor. As the current increases (which it's trying to do instantaneously), the strength of the magnetic field increases proportionally. Because the magnetic field is changing, it induces a back voltage that in turn induces a current in the coiled wire *in the opposite direction* to that of the current already flowing from the voltage source. The inductor seems to be trying to prevent the source current from changing too quickly; the effect is that the current doesn't increase instantaneously. This is why inductors are said to oppose changes in current.

The current induced in the coil reduces the strength of the expanding magnetic field a bit. As the source current keeps rising, the magnetic field continues to expand (but more and more slowly), and current opposing the source current continues to be induced (but it gets smaller and smaller). The cycle continues, until finally the overall current settles down to a steady DC. When current reaches a steady level, the magnetic field no longer changes — and the inductor ceases to affect the current in the circuit.

The overall effect is that it takes a finite amount of time for the current flowing through the inductor to reach a steady DC value. (The specific amount of time it takes depends on a few things, such as the characteristics of the inductor and the size of the resistor in the circuit. See the next section, "Calculating the RL time constant.") When this happens, the current flows freely through the inductor, which acts like a simple wire (commonly referred to as a *short circuit,* or simply a *short*), so V_L is nearly 0 V (a tiny voltage is dropped across the inductor due to the resistance of the coil's wire) and the steady-state current is determined by the source voltage and the resistor according to Ohm's Law $\left(I = \frac{V_{supply}}{R}\right)$.

If you then remove the DC voltage source and connect the resistor across the inductor, current will flow for a short period of time, with the inductor again opposing the sudden drop in current, until finally the current settles down to zero and the magnetic field disappears.

From an energy perspective, when you apply a DC source to an inductor, it stores electrical energy in a magnetic field. When you remove the DC source and connect a resistor across the inductor, the energy is transferred to the resistor, where it dissipates as heat. Inductors store electrical energy in magnetic fields. A real inductor — as opposed to a theoretical ideal inductor — exhibits a certain amount of resistance and capacitance in addition to inductance, due to the physical properties of its windings and core material as well as the nature of magnetic fields. Consequently, an inductor (unlike a capacitor) can't retain electrical energy for very long because energy is lost through heat dissipation.

To help you understand inductors, think about water flowing through a pipe with a turbine in it. When you first apply water pressure, the fins of the turbine obstruct the flow, applying back pressure on the water. As the fins start to turn, they apply less back pressure, so the water flows more easily. If you suddenly remove the water pressure, the fins will keep turning for a while, pulling water along with them, until eventually the fins stop turning and the water stops flowing.

Don't worry about the ins and outs of induced currents, expanding and contracting magnetic fields, and the like. Just remember a few things about inductors:

- An inductor opposes (resists) changes in current.
- An inductor acts like an open circuit when DC is first applied — that is, no current flows right away, and the entire source voltage is dropped across the inductor.
- An inductor eventually acts like a short in DC circuits — that is, when all the magnetic-field magic settles down, the voltage is zero, and the inductor allows the full DC current to pass through.

Calculating the RL time constant

You can calculate the amount of time (in seconds) it takes for the current flowing through an inductor in a resistor-inductor circuit to reach a steady DC value. To do this, you use the *RL time constant,* T, which tells you how long it takes for the inductor to conduct roughly two-thirds of the steady DC current that results from a voltage applied across the resistor-inductor series combination. The formula looks like this:

$$T = \frac{L}{R}$$

Just as the RC time constant in RC circuits (which you can read about in Chapter 7) gives you an idea of how long it takes a capacitor to charge to its full capacity, so the RL time constant helps you figure out how long it will take for an inductor to fully conduct a DC current: Direct current settles down to a steady value after roughly five RL time constants.

Keeping up with alternating current (or not!)

When you apply an AC voltage to a circuit containing an inductor, the inductor fights against any changes in the source current. If you keep varying the supply voltage up and down at a very high frequency, the inductor will keep opposing the sudden changes in current. At the extreme high end of the frequency spectrum, no current flows at all because the inductor simply can't react quickly enough to the change in current.

Picture yourself standing between two very tempting dessert platters. You want to go for each of them, but can't decide which to try first. You start out running toward one, but quickly change your mind and turn around and starting running toward the other. Then you change your mind again, so you turn and start racing toward the first one, and so forth. The faster you change your mind, the more you stay put in the middle — not getting anywhere (or any dessert). Those tempting desserts make you act like the electrons in an inductor when a high-frequency signal is applied to the circuit: Neither you nor the electrons make any progress.

Behaving differently depending on frequency

Like capacitors, inductors in an AC circuit act differently depending on the frequency of the voltage applied to them. Because the current passing through an inductor is affected by frequency, the voltage drops across the inductor and other components in the circuit are also affected by frequency. This frequency-dependent behavior forms the basis for useful functions, such as *filters,* which are circuits that allow some frequencies to pass through to another circuit stage while blocking other frequencies. When you adjust the bass and treble settings on your stereo system, you're using filters.

Here are some common types of filters:

- **Low-pass filters** are circuits that allow lower frequencies to pass from input to output while blocking frequencies above a particular *cutoff frequency*.
- **High-pass filters** are circuits that allow higher frequency signals to pass while blocking frequencies below the cutoff frequency.
- **Band-pass filters** are circuits that allow a band of frequencies — between a lower cutoff frequency and an upper cutoff frequency — to pass.
- **Band-stop (or band-rejection) filters** are circuits that allow every frequency except a specific band of frequencies to pass. You might use such a filter to filter out unwanted hum from a 60 Hz power line — as long as you know the frequency range of the hum.

Because inductors pass DC and block more and more AC as frequency increases, they are natural low-pass filters. In Chapter 7, you find out that capacitors block DC signals and allow AC signals to pass, so (as you might guess) they are natural high-pass filters. In electronic filter design — which is a complex topic beyond the scope of this book — components are carefully selected to precisely control which frequencies are allowed to pass through to the output.

Inductors are sort of the alter egos of capacitors. Capacitors oppose voltage changes; inductors oppose current changes. Capacitors block DC and pass more and more AC as frequency increases; inductors pass DC and block more and more AC as frequency increases.

Uses for Inductors

Inductors are used primarily in tuned circuits, to select or reject signals of specific frequencies, and to block (or choke) high-frequency signals, such as eliminating radio frequency (RF) interference in cable transmissions. In audio applications, inductors are also commonly used to remove the 60 Hz hum known as *noise* (often created by nearby power lines). Here's a list that explains some of the amazing things a simple coil of wire can do:

- **Filtering and tuning:** Like capacitors, inductors can be used to help select or reject certain electrical signals depending upon their frequency. Inductors are often used to tune in target frequencies in radio receiver systems.
- **AC motors:** In an AC induction motor, two pairs of coils are energized by an AC power supply (50 or 60 Hz), setting up a magnetic field that then induces a current in a rotor placed in the center of the magnetic field. That current then creates another magnetic field that opposes the original magnetic field, causing the rotor to spin. Induction motors are commonly used in fans and household appliances.

- **Blocking AC:** A *choke* is an inductor used to prevent a signal from passing through it to another part of a circuit. Chokes are often used in radio transmission systems to prevent the transmitted signal (an AC waveform) from shorting out through the power supply. By blocking the signal from traveling through a path to ground, a choke enables the signal to follow its intended path to the antenna, where it can be transmitted.
- **Contactless sensors:** Many traffic-light sensors use an inductor to trigger the light to change. Embedded in the street several yards before the intersection is an inductive loop consisting of several turns of a gigantic coil, roughly six feet in diameter. This loop is connected to a circuit that controls the traffic signal. As your car passes over the loop, the steel underbody of your car changes the magnetic flux of the loop. The circuit detects this change — and gives you the green light. Metal detectors use inductors in a similar way to pick up the presence of magnetic or metallic objects.
- **Smoothing out current:** Inductors can be used to reduce current swings (ripple) in a power supply. As the current changes, the magnetic flux around the coil changes and a back voltage is induced across the inductor, causing a current flow that opposes the main current.

Using Inductors in Circuits

The wire that makes up an inductor is often insulated to prevent unintended short circuits between the turns. Inductors may also be *shielded,* or encased in a nonferrous metal can (typically brass or aluminum), to prevent the magnetic lines of flux from infiltrating the neighborhood of other components in a circuit. You use a shielded inductor when you don't want to induce voltages or currents in other circuit elements. You use an unshielded inductor (or coil) when you do want to affect other circuit elements. I discuss the use of unshielded coils in circuits in the later section "Influencing the Coil Next Door: Transformers."

Reading inductance values

The value of an inductor is typically marked on its package using the same color-coding technique used for resistors, which you can read about in Chapter 5. You can often find the value of larger inductors printed directly on the components. Smaller-value inductors look a lot like low-wattage resistors; such inductors and resistors even have similar color-coding marks. Larger-value inductors come in a variety of sizes and shapes each offering trade-offs in terms of performance, cost, and other factors.

Inductors can be either fixed or variable. With either type, a length of wire is wound around a core. The number of turns of the wire, the core material, the wire's diameter, and length of the coil all determine the numerical value of the inductor. Fixed inductors have a constant value; variable inductors have adjustable values. The core of an inductor can be made of air, iron ferrite, or any number of other materials (including your car). Air and ferrite are the most common core materials.

Combining shielded inductors

Chances are, you won't use inductors in the basic electronic circuits you set up, but you may run across circuit diagrams for power supplies and other devices that include multiple inductors. Just in case you do, you should know how to calculate the equivalent inductance of combinations of shielded inductors so you can get a clear picture of how the circuit operates.

Inductors in series add up, just as resistors do:

$$L_{series} = L1 + L2 + L3 + \ldots$$

Like resistors, inductors in parallel combine by adding the reciprocals of each individual inductance, and then taking the reciprocal of that sum. (You may remember from math class that the reciprocal is the multiplicative inverse of a number, or the number that you multiply by so that the result equals 1. So for any integer x, 1/x is its reciprocal.)

$$L_{parallel} = \frac{1}{\frac{1}{L1} + \frac{1}{L2} + \frac{1}{L3} \ldots}$$

Another way to express the preceding equation is

$$\frac{1}{L_{parallel}} = \frac{1}{L1} + \frac{1}{L2} + \frac{1}{L3} \ldots$$

If you have just two inductors in parallel, you can simplify this equation as follows:

$$L_{parallel} = \frac{L1 \times L2}{L1 + L2}$$

Tuning in to Radio Broadcasts

Inductors are natural low-pass filters and capacitors are natural high-pass filters, so what happens when you put the two in the same circuit? As you might guess, inductors and capacitors are often used together in tuning circuits, to tune in a specific radio station's broadcast frequency.

Resonating with RLC circuits

Look at the RLC circuit in Figure 8-7. Now imagine what happens to the current, i, flowing through the circuit as you vary the frequency of the input signal, v_{in}. Because the capacitor blocks DC and allows more and more alternating current (AC) to flow as frequency increases, low frequency input signals tend to get squelched. Because the inductor passes DC and blocks more and more AC as frequency increases, high-frequency input signals are also squelched. What happens to the in-between frequencies? Well, some current is allowed to pass through, with the most current flowing when the input signal is at one specific frequency, known as the *resonant frequency*.

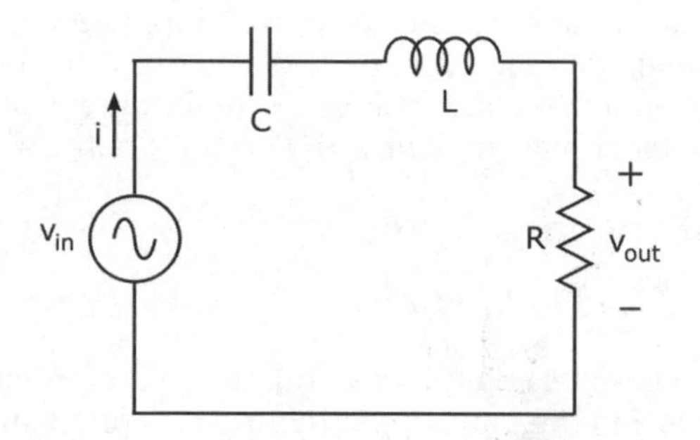

Figure 8-7: An RLC circuit has a resonant frequency, at which the maximum current flows.

The value of the resonant frequency, f_o, depends on the values of the inductance (L) and the capacitance (C), as follows:

$$f_0 = \frac{1}{2\pi\sqrt{LC}}$$

The circuit is said to *resonate* at that particular frequency, and so is known as a *resonant circuit.* Figure 8-8 shows a frequency plot of the current passing through the circuit; note that the current is highest at the resonant frequency.

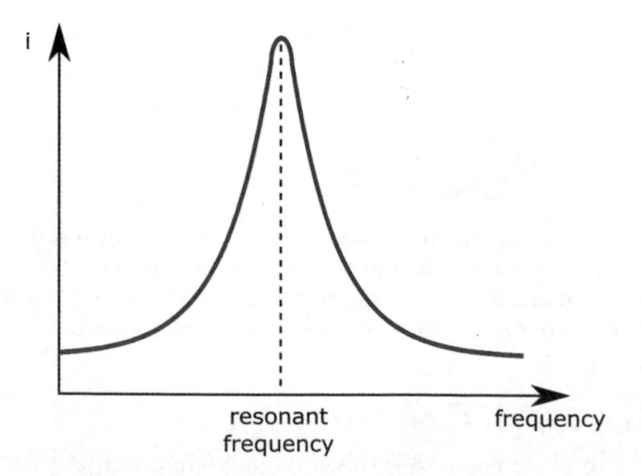

Figure 8-8: The current in a series RLC circuit is highest at the resonant frequency.

Analog radio receivers use RLC circuits to allow just one frequency to pass through the circuit. This process is known as *tuning in* to the frequency; used this way, the circuit is known as a *tuning circuit.* A variable capacitor is used to adjust the resonant frequency, so you can tune in different stations broadcasting at different frequencies. The knob that allows the capacitance to be changed is attached to the tuning-control knob on your radio.

A graphic equalizer uses a series of tuning circuits to separate an audio signal into several frequency bands. The slide controls on the equalizer allow you to adjust the gain (amplification) of each frequency band independently. In a later circuit stage, the individual bands are recombined into one audio signal — customized according to your taste — that is sent to your speakers.

By shifting components around a bit, you can create a variety of filter configurations. For instance, by placing the resistor, inductor, and capacitor in parallel with each other, you create a circuit that produces the *minimum* current at the resonant frequency. This sort of resonant circuit tunes out that frequency, allowing all others to pass, and is used to create band-stop filters. You might find such a circuit filtering out the 60 Hz hum that electronic equipment sometimes picks up from a nearby power line.

Ensuring rock-solid resonance with crystals

If you slice a quartz crystal in just the right way, mount two leads to it, and enclose it in a hermetically sealed package, you've created a single component that acts like an RLC combo in an RLC circuit, resonating at a particular frequency. *Quartz crystals,* or simply *crystals,* are used in circuits to generate an electrical signal at a very precise frequency. Figure 8-9 shows the circuit symbol for a crystal, which is labeled XTAL in schematics.

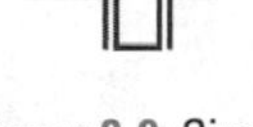

Figure 8-9: Circuit symbol for a crystal.

Crystals work because of something called the *piezoelectric effect:* If you apply a voltage in just the right way across a quartz crystal, it vibrates at a specific frequency, known as the resonant frequency. If you then remove the applied voltage, the crystal continues to vibrate until it settles back to its previous shape. As it vibrates, it generates a voltage at the resonant frequency.

You may be familiar with piezoelectric guitar pickups, which use crystals to convert the mechanical vibrations generated by guitar strings into electrical signals, which are then amplified. And if you predate compact disc (CD) technology, you may be interested to know that phonograph needles relied on the piezoelectric effect to convert the ups and downs of a vinyl record track into electrical energy.

The frequency at which a crystal resonates depends on its thickness and size, and you can find crystals with resonant frequencies ranging from a few tens of kilohertz to tens of megahertz. Crystals are more precise and more reliable than combinations of capacitors and inductors, but there's a catch: They're usually more expensive. You will find crystals used in circuits called *oscillators* to generate electric signals at a very precise frequency. Oscillators are responsible for the ticks and tocks that control quartz wristwatches and digital integrated circuits (which I discuss in Chapter 11), and for controlling the accuracy of radio equipment.

A quartz crystal is accurate to within roughly 0.001% of its stated resonant frequency. (That's why they're worth paying some extra bucks for!) You may also hear of ceramic resonators, which work the same way but cost less and are not as accurate as quartz. Ceramic resonators have a 0.5% frequency tolerance — meaning that the actual resonant frequency can vary by as much as 0.5% above or below its stated resonant frequency — and are used in many consumer-electronics devices, such as TVs, cameras, and toys.

Influencing the Coil Next Door: Transformers

Inductors used in tuning circuits are shielded so that the magnetic field they produce doesn't interact with other circuit components. Unshielded coils are sometimes placed close to one another for the express purpose of allowing their magnetic fields to interact. In this section, I describe how unshielded coils interact — and how you can exploit their interaction to do some useful things with an electronic device known as a *transformer.*

Letting unshielded coils interact

When you place two unshielded coils near each other, the varying magnetic field created as a result of passing AC through one coil induces a voltage in that coil *as well as in the other coil. Mutual inductance* is the term used to describe the effect of inducing a voltage in another coil, while *self-inductance* refers to the effect of inducing a voltage in the same coil that produced the varying magnetic field in the first place. The closer the coils, the stronger the interaction. Mutual inductance can add to or oppose the self-inductance of each coil, depending on how you match up the north and south poles of the inductors.

If you have an unshielded coil in one circuit, and place it close to an unshielded coil in another circuit, the coils will interact. By passing a current through one coil, you will cause a voltage to be induced in the neighboring coil — even though it is in a separate, unconnected circuit. This is known as *transformer action.*

A *transformer* is an electronic device that consists of two coils wound around the same core material in such a way that the mutual inductance is maximized. Current passing through one coil, known as the *primary,* induces a voltage in the other coil, known as the *secondary.* The job of a transformer is to transfer electrical energy from one circuit to another.

The circuit symbols for an air-core transformer and solid-core transformer, respectively, are shown in Figure 8-10.

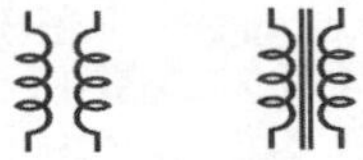

Figure 8-10: Circuit symbols for an air-core transformer and a solid-core transformer.

Isolating circuits from a power source

If the number of turns of wire in the primary winding of a transformer is the same as the number of turns in the secondary winding, theoretically, all of the voltage across the primary will be induced across the secondary. This is known as a *1:1 transformer,* because there is a 1:1 (read "one-to-one") relationship between the two coils. (In reality, no transformer is perfect, or *lossless,* and some of the electrical energy gets lost in the translation.)

A 1:1 transformer is also known as an *isolation transformer,* and is commonly used to electrically separate two circuits while allowing AC power or an AC signal from one to feed into the other. The first circuit typically contains the power source, and the second circuit contains the load. (You find out in Chapter 1 that the *load* is the destination for the electrical energy, or the thing you ultimately want to perform work on, such as a speaker diaphragm.) You may want to isolate circuits to reduce the risk of electrical shocks or to prevent one circuit from interfering with the other.

Stepping up, stepping down voltages

If the number of turns in the primary winding of a transformer is not the same as the number of turns in the secondary winding, the voltage induced in the secondary will be different from the voltage across the primary. The two voltages will be proportional to each other, with the proportion determined by the ratio of the number of turns in the secondary to the number of turns in the primary, as follows:

$$\frac{V_S}{V_P} = \frac{N_S}{N_P}$$

In this equation, V_S is the voltage induced in the secondary, V_P is the voltage across the primary, N_S is the number of turns in the secondary, and N_P is the number of turns in the primary.

Say, for instance, that the secondary consists of 200 turns of wire — twice as many as the primary, which consists of 100 turns of wire. If you apply an AC voltage with a peak value of 50 V to the primary, the peak voltage induced across the secondary will be 100 V, or twice the value of the peak voltage across the primary. This type of transformer is known as a *step-up transformer,* because it steps up the voltage from the primary to the secondary.

If, instead, the secondary consists of 50 turns of wire, and the primary consists of 100 turns, the same AC signal applied to the same primary has a different result: The peak voltage across the secondary would be 25 V, or half the primary's voltage. This is known as a *step-down transformer,* for obvious reasons.

In each case, the power applied to the primary winding is transferred to the secondary. Because power is the product of voltage and current $(P = V \times I)$, the *current* induced in the secondary winding is inversely proportional to the *voltage* induced in the secondary. So a step-up transformer steps up the voltage while stepping down the current; a step-down transformer steps down the voltage while stepping up the current.

Step-up and step-down transformers are used in electrical power-transmission systems. Electricity generated at a power plant is stepped up to voltages of 110 kV $(1\text{ kV} = 1{,}000\text{ V})$ or more, transported over long distances to a substation, and then stepped down to lower voltages for distribution to customers.

9

Diving into Diodes

In This Chapter

- Peeking inside a semiconductor
- Fusing two semiconductors to make a diode
- Letting current flow this way but not that way

Semiconductors are at the heart of nearly every major electronic system that exists today, from the programmable pacemaker to your smartphone to the space shuttle. It's amazing to think that teeny-tiny semiconductor devices are responsible for triggering enormous advances in modern medicine, space exploration, industrial automation, home entertainment systems, communications, and a slew of other industries.

The simplest type of semiconductor device, called a *diode,* can be made to conduct or block electric current and to allow current to flow in one direction but not the other — depending on how you control it electrically.

This chapter explains what semiconductors are, how to make them conduct current, and how to combine two semiconductors to create a diode. Then you get an eyewitness view of the valve-like behavior of diodes — and take a look at how to exploit that behavior to accomplish amazing things in circuits.

Are We Conducting or Aren't We?

Somewhere between insulators and conductors are materials that can't seem to make up their minds about whether to hold on to their electrons or let them roam freely. These *semiconductors* behave like conductors under some conditions and like insulators under other conditions, giving them unique capabilities.

With a device made from semiconductor materials, such as silicon or germanium, you can precisely control the flow of electrical charge carriers in one area of the device by adjusting a voltage in another area of the device. For instance, by adjusting the voltage across a two-terminal semiconductor diode (see Figure 9-1), you can allow current to flow in one direction while blocking its flow in the other direction, just like a check valve.

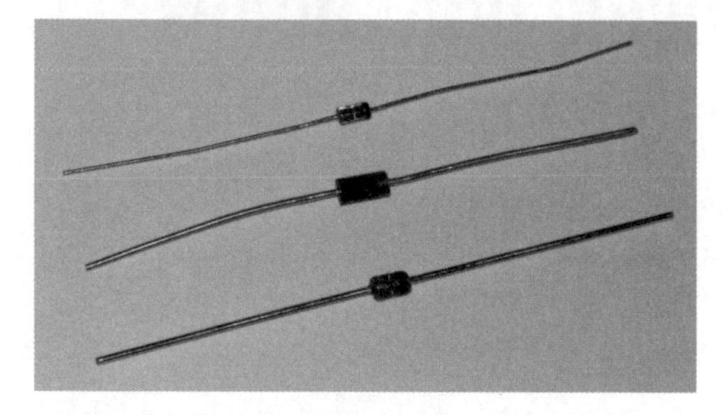

Figure 9-1: Diodes are two-terminal semiconductor devices similar in size and shape to resistors.

Sizing up semiconductors

The atoms of semiconductor materials align themselves in a structured way, forming a regular, three-dimensional pattern — a crystal — as shown in Figure 9-2. Atoms in the crystal are held together by a special bond, called a *covalent bond,* with each atom sharing its outermost electrons (known as *valence electrons*) with its neighbors.

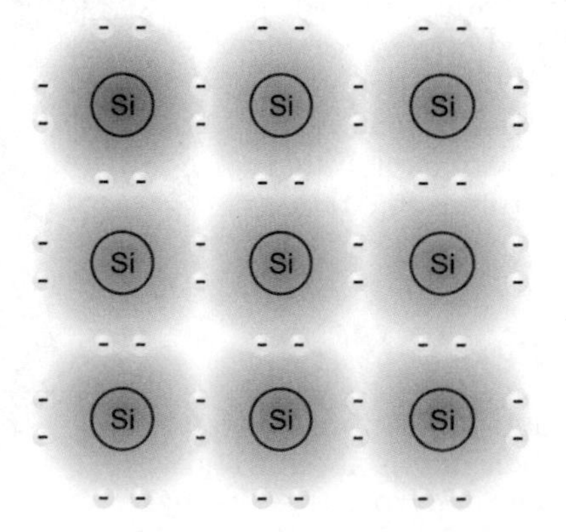

Figure 9-2: Silicon and other semiconductor materials contain strong covalent bonds that hold the atoms together in a crystalline structure. (Only the outermost electrons are shown.)

It's precisely because of this unique bonding and sharing of electrons that the semiconductor crystal acts like an insulator most of the time. Each atom thinks it has more valence electrons than it really has, and those electrons stay close to home. (This behavior is very different from a typical conductor atom, which often has just one valence electron that thinks it is free to roam around.)

Creating N-types and P-types

A pure semiconductor can be altered to change its electrical properties. The exact explanation of this process, which is called *doping* (relax; it has nothing to do with steroids), involves some fascinating physics that I won't go into it here. But the bottom line is this: Doping creates variants of semiconductor materials that (depending on the specific type of doping) either have *more* electrons or *fewer* electrons than the pure semiconductor material. These variants follow:

- **N-type semiconductors** have more electrons, which are treated like outsiders, unable to muscle in to the covalent bonds. N-types get their name from these spurned electrons (negative charge carriers), which move around within the crystal.
- **P-type semiconductors** have fewer electrons, leaving *holes* in the crystalline structure where the electrons used to be. Holes don't stay still long because neighboring electrons tend to occupy them, leaving new holes elsewhere, which are quickly occupied by more electrons, leaving more holes, and so on. The net result is that these holes appear to be moving within the crystal. Because holes represent an absence of electrons, you can think of them as positive charges. P-types are so named because of these positive charge carriers.

The doping process increases the conductivity of the semiconductors. If you apply a voltage source across either an N-type or a P-type semiconductor, electrons move through the material from the more negative voltage toward the more positive voltage. (For P-type semiconductors, this action is described as a movement of holes from the more positive voltage toward the more negative voltage.) In other words, N-type and P-type semiconductors are acting like conductors, allowing current to flow in response to an applied voltage.

Joining N-types and P-types to create components

Here's where things get interesting: When you fuse together an N-type and a P-type semiconductor, current can flow through the resulting *pn-junction* — but only in one direction.

Whether or not a current flows depends on which way you apply the voltage. If you connect the positive terminal of a battery to the P-type material and the negative terminal to the N-type material, current will flow (as long as the applied voltage exceeds a certain minimum and doesn't exceed a certain maximum). But if you reverse the battery, current will not flow (unless you apply a very large voltage).

Exactly how these N-type and P-type semiconductors are combined determines what sort of semiconductor device they become — and how they allow current to flow (or not) when voltage is applied. The pn-junction is the foundation for *solid-state* electronics, which involves electronic devices made of solid, non-moving materials, rather than the vacuum tubes of yore. Semiconductors have replaced the majority of vacuum tubes in electronics.

Forming a Junction Diode

A semiconductor *diode* is a two-terminal electronic device that consists of a single pn-junction. (See Figure 9-3.) Diodes act like one-way valves, allowing current to flow in only one direction when a voltage is applied to them. This capability is sometimes referred to as the *rectifying* property.

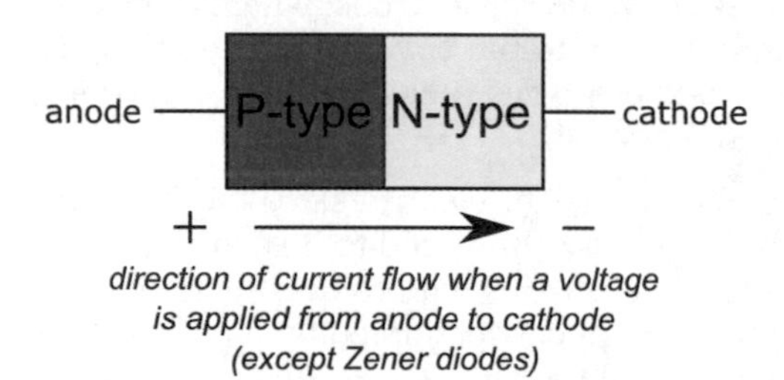

Figure 9-3: A diode is made up of a single pn-junction.

You refer to the P-side of the pn-junction in a diode as the *anode* and the N-side as the *cathode*. For most diodes, if you apply a more positive voltage to the anode and a more negative voltage to the cathode, current will flow from the anode to the cathode. If you reverse the voltage, the diode will not conduct current. (Zener diodes are an exception; for details, flip ahead in this chapter to "Regulating voltage with Zener diodes.")

In electronics, *current* refers to conventional current — which is just the opposite of real electron flow. So when you say that (conventional) current flows from the anode to the cathode, the electrons are moving from the cathode to the anode.

You can think of the junction within a diode as a hill and the current as a ball you are trying to move from one side of the hill to the other: Pushing the ball down the hill (from anode to cathode) is easy, but pushing the ball up the hill (from cathode to anode) is difficult.

Diodes are cylindrical, like resistors, but aren't quite as colorful as resistors. Most diodes sport a stripe or other mark at one end, signifying the cathode (refer to Figure 9-1). In the circuit symbols shown in Figure 9-4, the anode is on the left (broad end of the arrowhead) and the cathode is on the right.

For standard diodes and LEDs, the arrowhead is pointing in the direction of (conventional) current flow. (Current flows the other way in Zener diodes.)

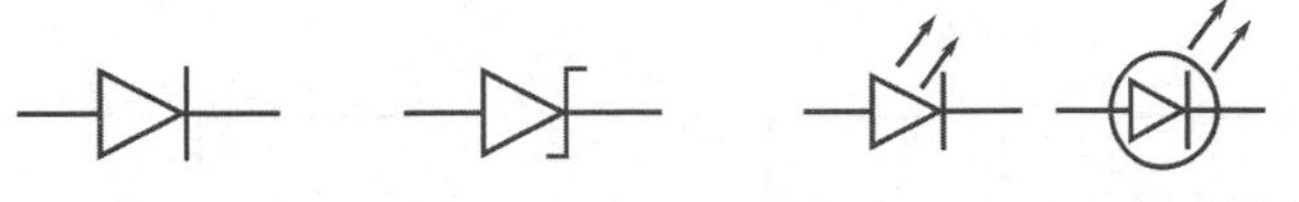

Figure 9-4: Circuit symbols for different types of semiconductor diodes.

Biasing the diode

When you bias a diode, you apply a voltage, known as the *bias voltage,* across the diode (from anode to cathode) so that the diode either allows current to flow from anode to cathode or blocks current from flowing. A standard diode has two basic operating modes:

- **Forward-bias (conducting):** When a high enough positive voltage is applied from anode to cathode, the diode turns on (conducts current).

  This minimum turn-on voltage is known as the *forward voltage,* and its value depends on the type of diode. A typical silicon diode has a forward voltage of about 0.6 V to 0.7 V, whereas forward voltages for light-emitting diodes (LEDs) range from about 1.5 V to 4.6 V (depending on the color). (Check the ratings on the particular diodes you use in circuits.)

  When the diode is forward-biased, current, known as *forward current,* flows easily across the pn-junction from anode to cathode. You can increase the amount of current flowing through the diode (up to the maximum current it can safely handle), but the forward voltage drop won't vary that much.

- **Reverse-bias (nonconducting):** When a *reverse voltage* (a negative voltage from anode to cathode) is applied across the diode, current is prohibited from flowing. (Actually, a small amount of current, in the μA range, will flow.)

  If the reverse-biased voltage exceeds a certain level (usually 50 V or more), the diode breaks down and *reverse current* starts flowing from cathode to anode. The reverse voltage at which the diode breaks down is known as the *peak reverse voltage* (PRV), or *peak inverse voltage* (PIV).

Figure 9-5 shows a forward-biased diode allowing current to flow through a lamp and a reverse-biased diode prohibiting current from flowing.

You usually don't purposely reverse-bias a diode (except if you're using a Zener diode, which I describe in the section, "Regulating voltage with Zener diodes," later in this chapter). You may accidentally reverse-bias a diode by orienting it incorrectly in a circuit (see the section "Which end is up?" later in this chapter), but don't worry: You won't harm the diode and you can simply reorient it. (But if you exceed the PRV, you may allow too much reverse current to flow, which can damage other circuit components.)

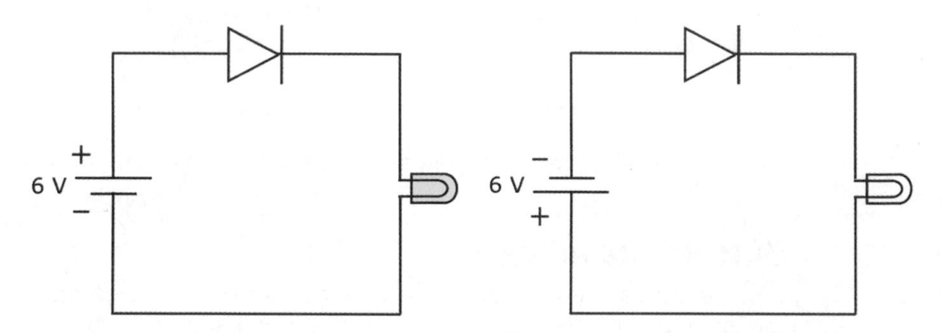

Figure 9-5: The battery on the left forward-biases the diode, allowing current to flow. Reversing the battery (right) reverse-biases the diode, preventing current from flowing.

If no voltage or a low voltage (less than the forward voltage) is applied across a diode, it is *unbiased*. (That doesn't mean the diode lacks prejudice; it just means that you haven't taken action on the diode yet.)

In electronics, the term *bias* refers to a steady DC voltage or current applied to an electronic device or circuit to get it to operate a certain way. Devices such as transistors (which I cover in Chapter 10) and diodes are nonlinear devices. That is, the relationship between voltage and current in these devices is not constant; it varies across different ranges of voltages and current. Diodes and transistors are not like resistors, which exhibit a linear (constant) relationship between voltage and current.

Conducting current through a diode

After current starts flowing through a diode, the forward voltage drop across the diode remains fairly constant — even if you increase the forward current. For instance, most silicon diodes have a forward voltage of between 0.6 V and 0.7 V over a wide range of forward currents. If you're analyzing a circuit that contains a silicon diode (such as the circuit on the left in Figure 9-5), you can assume that the voltage drop across the diode is about 0.7 V — even if you increase the source voltage from 6 V to 9 V. Increasing the source voltage increases the current through the circuit, but the diode voltage drop remains the same, so the increased source voltage is dropped across the lamp.

Of course, every electronic component has its limits. If you increase the current through a diode too much, you'll generate a lot of heat in the diode. At some point, the pn-junction will be damaged from all that heat, so you have to be careful not to turn up the source voltage too high.

Rating your diode

Most diodes don't have values like resistors and capacitors. A diode simply does its thing in controlling the on/off flow of electrons, without altering the shape or size of the electron flow. But that doesn't mean all diodes are the same. Standard diodes are rated according to two main criteria: peak reverse voltage (PRV) and current. These criteria guide you to choosing the right diode for a particular circuit, as follows:

- The **PRV rating** tells you the maximum reverse voltage the diode can handle before breaking down. For example, if the diode is rated at 100 V, you shouldn't use it in a circuit that applies more than 100 V to the diode. (Circuit designers build in considerable headroom above the PRV rating to accommodate voltage spikes and other conditions. For instance, it's common practice to use a 1,000 V PRV rectifier diode in power supply circuits running on 120 VAC.)
- The **current rating** tells you the maximum forward current the diode can withstand without sustaining damage. A diode rated for 3 A can't safely conduct more than 3 A without overheating and failing.

Identifying with diodes

Most diodes originating in North America are identified by five- or six-digit codes that are part of an industry-standard identification system. The first two digits are always 1N for diodes; the 1 specifies the number of pn-junctions, the N signifies semiconductor, and the remaining three or four digits indicate specific features of the diode. A classic example is the series of rectifier diodes identified as 1N40*xx*, where *xx* could be 00, 01, and so forth through 08. They are rated at 1 amp with PRV ratings ranging from 50 to 1,000 V, depending on the *xx* number. For instance, the 1N4001 rectifier diode is rated at 1 A and 50 V, and the 1N4007 is rated at 1 A and 1,000 V. Diodes in the 1N54*xx* series have a 3-A rating with PRV ratings from 50 to 1,000 V. You can readily find such information in any catalog of electronic components or cross-reference book of diode data, generally found online. (A *cross-reference book* tells you what parts can be substituted for other parts, in case a part specified in a circuit diagram is not available from your chosen source.)

Just to make things interesting (not to mention confusing), some diodes use the same color-coding scheme on their packaging as resistors, but instead of translating the code into a value (such as resistance), the color code simply gives you the semiconductor identification number for the diode. For instance, the color sequence "brown-orange-red" indicates the numerical sequence "1-3-2" so the diode is a 1N132 germanium diode. (Refer to the resistor color code chart in Chapter 5.)

Which end is up?

When you use a diode in a circuit, it's extremely important to orient the diode the right way (more about that in a minute). The stripe or other mark on the diode package corresponds to the line segment in the circuit symbol for a diode: Both indicate the cathode, or negative terminal, of the diode.

You can also determine which end is what by measuring the resistance of the diode (before you insert it in your circuit) with an ohmmeter or multimeter (which I discuss in Chapter 16). The diode has a low resistance when it's forward-biased, and a high resistance when it's reverse-biased. By applying the positive lead of your meter to the anode and the negative lead to the cathode, your meter is essentially forward-biasing the diode (because when used to measure resistance, a multimeter applies a small voltage across its leads). You can measure the resistance twice, applying the leads first one way and then the other way. The lower measurement result indicates the forward-biased condition.

Diodes are like one-way valves, allowing current to flow in one direction only. If you insert a diode backward in a circuit, either your circuit won't work because no current will flow or you may damage some components if you exceed the peak reverse voltage (PRV) and allow current to flow in reverse — which can damage components such as electrolytic capacitors. Always note the orientation of the diode when you use it in a circuit, double-checking to make sure you have it right!

Using Diodes in Circuits

You'll find several different flavors of semiconductor diodes designed for various applications in electronic circuits.

Rectifying AC

Figure 9-6 shows a circuit with a silicon diode, a resistor, and an AC power source. Note the orientation of the diode in the circuit: Its anode (positive end) is connected to the power source. The diode conducts current when it's forward-biased, but not if it's reverse-biased. When the AC source is positive (and provides at least 0.7 V to forward-bias the silicon diode), the diode conducts current; when the AC source is less than 0.7 V, the diode does not conduct current. The output voltage is a clipped version of the input voltage; only the portion of the input signal that is greater than 0.7 V passes through to output.

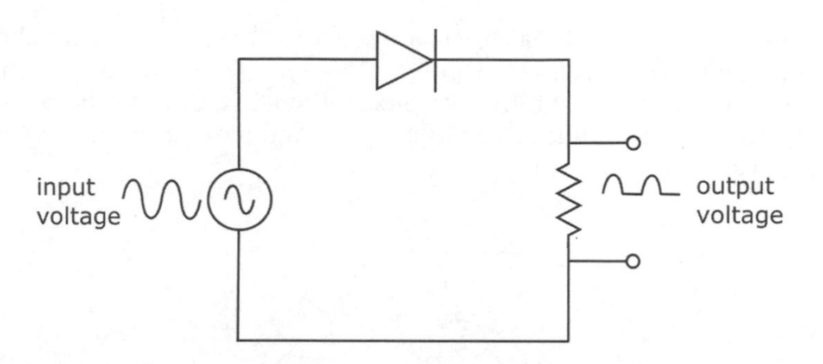

Figure 9-6: The diode in this circuit clips off the lower portion of the AC source voltage.

If the diode orientation is reversed in the circuit, the opposite happens. Only the negative part of the input voltage is passed through to the output:

- When the input voltage is positive, the diode is reverse-biased and no current flows.
- When the input is sufficiently negative (at least –0.7 V), the diode is forward-biased and current flows.

Diodes used this way — to convert AC current into varying DC current (it's DC because the current is flowing in one direction only, but it isn't a constant current) — are called *rectifier diodes,* or just *rectifiers*. Rectifier diodes are designed to handle currents ranging from several hundred milliamps to a few amps — much higher strengths than general-purpose *signal diodes* are designed to handle (those currents go up to only about 100 mA). You'll see rectifiers used in two major ways:

- **Half-wave rectification:** Using a single rectifier diode to clip an AC signal is known as *half-wave rectification* because it converts half the AC signal into DC.
- **Full-wave rectification:** By arranging four diodes in a circuit known as a *bridge rectifier,* you can convert both the ups and the downs (relative to 0 V) of an AC voltage into just ups (see Figure 9-7). This *full-wave rectifier* is the first stage of circuitry in a *linear power supply,* which converts AC power into a steady DC power supply.

  Bridge rectifiers are so popular, you can purchase them as a single four-terminal part, with two leads for the AC input and two leads for the DC output.

Regulating voltage with Zener diodes

Zener diodes are special diodes that are meant to break down. They are really just heavily doped diodes that break down at much lower voltages than standard diodes. When you reverse-bias a Zener diode, and the voltage across it reaches or exceeds its breakdown voltage, the Zener diode suddenly starts conducting current backward through the diode (from cathode to anode). As you continue to increase the reverse-biased voltage beyond the breakdown point, the Zener continues to conduct more and more current — while maintaining a steady voltage drop.

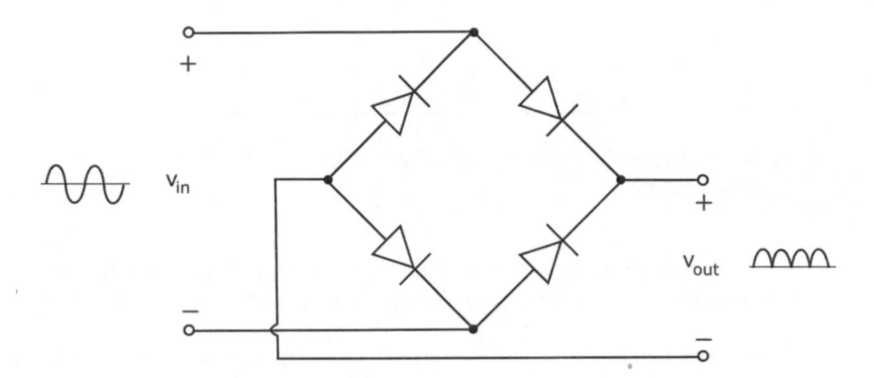

Figure 9-7: In a bridge rectifier, four diodes transform an AC voltage or current into a purely DC voltage or current.

Keep in mind these two important ratings for Zener diodes:

- The **breakdown voltage,** commonly called the *Zener voltage,* is the reverse-biased voltage that causes the diode to break down and conduct current. Breakdown voltages, which are controlled by the semiconductor doping process, range from 2.4 V to hundreds of volts.
- The **power rating** tells you the maximum power (voltage × current) the Zener diode can handle. (Even diodes designed to break down can *really* break down if you exceed their power ratings.)

To see the circuit symbol for a Zener diode, refer to Figure 9-4.

Because Zener diodes are so good at maintaining a constant reverse-biased voltage, even as current varies, they're used to regulate voltage in circuits. In the circuit in Figure 9-8, for example, a 9 V DC supply is being used to power a load, and a Zener diode is placed so that the DC supply exceeds the breakdown voltage of 6.8 V. (Note that this voltage is reverse-biasing the diode.) Because the load is in parallel with the Zener diode, the voltage drop across the load is the same as the Zener voltage, which is 6.8 V. The remaining supply voltage is dropped across the resistor (which is there to limit the current through the diode so the power rating is not exceeded).

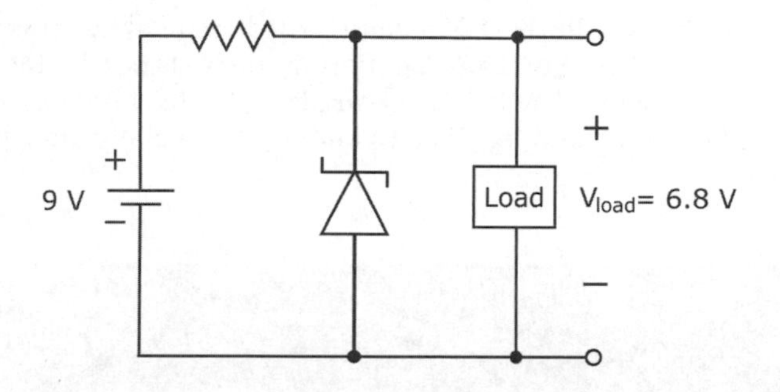

Figure 9-8: The Zener diode stabilizes the voltage drop across the load in this circuit.

Here's the important thing: If the supply voltage varies up or down around its nominal 9 V value, the current in the circuit will fluctuate *but the voltage across the load will remain the same:* a constant 6.8 V. The Zener diode allows current fluctuations while stabilizing the voltage, whereas the resistor voltage varies as the current fluctuates.

Seeing the light with LEDs

All diodes release energy in the form of light when forward-biased. The light released by standard silicon diodes is in the infrared range, which is not visible to the human eye. *Infrared light-emitting diodes (IR LEDs)* are commonly used in remote-control devices to send secret (well, okay, invisible) messages to other electronic devices, such as your TV or DVD player.

Diodes known as *visible LEDs* (or just *LEDs*) are specially made to emit copious amounts of visible light. By varying the semiconductor materials used, diodes can be engineered to emit light in many colors, including red, orange, yellow, green, blue, white, and even pink. (The 2014 Nobel Prize in Physics was awarded to researchers Akasaki, Amano, and Nakamura for their invention of the blue LED in the 1990s.) Bi-color and tri-color LEDs contain two or three different diodes in one package.

Refer to Figure 9-4 for two commonly used circuit symbols for an LED. Note that the outward-pointing arrows represent the visible light emitted by the diode.

The diode in an LED is housed in a plastic bulb designed to focus the light in a particular direction. The lead connected to the cathode is shorter than the lead connected to the anode. Compared to standard incandescent light bulbs, LEDs are more durable and efficient, run cooler, achieve full brightness much faster, and last much longer. LEDs are commonly used as indicator lights, task lights, and holiday lights, and in automobile headlights, displays (such as alarm clocks), and high-definition TVs.

Figure 9-9 shows a single-color LED. The shorter lead is usually attached to the cathode (negative side). You can also identify the cathode by looking through the plastic housing: The larger metal plate inside is the cathode and the smaller plate is the anode. (Nice to know — especially after you've clipped the leads.)

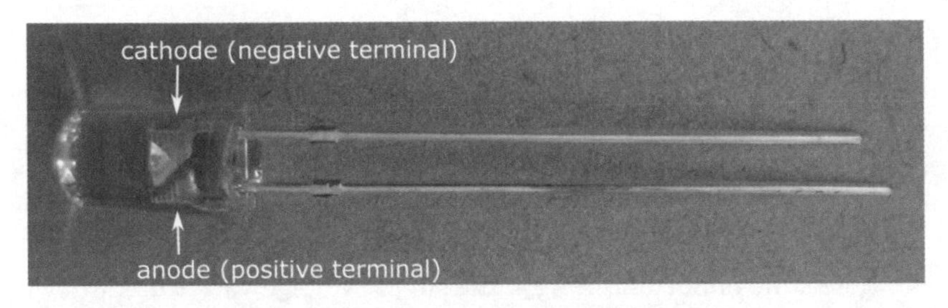

Figure 9-9: The cathode (negative side) of a typical single-color LED is attached to the larger metal plate inside the plastic housing and to the shorter lead (until you clip the leads).

LEDs carry the same specifications as standard diodes, but they usually have fairly low current and PRV ratings. A typical LED has a PRV rating of about 5 V with a maximum current rating of under 50 mA. If more current passes through an LED than its maximum rating specifies, the LED burns up like a marshmallow in a campfire. Forward voltages vary, depending on the type of LED; they range from 1.5 V for IR LEDs up to 3.4 V for blue LEDs. Red, yellow, and green LEDs typically have a forward voltage of about 2.0 V. Be sure to check the specifications of any LEDs you use in circuits.

The maximum current rating for an LED is usually referred to as the maximum *forward current,* which is different from another LED rating, known as the *peak current* or *pulse current*. The peak or pulse current, which is higher than the maximum forward current, is the absolute maximum current that you can pass through the LED for a very short period of time. Here, short means *short* — on the order of milliseconds. If you confuse forward current with peak current, you may wreck your LED.

You should never connect an LED directly to a power source, or you may fry the LED instantly. Instead, use a resistor in series with the LED to limit the forward current. For instance, in the circuit in Figure 9-10, a 9-volt battery is used to power a red LED. The LED has a forward voltage drop of 2.0 V and a maximum current rating of 24 mA. The voltage drop across the resistor is the difference between the source voltage and the LED forward voltage, or $9\text{ V} - 2\text{ V} = 7\text{ V}$. The question is, how big should the resistor be to limit the current to 24 mA (that's 0.024 A) *or less* when the voltage dropped across the resistor is 7 V? You apply Ohm's Law (which I discuss in Chapter 6) to calculate the *minimum* value of resistance required to keep the current below the maximum current rating as follows:

$$
\begin{aligned}
R &= \frac{V_R}{I_{max}} \\
&= \frac{7\text{ V}}{0.024\text{ mA}} \\
&\approx 292\ \Omega
\end{aligned}
$$

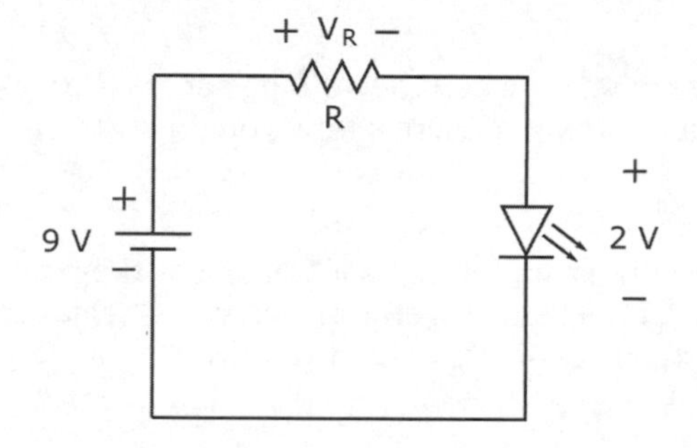

Figure 9-10: Be sure to insert a resistor in series with an LED to limit current to the LED.

Chances are you won't find a resistor with the exact value you calculated, so choose a standard resistor with a *higher* value (such as 330 Ω or 390 Ω) to limit the current a bit more. If you choose a lower value (such as 270 Ω), the current will exceed the maximum current rating.

Turning on an LED

The circuit in Figure 9-11 is designed to demonstrate the on/off operation of an LED, and how increasing current strengthens the light emitted by the diode.

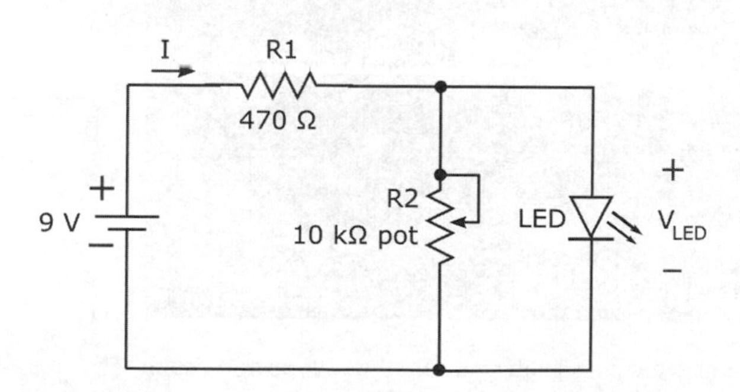

Figure 9-11: Use this circuit to turn an LED on and off, and to vary the intensity of the light.

Here's what you need to build this circuit:

- One 9-volt battery and battery clip
- One 470 Ω 1/4 W (minimum) resistor (yellow-violet-brown)
- One 10 kΩ potentiometer with attached lead wires
- One red, yellow, or green LED (any size)
- At least one jumper wire
- One solderless breadboard

You can witness the LED turning on when the voltage across it reaches its forward voltage by using a multimeter. Here's how you do it:

1. **Zero the 10 kΩ pot.**

   Set your multimeter to ohms and place its leads across the potentiometer. Make sure you tie two of the leads together when you do this. Turn the pot knob all the way to one end or the other until you get a reading of 0 Ω (or just about 0 Ω).

2. **Build the circuit, using Figure 9-12 as your guide.**

   Keep the same two pot leads together so that the pot is providing 0 ohms of resistance to the circuit. Make sure you orient the LED properly, with the cathode (negative side) connected to the negative battery terminal. (Remember that the cathode has a shorter lead and a larger plate inside the plastic housing.)

Figure 9-12: By turning the pot's knob, you vary the voltage across the LED. When the LED voltage exceeds about 2 V, the LED turns on.

3. **Set your multimeter to DC volts with a range of 10 V and place the leads across the LED (red lead at anode; black lead at cathode).**

   Is the LED lit? What voltage reading do you get? The LED voltage should be just a few millivolts, which is not enough to turn on the diode.

4. **Dial the pot up slowly, keeping your eye on the LED. When the LED turns on, stop dialing the pot.**

   Observe the reading on the multimeter. Is the LED voltage close to 2 V?

5. **Continue to dial the pot up as you watch the LED.**

   What is happening to the light? (It should be glowing brighter.)

6. **Dial the pot all the way up to 10 kΩ and observe the intensity of the LED. Note the voltage reading on your multimeter.**

   Did the LED voltage change much as the light intensity increased?

To understand why the LED was off when the pot was set to 0 Ω and then turned on as you increased the pot resistance, consider the circuit if you remove the LED. The circuit is a voltage divider (described in Chapter 6), and the voltage across the pot (resistor *R2*) — which is the same as the LED voltage — is given by a ratio of resistance times the supply voltage:

$$V_{LED} = \frac{R2}{R1 + R2} \times 9\text{ V}$$
$$= \frac{R2}{470 + R2} \times 9\text{ V}$$

If the resistance of the potentiometer is 0 Ω, the voltage across the LED is 0 V. As you increase *R2* (that is, the resistance of the pot), the voltage across the LED increases. When *R2* reaches a certain point, the voltage across the LED rises enough to turn on the LED. V_{LED} climbs to about 2.0 V when *R2* rises to about 134 Ω. (Plug in 134 for *R2* in the preceding equation and see for yourself!)

Of course, your particular LED may turn on at a slightly different voltage, say in the range of 1.7 V to 2.2 V. If you measure the resistance of the pot (after removing it from the circuit) at the point at which your LED turns on, you may see a somewhat lower or higher resistance than 134 Ω.

You can also observe the current flowing through the LED by following these steps:

1. **Break the circuit between the cathode (negative side) of the LED and the negative battery terminal.**

2. **Set your multimeter to DC amps and insert it in series where you broke the circuit.**

   Make sure the red lead of your multimeter is connected to the cathode of the LED and the black lead of your multimeter is connected to the negative battery terminal so that you are measuring positive current flow.

3. **Start with the pot turned all the way down to 0 Ω. As you dial the pot up, observe the current reading.**

   Note the reading when the LED first turns on. Then continue to dial the pot up and observe the current readings. You should see the current increase to over 14 mA as the light intensifies.

If you have two multimeters, try measuring the voltage across the LED with one multimeter (set to DC volts) and the current flowing through the LED with the other multimeter (set to DC amps) simultaneously. You should notice that the LED turns on when its voltage approaches 2.0 V, with just a tiny current passing through it at this point. As you increase the current through the LED, the light brightens but the voltage across it remains fairly steady.

Other uses of diodes

Among the many other uses of diodes in electronic circuits are the following:

- **Overvoltage protection:** Diodes placed in parallel with a piece of sensitive electronic equipment protect the equipment from large voltage spikes. The diode is placed backward so that it's normally reverse-biased, acting like an open circuit and not playing any part in the normal operation of the circuit. However, under abnormal circuit conditions, if a large voltage spike occurs, the diode becomes forward-biased — which limits the voltage across the sensitive component and shunts excess current to ground to prevent harm to the component. (The diode may not be so lucky.)
- **Construction of logic gates:** Diodes are the building blocks of specialized circuits known as *logic circuits,* which process signals consisting solely of two voltage levels that are used to represent binary information (such as on/off, high/low, or 1/0) in digital systems. I discuss logic a bit more in Chapter 11.
- **Current steering:** Diodes are sometimes used in uninterruptible power supplies (UPSs) to prevent current from being drawn out of a backup battery under normal circumstances, while allowing current to be drawn from the battery during a power outage.

10

Tremendously Talented Transistors

In This Chapter

- Revolutionizing electronics with the tiny transistor
- Understanding transistor action
- Using transistors as teeny-tiny switches
- Giving signals a boost with transistors

Imagine the world without the amazing electronics building block known as the transistor. Your cellphone would be the size of a washing machine, your laptop wouldn't fit on your lap (or in a single room), your iPod would be just another pie-in-the-sky idea — and your Apple stock would be worthless.

Transistors are the heart of nearly every electronic device in the world, quietly working away without taking up much space, generating a lot of heat, or breaking down every so often. Generally regarded as the most important technological innovation of the 20th century, transistors were developed as an alternative to the vacuum tube, which drove the development of electronic systems ranging from radio broadcasting to computers but exhibited some undesirable characteristics. The solid-state transistor enabled the miniaturization of electronics, leading to the development of cellphones, iPods, GPS systems, implantable pacemakers — and much more.

In this chapter, you find out what transistors are made of and the secrets of their success. You discover how transistors amplify tiny signals and how to use transistors as microscopic switches. Finally, you observe transistor action firsthand by building a couple of simple transistor circuits.

Transistors: Masters of Switching and Amplifying

Transistors basically do just two things in electronic circuits: switch and amplify. But those two jobs are the key to getting really interesting things done. Here's why those functions are so important:

- **Switching:** If you can switch electron flow on and off, you have control over the flow, and you can build involved circuits by incorporating lots of switches in the right places.

  Consider, for instance, the telephone-switching system: By dialing a 10-digit number, you can connect with any one of millions of people around the world. Or look at the Internet: Switching enables you to access a website hosted in, say, Sheboygan while you're sitting on a train in, say, London. Other systems that rely on switching are computers, traffic lights, the electric power grid — well, you get the idea. Switching is pretty darn important.

- **Amplifying:** If you can amplify an electrical signal, you can store and transmit tiny signals and boost them when you need them to make something happen.

  For instance, radio waves carry tiny audio signals over long distances, and it's up to the amplifier in your stereo system to magnify the signal so it can move the diaphragm of a speaker so you can hear the sound.

Before the invention of the transistor, vacuum tubes did all the switching and amplifying. In fact, the vacuum tube was widely regarded as the greatest marvel in electricity in the early 20th century. Then, Bardeen, Brattain, and Shockley showed the world that tiny semiconductor transistors could do the same job — only better (and for less money). The trio was awarded the 1956 Nobel Prize in Physics for their invention of the transistor.

Transistors these days are microscopically small, have no moving parts, are reliable, and dissipate a heck of a lot less power than their vacuum-tube predecessors. (However, many audiophiles believe tubes offer superior sound quality compared to solid-state transistor technology.)

The two most common types of transistors are

- Bipolar junction transistors
- Field-effect transistors

Figure 10-1 shows the circuit symbols commonly used for various types of transistors. The sections that follow provide a closer look at bipolar transistors and field-effect transistors.

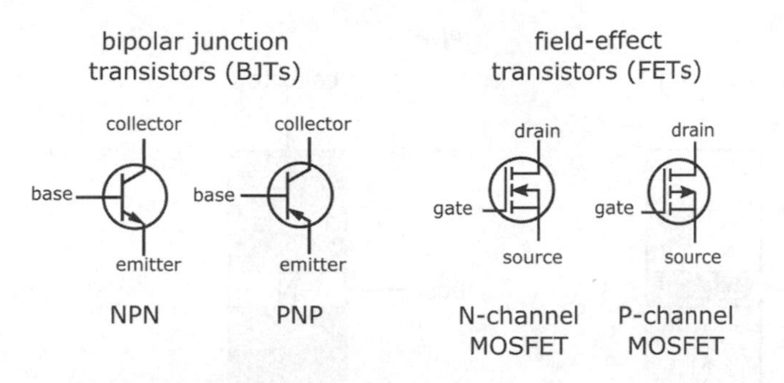

Figure 10-1: Circuit symbols for bipolar junction transistors and field-effect transistors, with labeled leads.

Bipolar junction transistors

One of the first transistors to be invented was the *bipolar junction transistor* (BJT), and BJTs are what most hobbyists use in home-brewed circuits. BJTs consist of two pn-junctions fused together to form a three-layer sandwich-like structure. As Chapter 9 explains, a *pn-junction* is the boundary between two different types of semiconductors: a P-type semiconductor, which contains positive charge carriers (known as *holes*), and an N-type semiconductor, which contains negative charge carriers (electrons).

Leads are attached to each section of the transistor, and are labeled the *base, collector,* and *emitter.* There are two types of bipolar transistors (see Figure 10-2):

- **NPN transistor:** A thin piece of P-type semiconductor is sandwiched between two thicker pieces of N-type semiconductor, and leads are attached to each of the three sections.
- **PNP transistor:** A thin piece of N-type semiconductor is sandwiched between two thicker pieces of P-type semiconductor, and leads are attached to each section.

Bipolar transistors essentially contain two pn-junctions: the base-emitter junction and the base-collector junction. By controlling the voltage applied to the base-emitter junction, you control how that junction is biased (forward or reverse), ultimately controlling the flow of electrical current through the transistor. (In Chapter 9, I explain that a small positive voltage *forward-biases* a pn-junction, allowing current to flow, and that a negative voltage *reverse-biases* a pn-junction, prohibiting current from flowing.)

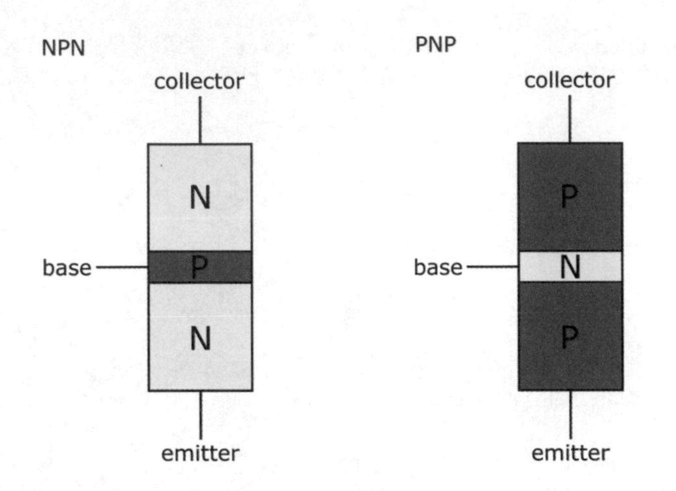

Figure 10-2: Bipolar junction transistors contain two pn-junctions: the base-emitter junction and the base-collector junction.

Field-effect transistors

A *field-effect transistor (FET)* consists of a channel of N- or P-type semiconductor material through which current can flow, with a different material (laid across a section of the channel) controlling the conductivity of the channel (see Figure 10-3).

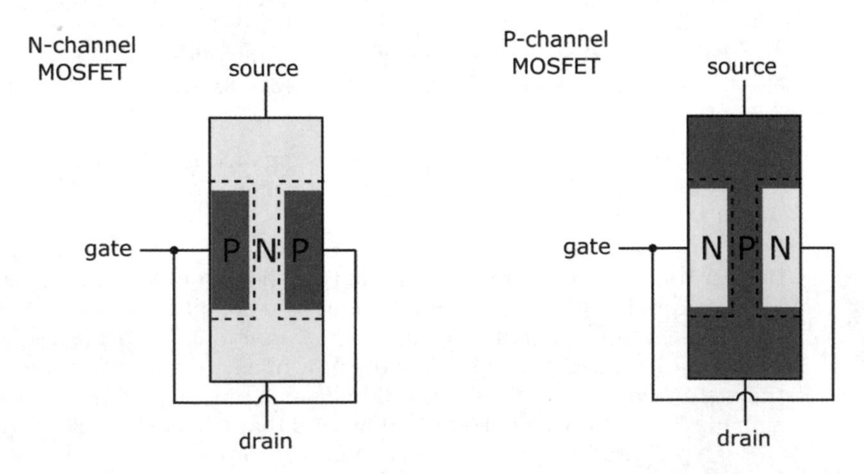

Figure 10-3: In a field-effect transistor (FET), voltage applied to the gate controls the flow of current through a channel from the source to the drain.

One end of the channel is known as the *source,* the other end of the channel is called the *drain,* and the control mechanism is called the *gate.* By applying a voltage to the gate, you control the flow of current from the source to the drain. Leads are attached to the source, drain, and gate. Some FETs include a fourth lead so you can ground part of the FET to the chassis of the circuit. (But don't confuse these four-legged creatures with *dual-gate MOSFETs,* which also have four leads.)

FETs (pronounced "fetts") come in two flavors — N-channel and P-channel — depending on the type of semiconductor material (N-type or P-type, respectively) through which current flows. There are two major subtypes of FET: *MOSFET (metal-oxide-semiconductor FET)* and *JFET (junction FET).* Which is which depends on how the gate is constructed — which results, in turn, in different electrical properties and different uses for each type. The details of gate construction are beyond the scope of this book, but you should be aware of the names of the two major types of FETs.

FETs (particularly MOSFETs) have become much more popular than bipolar transistors for use in *integrated circuits (ICs*), which I discuss in Chapter 11, where thousands of transistors work together to perform a task. That's because they're low-power devices whose structure allows thousands of N- and P-channel MOSFETs to be crammed like sardines on a single piece of silicon (that is, semiconductor material).

Electrostatic discharge (ESD) can damage FETs. If you purchase FETs, be sure to keep them in an antistatic bag or tube — and leave them there until you're ready to use them. You can read more about the harmful effects of ESD in Chapter 13.

Recognizing a transistor when you see one

The semiconductor material in a transistor is the size of a grain of sand or even smaller, so manufacturers put these teensy components in a metal or plastic case with leads sticking out so you can connect them in your circuits. You can find literally dozens upon dozens of different shapes and sizes of transistors, some of which are shown in Figure 10-4.

The smaller packages generally house *signal transistors,* which are rated to handle smaller currents. Larger packages contain *power transistors,* which are designed to handle larger currents. Most signal transistors come in plastic cases, but some precision applications require signal transistors housed in metal cases to reduce the likelihood of stray radio-frequency (RF) interference.

Bipolar transistors typically have three wire leads so you can access the base, collector, and emitter of the transistor. One exception to this is a *phototransistor* (which I discuss in Chapter 12), which is packaged in a clear case and has just two leads (collector and emitter) — light is used to bias the transistor, so you don't have to apply a voltage to the base. All FETs have leads for the source, drain, and gate, and some include a fourth lead so you can ground the transistor's case to the chassis of your circuit, or for the second gate of a dual-gate MOSFET.

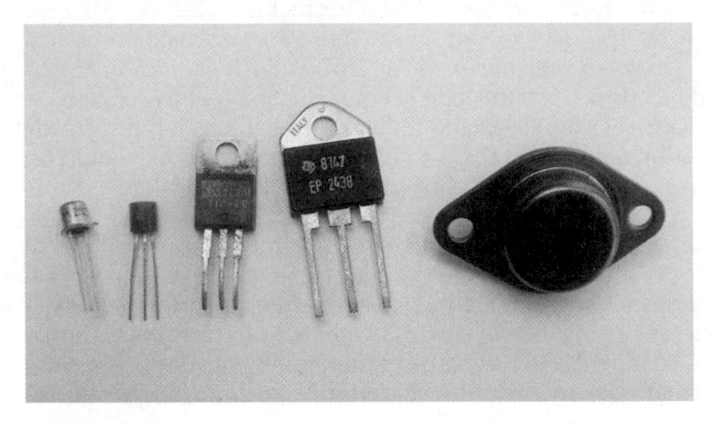

Figure 10-4: The drab, dull packaging of the average transistor is just a cover for its exciting, game-changing interior.

To figure out which package lead is which, consult the documentation for the specific transistor. Be careful how you interpret the documentation: Transistor connections are often (though not always) shown from the underside of the case, as if you've turned the transistor over and are gazing at it from the bottom.

You absolutely must install transistors the proper way in your circuits. Switching the connections around can damage a transistor and may even damage other circuit components.

Making all kinds of components possible

Transistors can be combined in all sorts of ways to make lots of incredible things happen. Because the semiconductor material that makes up a transistor is so small, it's possible to create a circuit containing hundreds or thousands of transistors (along with resistors and other components) and plop the entire circuit into a single component that fits easily into the palm of your hand. These amazing creations, known as *integrated circuits (ICs)*, enable you to build *really* complex circuits with just a few parts. The next chapter takes a look at some of the ICs that are available today as a result of the semiconductor revolution.

Examining How Transistors Work

BJTs and FETs work basically the same way. The voltage you apply to the input (*base*, for a BJT, or *gate*, for a FET) determines whether or not current flows through the transistor (from collector to emitter for a BJT, and from source to drain for a FET).

To get an idea of how a transistor works (specifically, a FET), think of a pipe connecting a source of water to a drain with a controllable valve across a section of the pipe, as shown in Figure 10-5. By controlling whether the valve is fully closed, fully open, or partially open, you control the flow of water from the source to the drain.

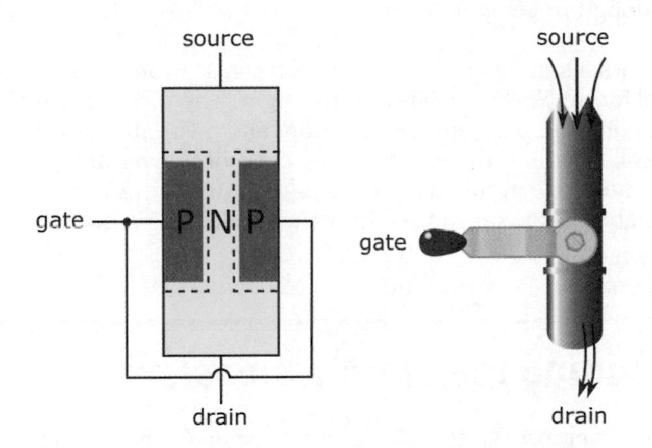

Figure 10-5: Like a valve, a transistor can be off (no current), fully on (maximum current), or partially on (amount of current depends on how wide the gate is open, so to speak).

You can set up the control mechanism for your valve in two ways. The control mechanism can

- **Act like an on/off switch,** either fully opening or fully closing with nothing in between.

- **Open partially,** depending on how much force you exert on it. When it's partially open, you can adjust the valve a little to allow more or less water to flow from source to drain; small changes in the force you exert on the valve create similar, yet larger, changes in the flow of water. That's how a transistor acts as an amplifier.

Bipolar transistors work in a similar way: The base acts like the controllable valve in Figure 10-5, controlling the flow of electrons from the emitter to the collector (or, in circuit-speak, the flow of conventional current from the collector to the emitter). By controlling the base, you can turn the transistor fully on or fully off, or you can allow small changes at the base to control large changes in current from collector to emitter.

Using a model to understand transistors

Exactly how a transistor works internally involves technical details about free electrons, moving holes, pn-junctions, and biasing. You don't have to know all that technical stuff by chapter and verse to use transistors in circuits. Instead, familiarize yourself with a functional model of a transistor, and you'll know enough to get going.

Figure 10-6 shows a simple model of an NPN transistor on the left, and the circuit symbol for an NPN transistor on the right. The model includes a diode representing the base-emitter junction, and a variable resistance, R_{CE}, between the collector and the emitter. The value of the variable resistance is controlled by the diode in this model. Voltages, currents, and transistor terminals are labeled so you can see how the model corresponds to the actual device.

Decoding the word *transistor*

So why are transistors called transistors? Well, the word *transistor* is a combination of two word parts: *trans* and re*sistor.*

The *trans* part of the name conveys the fact that by placing a forward-biased voltage on the base-emitter junction, you cause electrons to flow in another part of the component, from emitter to collector. You *transfer* the action from one part of the component to another. This is known as *transistor action.*

Because fluctuations in base current result in proportional fluctuations in collector/emitter current, you can think of the transistor as a sort of variable resistor: When you turn the dial (by varying the base current), the resistance changes, producing a proportionally varying collector/emitter current. That's where the *sistor* part of the name comes from.

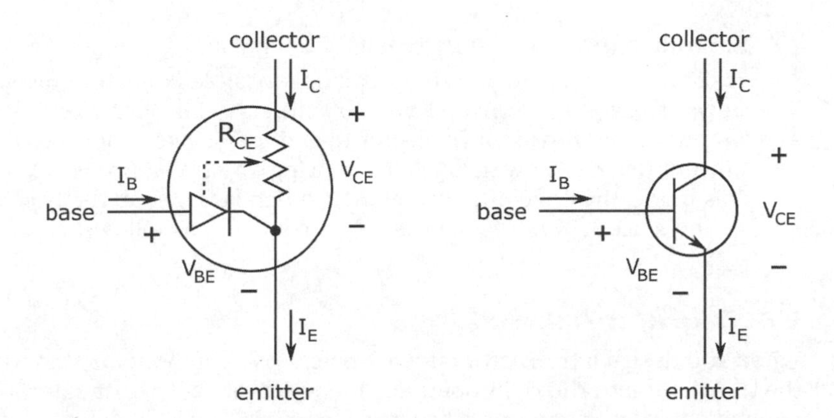

Figure 10-6: A transistor works as a switch or an amplifier, depending on what you input to the base.

Here's what the voltages and currents represent:

- V_{BE} is the voltage across the base-emitter junction, which is a pn-junction, just like a diode.
- I_B (the base current) is the current flowing into the base of the transistor.
- V_{CE} is the voltage from the collector to the emitter. This voltage will vary depending on what is going on at the base.
- I_C (the collector current) is the current flowing into the collector.
- I_E (the emitter current) is the current flowing out of the emitter. The emitter current is the sum of the collector current and the base current: $I_E = I_C + I_B$.

The transistor has three *operating modes,* or possibilities for how it operates:

- **Cutoff (transistor off):** If $V_{BE} \leq 0.7$ V, the diode is off, so $I_B = 0$. This makes the resistance R_{CE} infinite, which means $I_C = 0$. The output of the transistor (collector-to-emitter) is like an open switch: No current is flowing. This mode of operation is called *cutoff.*
- **Active (transistor partially on):** If $V_{BE} \geq 0.7$ V, the diode is on, so base current flows. If I_B is small, the resistance R_{CE} is reduced and some collector current, I_C, flows. I_C is directly proportional to I_B, with a *current gain,* h_{FE}, equal to I_C/I_B, and the transistor is functioning as a current amplifier — that is, operating in *active* mode.

- **Saturation (transistor fully on):** If $V_{BE} \geq 0.7$ V and I_B is increased a lot, the resistance R_{CE} is zero and the maximum possible collector current, I_C, flows. The voltage from collector to emitter, V_{CE}, is nearly zero, so the output of the transistor (collector-to-emitter) is like a closed switch: All current that can flow through it is flowing; the transistor is *saturated.* In this mode, the collector current, I_C, is much larger than the base current, I_B, and since $I_E = I_C + I_B$, you can approximate I_E as follows: $I_E \approx I_C$.

Operating a transistor

When you design a transistor circuit, you choose components that will put the transistor into the right operating mode (cutoff, active, or saturation), depending on what you want the transistor to do. Here's how:

- **Transistor amplifier:** If you want to use the transistor as an amplifier (active mode), you select supply voltages and resistors to connect to the transistor so that you forward-bias the base-emitter junction and allow just enough base current to flow — but not so much that the transistor becomes saturated. This selection process is known as *biasing* the transistor.
- **Transistor switch:** If you want the transistor to act like an on/off switch, you choose values of supply voltages and resistors so that the base-emitter junction is either nonconducting (the voltage across it is less than 0.7 V) or fully conducting — with nothing in between. When the base-emitter junction is nonconducting, the transistor is in cutoff mode and the switch is off. When the base-emitter junction is fully conducting, the transistor is in saturation mode and the switch is on.

Choosing the right switch

You may wonder why you would use a transistor as a switch when there are so many other types of switches and relays available (as Chapter 4 describes). Well, transistors have several advantages over other types of switches, and so are used when they're the best choice. Transistors use very little power, can switch on and off several billion times per second, and can be made microscopically small, so integrated circuits (which I discuss in Chapter 11) use thousands of transistors to switch signals around on a single tiny chip. Mechanical switches and relays have their uses, too, in situations where transistors just can't handle the load, such as switching currents bigger than about 5 A or switching higher voltages (as in electrical power systems).

Amplifying Signals with a Transistor

Transistors are commonly used to amplify small signals. (See the sidebar titled "Deciphering electrical signals" for details on what signals are.)

Suppose that you produce an audio signal as the output of one stage of an electronic circuit, and you'd like to amplify it before shipping it off to another stage of electronics, such as a speaker. You use a transistor, as shown in Figure 10-7, to amplify the small up-and-down fluctuations in the audio signal (v_{in}), which you input to the base of the transistor (labeled Q in the schematic). The transistor transforms them into *large* signal fluctuations (v_{out}), which appear at the output (collector) of the transistor. Then you take the transistor output and apply it to the input of your speakers.

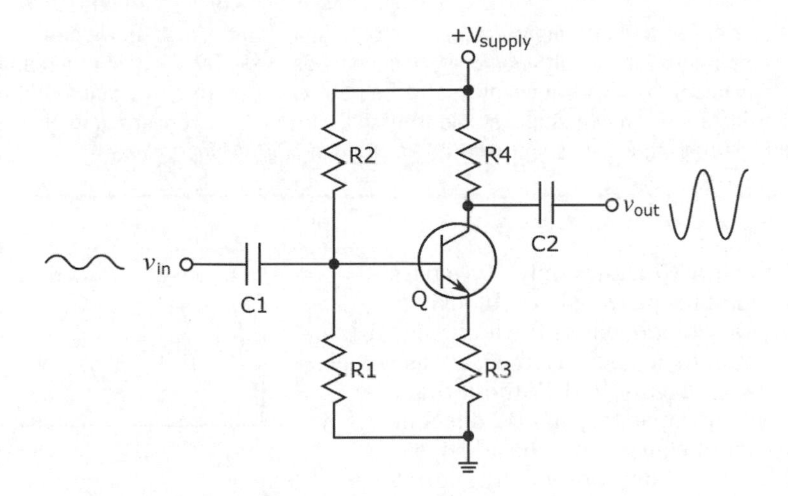

Figure 10-7: By strategically positioning a few resistors in a transistor circuit, you can properly bias a transistor and control the gain of this amplifier circuit.

Biasing the transistor so it acts like an amplifier

A transistor must be partially on to work as an amplifier. To put a transistor in this state, you bias it by applying a small voltage to the base. In the example that was shown in Figure 10-7, resistors *R1* and *R2* are connected to the base of the transistor and configured as a voltage divider (for more about how a voltage divider works, see Chapter 6), dividing the supply voltage, V_{supply}. The output of this voltage divider, $\frac{R1}{R1+R2} \times V_{supply}$, supplies enough voltage to the base to turn the transistor on and allow current to flow through it, biasing the transistor so it's in the active mode (that is, partially on).

Deciphering electrical signals

Transistors are commonly used to amplify signals. An *electrical signal* is the pattern over time of an electrical current. Often, the way an electrical signal changes its shape conveys information about something physical, such as the intensity of light, heat, or sound, or the position of an object, such as the diaphragm in a microphone or the shaft of a motor. Think of an electrical signal as a code, somewhat like Morse code, sending and receiving secret messages that you can figure out — if you know the key.

An *analog electrical signal,* or simply *analog signal,* is so named because it is an *analog,* or one-to-one mapping, of the physical quantity it represents. For instance, when a sound studio records a song, fluctuations in air pressure (that's what sound is) move the diaphragm of a microphone, which produces corresponding variations in electrical current. That fluctuating current is a representation of the original sound, or an audio signal.

Digital systems, like computers, can't handle continuous analog signals, so electrical signals must be converted into digital format before entering the depths of a digital system. *Digital format* is just another coding scheme, one that uses only the binary values 1 and 0 to represent information, much like Morse code uses dots and dashes. A *digital signal* is created by sampling the value of an analog signal at regular intervals in time and converting each value into a string of *bits,* or *binary digits.*

Capacitor *C1* allows only AC to pass through to the transistor, blocking any DC component of the input signal (an effect known as a *DC offset*), as shown in Figure 10-8. Without that blocking capacitor, any DC offset in the input signal would be added to the bias voltage, potentially saturating the transistor or shutting it off (cutoff) so it no longer acts like an amplifier.

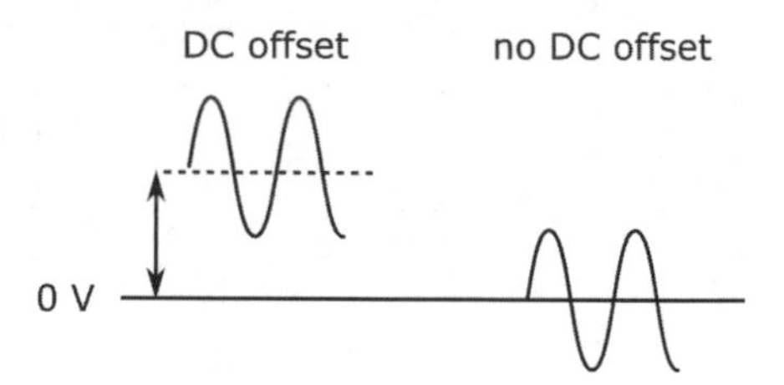

Figure 10-8: Blocking capacitor *C1* helps maintain the bias of the transistor by filtering out DC offsets in the input signal before the signal gets to the transistor.

Controlling the voltage gain

With the transistor in Figure 10-7 partially on, fluctuations in the base current caused by the AC input signal get amplified by the transistor. Because the current gain, h_{FE}, of any transistor you happen to choose can be somewhat variable (schizophrenic, in fact), you design your amplifier circuit to eliminate any dependency on the flaky current gain. You'll give up some strength of amplification, but you'll get stability and predictability in return.

By placing resistors *R3* and *R4* in the circuit, you can control the *voltage gain,* or how much the input signal is amplified — without worrying about the exact current gain of the specific transistor at the heart of your circuit. (This is truly amazing stuff!) The AC voltage gain of a transistor circuit with resistors as shown in Figure 10-7 is –*R4*/*R3*. The negative sign just means that the input signal is *inverted:* As the input voltage varies up and then down, the output voltage varies down and then up, as shown by the input and output signal waveforms in Figure 10-7. Before sending the output signal to, say, a speaker, you pass it through another blocking capacitor (*C2*) to remove any DC offset.

Configuring transistor amplifier circuits

The type of transistor setup I discuss in the preceding section is known as a *common-emitter amplifier* (so-named because the emitter is tied to common ground); this circuit is just one of many ways to configure transistor circuits for use as amplifiers. You use different configurations to achieve different goals, such as high power gain versus high voltage gain. How the circuit behaves depends on

- How you connect the transistor to the power supplies
- The location of the load
- What other circuit components (such as resistors, capacitors, and other transistors) you add to the circuit
- Where you add other components in the circuit

For instance, you can piggyback two bipolar transistors in a setup known as a *Darlington pair* to produce multiple stages of amplification. (In the section "Gaining Experience with Transistors," later in this chapter, you find out how to configure a simple Darlington pair.) Or you can get the same result the easy way: Purchase a three-lead component called a *Darlington transistor,* which includes a Darlington pair already hooked up.

Designing transistor amplifier circuits is a field of study on its own, and many excellent books have been written on the subject. If you're interested in learning more about transistors and how to design amplifier circuits using transistors, try getting your hands on a good electronics design book, such as *The Art of Electronics,* 3rd Edition by Paul Horowitz and Winfield Hill (Cambridge University Press). It isn't cheap, but it's a classic.

Switching Signals with a Transistor

You can also use a transistor as an electrically operated switch. The base lead of the transistor works like the toggle on a mechanical switch as follows:

- The transistor switch is off when no current flows into the base (in cutoff), and the transistor acts like an open circuit — even if there is a voltage difference from collector to emitter.
- The transistor switch is on when current flows into the base (in saturation), and the transistor acts like a closed switch, fully conducting current from collector to emitter — and out to whatever load you want to turn on.

How do you get this on/off thing to work? Say you use an electronic gadget to scatter chicken feed automatically at dawn. You can use a *photodiode,* which conducts current when exposed to light, to control the input to a transistor switch that delivers current to your gadget (the load). At night, the photodiode doesn't generate any current, so the transistor is off. When the sun rises, the photodiode generates current, turning the transistor on and allowing current to flow to your gadget. The gadget then starts scattering chicken feed — keeping the chickens happy while you continue to snooze.

Are you're wondering why you don't just supply the current from the photodiode to the gadget? Your gadget might need a larger current than can be supplied by the photodiode. The small photodiode current controls the on/off action of the transistor, which acts like a switch to allow a larger current from a battery to power your gadget.

One of the reasons transistors are so popular for switching is that they don't dissipate a lot of power. Remember that power is the product of current and voltage. When a transistor is off, no current flows, so the power dissipated is zero. When a transistor is fully on, V_{CE} is nearly zero, so the power dissipated is nearly zero.

Choosing Transistors

Transistors have become so popular that thousands upon thousands of different transistors are available. So how do you choose one for your circuit, and how do you make sense of all the choices on the market?

If you're designing a transistor circuit, you need to understand how your circuit will operate under various conditions. What is the maximum amount of collector current your transistor will have to handle? What is the minimum current gain you need to amplify an input signal? How much power could possibly be dissipated in your transistor under extreme operating conditions (for instance, when the transistor is off and the entire power supply voltage may be dropped across the collector-emitter)?

After you understand the ins-and-outs of how your circuit will operate, you can start looking up transistor specifications to find one that meets your needs.

Important transistor ratings

Loads of parameters are used to describe the loads of different transistors, but you need to know only a few parameters to choose the right transistor for your circuit. For bipolar (NPN or PNP) transistors, here's what you need to know:

- **Maximum collector current (I_{Cmax}):** The maximum DC current that the transistor can handle. When designing a circuit, make sure you use a resistor to limit the collector current so it doesn't exceed this value.
- **DC current gain (h_{FE} or β):** The ratio of collector current to base current (that is, I_C/I_B), which provides an indication of the amplifying capability of the transistor. Typical values are 50 to 200. Because the current gain can vary — even among transistors of the same type — you need to know the guaranteed minimum value of h_{FE}, and that's what this parameter tells you. The h_{FE} value also varies for different values of I_C, so sometimes h_{FE} is given for a specific value of I_C, such as 20 mA.
- **Maximum collector-to-emitter voltage (V_{CEmax}):** The maximum voltage across the collector and emitter; this value is usually at least 30 V. If you're working with low-power applications such as hobby electronics circuits, don't worry about this value.
- **Total power dissipation (P_{total}):** The total power that the transistor can dissipate; this value is roughly $V_{CE} \times I_{Cmax}$. No need to worry about this rating if you're using the transistor as a switch because power dissipation is nearly zero anyway. If you're using the transistor as an amplifier, however, you need to be aware of this rating.

If you think your circuit will approach the total power dissipation value, be sure you attach a heat sink to the transistor.

To determine these characteristics, consult the component's *data sheet,* which provides technical specifications for the component, on the manufacturer's website. If you're building a circuit someone else designed, you don't have to worry too much about the specs; you can simply use the transistor specified by the designer or consult a cross-reference to find a similar model to substitute.

Identifying transistors

Many bipolar transistors originating in North America are identified by a five- or six-digit code that is part of an industry-standard semiconductor identification system. The first two digits are always 2N for transistors, with the 2 specifying the number of pn-junctions and the N signifying a semiconductor.

The remaining three or four digits indicate the specific features of the transistor. However, different manufacturers may use different coding schemes, so your best bet is to consult the appropriate website, catalog, or specification sheet to make sure you're getting what you need for your circuit.

Many suppliers categorize transistors according to the type of application they are used in, such as low-power, medium-power, high-power, audio (low-noise), or general-purpose. Knowing the category that describes your project can help guide you to choose the right transistor for your particular circuit.

Gaining Experience with Transistors

In this section, you see how tiny transistors control the current in one circuit (at the output of the transistor) using electronic components in another circuit (at the input of the transistor). That's what transistor action is all about!

Amplifying current

You can use the circuit in Figure 10-9 to demonstrate the amplification capabilities of a transistor.

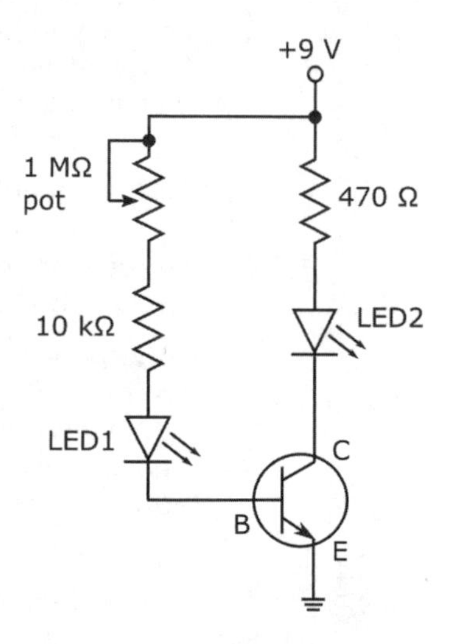

Figure 10-9: A pair of LEDs helps you visualize the amplification capabilities of a transistor.

Here are the parts you need to build the circuit:

- One 9-volt battery with battery clip
- One 2N3904 or BC548 (or any general-purpose) NPN bipolar transistor
- One 470 Ω resistor (yellow-violet-brown)
- One 10 kΩ resistor (brown-black-orange)
- One 1 MΩ potentiometer
- Two LEDs (any size, any color)
- One solderless breadboard and jumpers

Refer to Chapter 2 or Chapter 19 for information on where to get parts.

Using Figure 10-10 as your guide, follow these steps:

1. **Build the circuit using a general-purpose NPN bipolar transistor, such as a 2N3904 or a BC548.**

   Be careful to connect the base, collector, and emitter leads properly (consult the transistor package or datasheet), and to orient the LEDs correctly (as shown in Chapter 9)

2. **Dial the potentiometer all the way up so that the resistance is 1 MΩ.**

   You'll probably see a tiny glow from LED2, but you may not see any light coming from LED1 — although there *is* a teeny current passing through LED1.

3. **Now slowly dial the pot down and observe the LEDs.**

   You should see LED2 getting steadily brighter as you dial the pot down. At some point, you'll start to see light from LED1 as well. As you continue to dial the pot down, both LEDs will glow brighter, but LED2 will be clearly much brighter than LED1.

Figure 10-10: The small base current barely lights the red LED, while the larger collector current brightly illuminates the green LED.

You're witnessing transistor action: The tiny base current passing through LED1 is amplified by the transistor, which allows a much larger current to flow through LED2. You see a dim glow from LED1 due to the tiny base current, and a bright glow from LED2 due to the stronger collector current. You can measure each current, if you'd like. (See the sidebar "Measuring teeny tiny currents" for a tip on how to measure the small base current.)

Measuring teeny tiny currents

The base current of the bipolar transistor in Figure 10-10, which passes through LED1, is very small, especially when the pot is set to its maximum resistance. If you'd like to measure this teeny current, you can do it in these different ways:

- Make the measurement directly, by setting your multimeter to DC amps, breaking the circuit on one side of LED1, and inserting your multimeter in series with LED1. (The current is so small that it may not register on your meter.)
- Measure the current indirectly, using Ohm's Law to help you. The same current that passes through LED1 and into the base of the transistor also passes through two resistors: the 10 kΩ resistor and the potentiometer. You can measure the voltage drop across either resistor and divide the voltage reading by the resistance. (In Chapter 6, you find out that Ohm's Law tells you that the current passing through a resistor is equal to the voltage across the resistor divided by the resistance.)
- If you want an exact measurement, power the circuit down, pull the resistor out of the circuit, and measure its exact resistance with your multimeter. Then perform the current calculation. Using this method, I measured a base current of 6.7 μA (that's 0.0000067 A).

With the pot set to 1 MΩ, I measured a base current of 6.7 μA (that's 0.0000067 A) and a collector current of 0.94 mA. Dividing the collector current by the base current, I found that the current gain of this transistor circuit is 140. With the pot set to 0 Ω, I measured a base current of 0.65 mA and a collector current of 14 mA, for a current gain of approximately 21.5. Pretty intense!

The switch is on!

The circuit in Figure 10-11 is a touch switch. It uses a pair of NPN transistors to amplify a really teeny base current enough to light the LED. This piggyback configuration of two bipolar transistors, with their collectors connected and the emitter of one feeding into the base of the other, is known as a *Darlington pair.* (The letter Q is used to label transistors in schematics.)

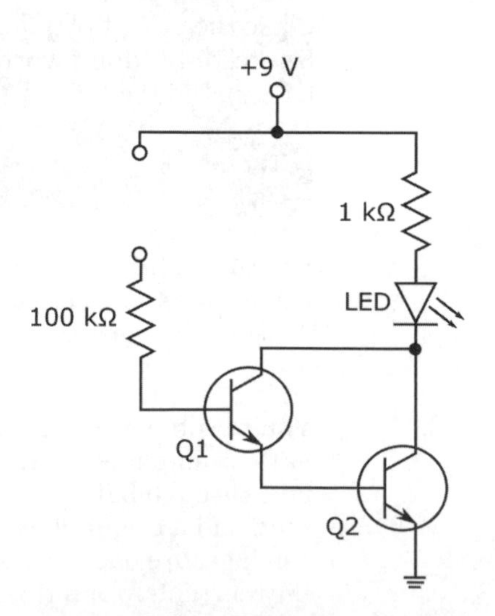

Figure 10-11: A Darlington pair can be used as a touch switch.

To test it, set up the circuit as shown in Figure 10-12 using the following parts:

- One 9-volt battery with battery clip
- One 100 kΩ resistor (brown-black-yellow)

- One 1 kΩ resistor (brown-black-red)
- One LED (any size, any color)
- Two 2N3904 or BC548 (or any general-purpose) NPN bipolar transistors
- One solderless breadboard and jumper wires

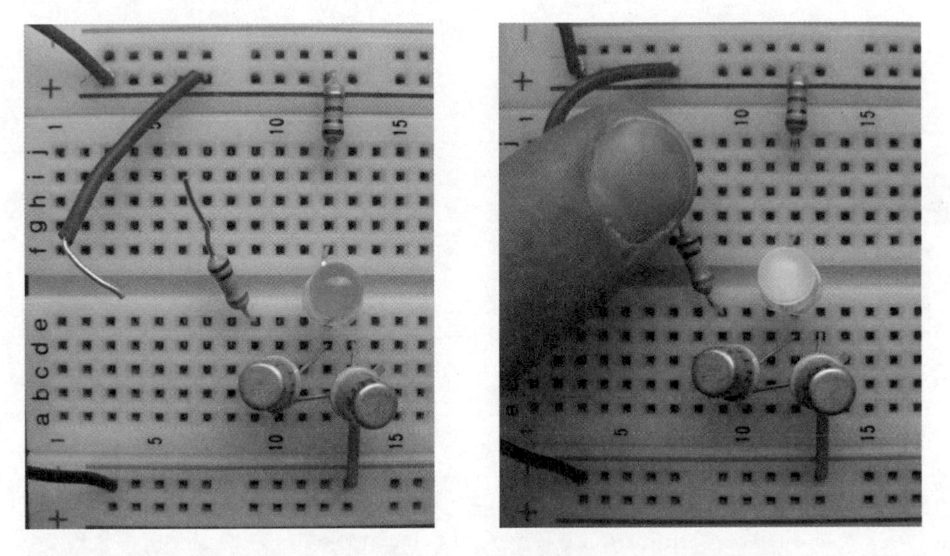

Figure 10-12: One way to set up the touch switch transistor circuit. You can turn on the LED by placing your finger across the red wire and the 100 kΩ resistor.

Close the circuit by placing your finger across the open circuit, as shown in Figure 10-12 (don't worry, my ten-year-old son didn't get hurt and you won't either). Did the LED turn on? When you close the circuit, your skin conducts a teeny tiny current (a few microamps), which is amplified by the pair of transistors, lighting the LED.

If you touch the unconnected lead of the 100 kΩ resistor (without closing the circuit), you may see the LED light up briefly and go out (especially if you shuffle your feet on a carpet first). That's because you've built up a minute amount of charge on your finger, and when you touch the resistor, the charge flows into the base of the first transistor and is amplified enough by the Darlington pair to light the LED. (If the nominal gain, h_{FE}, of your transistors is 100, the total gain of the Darlington pair is $100 \times 100 = 10{,}000$.) After the charge has dissipated, the LED goes out. (Note that if you use an antistatic wrist strap when you touch the resistor, the LED will not light.)

11

Innovating with Integrated Circuits

In This Chapter

- Corralling components onto a chip
- Speaking the language of bits
- Thinking logically about gates
- Reading into IC packages
- Pondering IC pinouts
- Boosting signals with op amps
- Timing, counting, and controlling everything in sight

Space exploration, programmable pacemakers, consumer electronics, and much more would be nothing but the idle dreams of creative minds were it not for the integrated circuit (IC). This incredible innovation — really, a series of incredible innovations — makes possible your smartphone, tablet, iPod, and GPS navigation system, as well as newer disruptive technologies, such as 3D printing and autonomous (self-driving) vehicles. Who knows? One day ICs may enable you to 3D-print your own driverless car!

An *integrated circuit* incorporates anywhere from a few dozen to many billions (yes — billions!) of circuit components into a single device that fits easily into the palm of your hand. Each IC contains an intricate mesh of tiny transistor-based taskmasters, with access to the outside world provided by a finite number of inputs and outputs.

This chapter explores how integrated circuits came to be, identifies the three major IC flavors, and dissects the inner workings of one variety — digital ICs. You get a look at how computers and other digital devices manipulate two distinct voltage levels to process information using special rules known as logic. Next up is an explanation of how to read an IC to understand what the heck it does (because you can't tell by its cover) and how to connect it for use in circuits. Finally, you get a closer look at three best-selling ICs, what they do, and how you can use them to create your own innovative circuits.

Why ICs?

The integrated circuit (IC) was invented in 1958 (see the sidebar, "The birth of the IC") to solve the problems inherent in manually assembling mass quantities of tiny transistors. Also called *chips,* integrated circuits are miniaturized circuits produced on a single piece of semiconductor. A typical integrated circuit contains hundreds of transistors, resistors, diodes, and capacitors; the most advanced ICs contain several billion components. Because of this circuit efficiency, you can build complex circuits with just a few parts. ICs are the building blocks of larger circuits. You string them together to form just about any electronic device you can think up.

The birth of the IC

With the invention of the transistor in 1947, the focus of electronic design shifted away from bulky vacuum tubes to this newer, smaller, more reliable device. Because size was no longer an obstacle, engineers worked to build more and more advanced circuits. Their success in creating advanced designs, however, led to some practical problems: Interconnecting hundreds of components inevitably resulted in errors that were extremely difficult to isolate. Additionally, complex circuits often failed to meet speed requirements (because it does take some time for electrons to travel through a maze of wires and components). Throughout the 1950s, a major focus in the electronics industry was figuring out how to make circuits smaller and more reliable.

In 1952, a British engineer named Geoffrey Dummer publicly presented his idea for combining multiple circuit elements onto a single piece of semiconductor material with no connecting wires. He reasoned this would eliminate the faulty wiring and the cumbersome manual assembly of discrete components. Although Dummer never actually built an IC, he is widely regarded as the "Prophet of the Integrated Circuit."

Then, in the summer of 1958, Jack Kilby, a newly employed engineer at Texas Instruments working alone in a lab (while his colleagues were on vacation) was able to build multiple circuit components out of a single, monolithic piece of germanium (a semiconductor material), and lay metal connectors in patterns on top of it. Kilby's crude design was the first successful demonstration of the integrated circuit. Six months later, Robert Noyce of Fairchild Semiconductor (who also co-founded Intel) invented his own version of the IC, which solved many of the practical problems inherent in Kilby's design and led to the mass production of ICs. Together, Kilby and Noyce are credited with the invention of the integrated circuit. (Kilby was awarded the Nobel Prize in Physics for his contributions to the invention of the integrated circuit — but not until 42 years later — and stated that had Noyce been alive when the Prize was awarded, he surely would have shared it.)

A lot has happened since 1958. All those smart, creative people continued to plug away at their work, and many more innovations took place. As a result, the electronics industry has exploded as *chip densities* (a measure of how closely packed the transistors are) have increased exponentially. Today, semiconductor manufacturers routinely carve millions of transistors into a piece of silicon smaller than the size of a dime, and microprocessors containing more than one billion transistors have been around since 2006. (Kinda makes your head spin, doesn't it?)

Linear, Digital, or Combination Plate?

Over the years, chip makers have come out with lots of different ICs, each of which performs a specific function depending on how the components inside are wired. Many of the integrated circuits you encounter are so popular they have become standardized, and you can find a wealth of information about them online and in books. A lot of different chip makers offer these standardized ICs, and manufacturers and hobbyists the world over buy and use them in various projects. Other special-purpose ICs are designed to accomplish a unique task. More often than not, only a single company sells a particular special-purpose chip.

Whether standardized or special-purpose, you can separate ICs into three main categories: *linear (analog), digital,* and *mixed signal.* These terms relate to the kinds of electrical signals (more about those Chapter 10) that work in the circuit:

- **Linear (analog) ICs:** These ICs contain circuits that process *analog signals,* which consist of continuously varying voltages and currents. Such circuits are known as *analog circuits.* Examples of analog ICs are power-management circuits, sensors, amplifiers, and filters.
- **Digital ICs:** These ICs contain circuits that process *digital signals,* which are patterns consisting of just two voltage (or current) levels representing binary digital data, for instance, on/off, high/low, or 1/0. (I discuss digital data a bit more in the next section.) Such circuits are known as *digital circuits.* Some digital ICs, such as microprocessors, contain millions — or billions — of tiny circuits in just a few square millimeters.
- **Mixed signal ICs:** These ICs contain a combination of analog and digital circuits and are commonly used to convert analog signals to digital signals for use in digital devices. Analog-to-digital converters (ADCs), digital-to-analog converters (DACs), and digital radio processors are examples of mixed signal chips.

Making Decisions with Logic

When you first learned to add numbers, you memorized facts such as "2 + 2 = 4" and "3 + 6 = 9." Then, when you learned to add multi-digit numbers, you used those simple facts as well as a new one — "carrying" numbers to us older folks, "regrouping" numbers to the younger generation. By applying a few simple addition facts and one simple rule, you can add two large numbers together fairly easily.

The microprocessor in your computer works in much the same way. It uses lots of teeny digital circuits — known among computer types as *digital logic* — to process simple functions similar to "2 + 2 = 4." Then the logic combines the outputs of those functions by applying rules similar to carrying/regrouping to get an answer. By piggybacking lots of these "answers" together in a complex web of circuitry, the microprocessor can perform some complicated mathematical tasks. Deep down inside, though, there's just a bunch of logic applying simple little rules.

In this section, you get a look at how digital logic circuits work.

Beginning with bits

When you add two digits together, you have ten choices for each digit (0 through 9) because that's how our numbering system (known as a *base 10,* or *decimal, system*) works. When a computer adds two digits together, it uses only two possible digits: 0 and 1 (this is known as a *base 2* or *binary system*). Because there are only two, these digits are known as binary digits, or *bits.* Bits can be strung together to represent letters or numbers — for instance, the bit string 1101 represents the number 13. The sidebar "Adding bits of numbers" offers a glimpse of how this process works.

In addition to representing numbers and letters, bits can be used also to carry information. As information carriers, data bits are versatile. They can represent many *two-state* (binary) things: a screen's pixel is either on or off; a CTRL key is either up or down; a laser pit is either present or absent on a DVD surface; an ATM transaction is either authorized or not; and much more. By assigning logical values of 1 and 0 to a particular on/off choice, you can use bits to carry information about physical events — and allow that information to control other things by processing the bits in a digital circuit.

Logical 1 and logical 0 are also referred to as true and false, or high and low. But what exactly *are* these ones and zeros in a digital circuit? They are simply high or low currents or voltages that are controlled and processed by transistors. (In Chapter 10, I discuss how transistors work and how they can be used as on/off switches.) Common voltage levels used to represent digital data are 0 volts for logical 0 (low), and (often) 5 volts for logical 1 (high).

A *byte,* which you've probably heard about quite a bit, is a grouping of eight bits used as a basic unit of information for storage in computer systems. Computer memory, such as random access memory (RAM), and storage devices such as CDs and flash drives, use bytes to organize gobs of data. Just as banks pack 40 quarters into a quarter roll, 50 dimes into a dime roll, and so forth to simplify the process of supplying merchants with change for their cash registers, so computer systems pack data bits together in bytes to simplify the storage of information.

TECHNICAL STUFF

Adding bits of numbers

In the decimal *(base 10)* system, if you want to express a number greater than 9, you need to use more than one digit. Each position, or place, in a decimal number represents a *power of ten* (10^0, 10^1, 10^2, 10^3, and so forth), and the value of the digit (0–9) sitting in that position is a multiplier for that power of ten. With powers of ten, the *exponent* (that tiny number raised up next to the 10) tells you how many times to multiply 10 times itself, so 10^1 equals 10, 10^2 equals 10×10 which is 100, 10^3 equals $10 \times 10 \times 10$ which is 1,000, and so on. As for 10^0, it just equals 1 because *any* number raised to the zeroth power equals 1. So the positions in a decimal number, starting from the rightmost position, represent 1, 10, 100, 1,000, and so forth. These are also known as *place values* (ones or units, tens, hundreds, thousands, and so forth). The digit (0–9) sitting in that position (or place) tells you how many ones, tens, hundreds, thousands, and so forth are contained in that decimal number.

For example, the number 9,452 can be written in *expanded notation* as

$$(9\times1{,}000)+(4\times100)+(5\times10)+(2\times1)$$

Our entire mathematics system is based on the number 10 (but if humans had only eight fingers, we might be using a *base 8* system), so your brain has been trained to automatically think in decimal format (it's like a math language). When you add two digits together, such as 6 and 7, you automatically interpret the result, 13, as "1 group of 10 plus 3 groups of 1." It's ingrained in your brain, no less than your native language is.

Well, the binary system is like another language: It uses the same methodology, but it's based on the number 2. If you want to represent a number greater than 1, you need more than one digit and each position in your number represents a *power of two:* 2^0, 2^1, 2^2, 2^3, 2^4, and so forth, which is the same as 1, 2, 4, 8, 16, and so forth. The *bit* (a bit is a binary digit, just 0 or 1) that sits in that position in your number is a multiplier for that power of two. For example, the *binary number* 1101 can be written in expanded notation as

$$(1\times2^3)+(1\times2^2)+(0\times2^1)+(1\times2^0)$$

By translating this into decimal format, you can see what numerical quantity the bit string 1101 represents:

$$\begin{aligned}&(1\times8)+(1\times4)+(0\times2)+(1\times1)\\&=8+4+0+1\\&=13\end{aligned}$$

So the binary number 1101 is the same as the decimal number 13. They're just two different ways of representing the same physical quantity. This is analogous to saying "bonjour" or "buenos días" rather than "good day." They're just different words for the same greeting.

When you add two binary numbers, you use the same methodology that is used in the decimal system, but using 2 as a base. In the decimal system, $1+1=2$, but in the binary system, $1+1=10$ (remember, the binary number 10 represents the same quantity as the decimal number 2). Computers use the binary system for arithmetic operations because the electronic circuits inside computers can work easily with bits, which are just high or low voltages (or currents) to them. The circuit that performs addition inside a computer contains several transistors arranged in just the right way so that when high or low signals representing the bits of two numbers are applied to the transistor inputs, the circuit produces the right combination of high or low outputs to represent the bits of the numerical sum. Exactly how this is done is beyond the scope of this book, but I hope that you now have an idea of how this type of thing works.

Processing data with gates

Logic gates, or simply *gates,* are tiny digital circuits that accept one or more data bits as inputs and produce a single output bit whose value (1 or 0) is based on a specific rule. In the same way that different arithmetic operators produce different outputs for the same two inputs (for instance, three *plus* two produces five, while three *minus* two produces one), different types of logic gates produce different outputs for the same inputs:

- **NOT gate (inverter):** This single-input gate produces an output that inverts the input. An input of 1 generates an output of 0, and an input of 0 generates an output of 1. A more common name for a NOT gate is an *inverter.*
- **AND gate:** The output is 1 only if both inputs are 1. If either input is 0, the output is 0. A standard AND gate has two inputs, but you can also find three-, four-, and eight-input AND gates. For those gates, the output is 1 only if *all* inputs are 1.
- **NAND gate:** This function behaves like an AND gate followed by an inverter (hence the NAND, which means NOT AND). It produces an output of 0 only if all of its inputs are 1. If any input is 0, the output is 1.
- **OR gate:** The output is 1 if at least one of its inputs is 1. It produces an output of 0 only if both inputs are 0. A standard OR gate has two inputs, but three- and four-input OR gates are also available. For these gates, an output of 0 is generated only when all inputs are 0; if one or more inputs is 1, the output is 1.
- **NOR gate:** This behaves like an OR gate followed by a NOT gate. It produces an output of 0 if one or more of its inputs are 1, and generates an output of 1 only if all inputs are 0.
- **XOR gate:** The exclusive OR gate produces an output of 1 if either one — but not both — of the inputs is 1; otherwise, it produces an output of 0. All XOR gates have two inputs, but multiple XOR gates can be cascaded together to create the effect of XORing multiple inputs.
- **XNOR gate:** The exclusive NOR gate produces an output of 0 if either one — but not both — of the inputs is 1. All XNOR gates have two inputs.

Figure 11-1 shows the circuit symbols for these common logic gates.

Most logic gates are built using diodes and transistors, which I discuss in Chapters 9 and 10. Inside each logic gate is a circuit that arranges these components in just the right way, so that when you apply input voltages (or currents) representing a specific combination of input bits, you get an output voltage (or current) that represents the appropriate output bit. The circuitry is built into a single chip with leads, known as *pins,* providing access to the inputs, outputs, and power connections in the circuit.

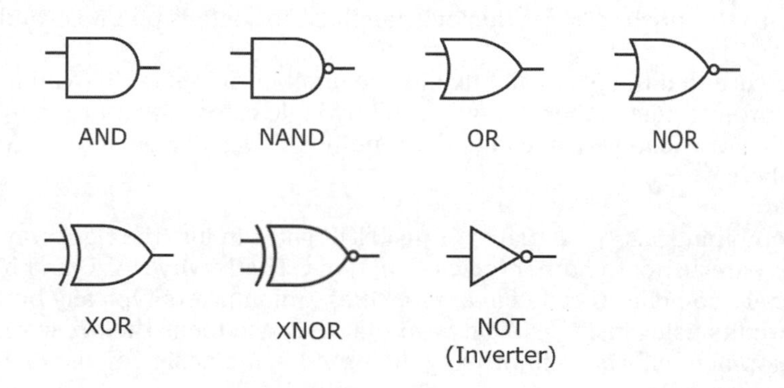

Figure 11-1: Circuit symbols for logic gates.

You usually find multiple logic gates sold in integrated circuits, such as an IC containing four two-input NAND gates (called a *quad 2-input* NAND gate), as shown in Figure 11-2. The package sports pins that connect to each gate's inputs and output, as well as other pins that connect a power supply to the circuitry. Look on the website of the IC's manufacturer for a datasheet that tells you which pins are inputs, outputs, V+ (voltage), and ground. (A *datasheet* is like a user's manual; it provides technical specifications and performance information about the chip.)

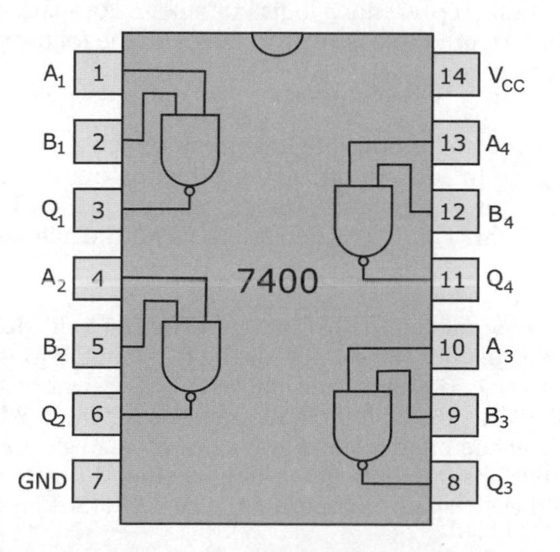

Figure 11-2: Functional diagram of the industry-standard 7400 quad 2-input NAND gate IC.

You may see labels in addition to V+ for the positive power supply pin on an IC. Different chip manufacturers use different conventions, but it's common to see V_{DD} used in CMOS chips and V_{CC} used in TTL chips (see Figure 11-2). Refer to the sidebar "Meet the logic families" for details on CMOS and TTL.

Make sure that the part you buy has the number of inputs that you need for your project. Remember that you can buy logic gates with more than two inputs. For example, you can find a 3-input NAND gate from most electronics suppliers.

By combining just NAND gates or just NOR gates in just the right way, you can create any of the other logical functions. That's why NAND and NOR gates are sometimes called *universal gates.* Chip makers typically build digital circuits using just NAND gates so that they can focus their research and development efforts on improving the process and design of just one basic logic gate.

Simplifying gates with truth tables

Tracking all the 0 and 1 inputs to logic gates and the outputs they produce can get a bit confusing — especially for gates with more than two inputs — so designers use a tool called a *truth table* to keep things organized. This table lists all the possible combinations of inputs and corresponding outputs for a given logical function.

Figure 11-3 shows the truth tables for the NOT (inverter), AND, NAND, OR, NOR, XOR, and XNOR logic gates, along with labeled symbols for each gate. Each row in the truth table represents a logic statement. For instance, the second row in the NAND truth table is really telling you the following:

$$0 \text{ NAND } 1 = 1$$

You can use truth tables also for other digital circuits, such as a *half-adder* circuit, which is designed to add two bits and produce an output consisting of a sum bit and a carry bit. For instance, for the binary equation 1 + 1 = 10, the sum bit is 0 and the carry bit is 1. The truth table for the half adder is shown in Figure 11-4.

If you look at the carry-bit column in the truth table for the half adder, you may notice that it looks just like the output for the two-input AND gate shown in Figure 11-3: That is, the carry bit is the same as A AND B, where A and B are the two input bits. Similarly, the sum bit is the same as A XOR B. What's the significance of this? You can build a half adder using an AND gate and an XOR gate. You feed the input bits into both gates, and use the AND gate to generate the carry bit and the XOR gate to generate the sum bit. (See Figure 11-5.)

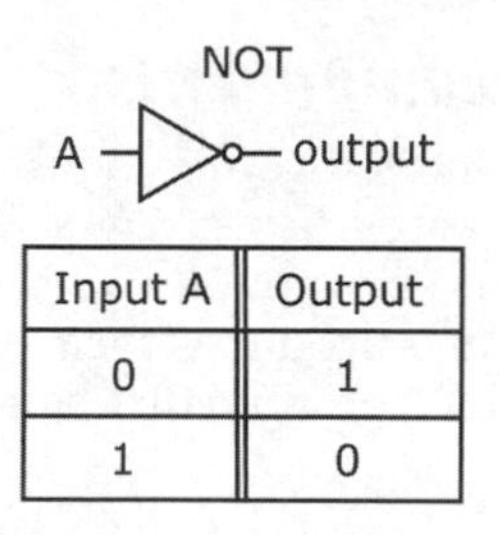

Input A	Output
0	1
1	0

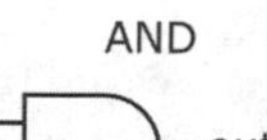
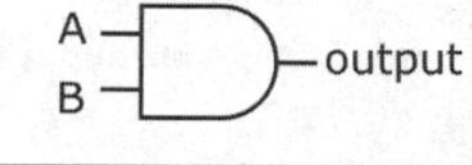

Input A	Input B	Output
0	0	0
0	1	0
1	0	0
1	1	1

Input A	Input B	Output
0	0	1
0	1	1
1	0	1
1	1	0

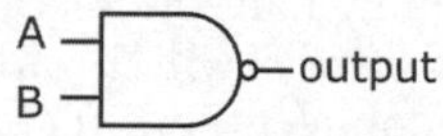

OR

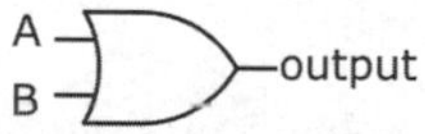

A
B
output

Input A	Input B	Output
0	0	0
0	1	1
1	0	1
1	1	1

NOR

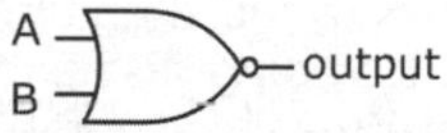

A
B
output

Input A	Input B	Output
0	0	1
0	1	0
1	0	0
1	1	0

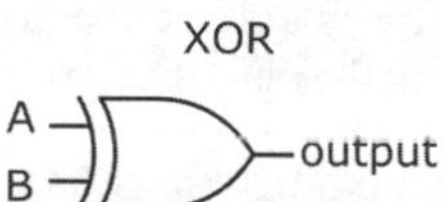

Input A	Input B	Output
0	0	0
0	1	1
1	0	1
1	1	0

XNOR

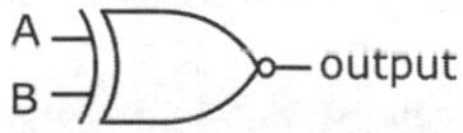

A
B
output

Input A	Input B	Output
0	0	1
0	1	0
1	0	0
1	1	1

Figure 11-3: Truth tables tell you the outputs of a logic gate for each combination of inputs.

Creating logical components

By connecting several adders together in just the right way, you can create a larger digital circuit that takes two multi-bit inputs, such as 10110110 and 00110011, and produces their sum, 11101001. (In decimal notation, that sum is 182+51=233.)

Inputs		Outputs	
A	B	Carry	Sum
0	0	0	0
0	1	0	1
1	0	0	1
1	1	1	0

Figure 11-4: The truth table for the half-adder circuit.

You can create loads of other complex functions by combining multiple AND, OR, and NOT gates. It's all a matter of which gates you use and how you interconnect them. Think about forming words from letters. With just 26 different choices, you can create millions of words. Likewise, you can create circuits that perform math functions (such as adders, multipliers, and many others) by connecting a whole bunch of gates in the right combination.

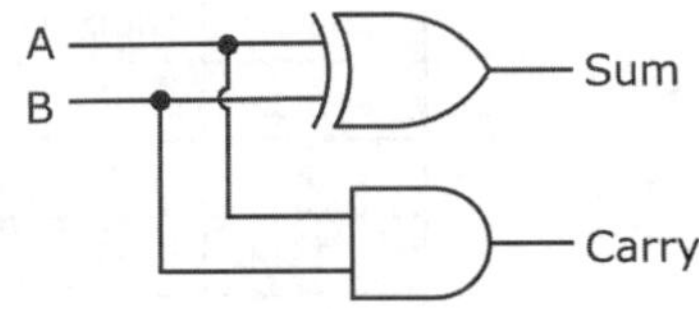

Figure 11-5: The half-adder circuit consists of an AND gate and an XOR gate.

Over the years, digital circuit designers have perfected the design of adders and other commonly used digital circuits, figuring out clever ways to speed up the computation time, reduce power dissipation, and ensure that the results are accurate, even under harsh circuit conditions such as extreme temperatures. Tried-and-true digital circuit designs are commonly turned into standardized IC product offerings — so you and other circuit builders don't have to re-create the wheel over and over again.

Storing bits in registers

Connecting dozens of gates in a complex web of circuitry presents a bit of a timing problem. As the gate inputs in one stage of logic change, the gate outputs will change — but not instantaneously. (It takes some time, albeit a very short time, for each gate to react.) Those outputs then feed the inputs to another stage of logic, and so on.

Complex logic devices use special circuits called *registers* between stages of logic to hold (or store) the output bits of one stage for a short time before allowing them to be applied to the next logic stage. Registers ship their contents out and accept new contents in upon receiving a signal known as a *clock tick,* which gives every gate enough time to compute its output. Clock signals are produced by special precision timing circuits. (See the section on the 555 timer later in this chapter for more information on how to create clocks and registers.)

Using ICs

Integrated circuits are nothing like discrete components — such as individual resistors, capacitors, and transistors — which have two or more leads connected directly to the component inside the package. The miniature prebuilt components inside an IC are already interconnected in one big happy circuit, ready to perform a specific task. You just have to add a few ingredients — say, power and one or more input signals — and the IC will do its thing. Sounds simple, right? Well, it is. You just have to know how to read IC packages — because they all look like black multilegged critters — so you know how to make the right connections.

Identifying ICs with part numbers

Every IC has a unique code, such as 7400 or 4017, to identify the type of device — really, the circuit — that's inside. You can use this code, also known as a *part number,* to look up specifications and parameters about an IC in an online resource. The code is printed on the top of the chip.

Many ICs also contain other information, including the manufacturer's catalog number and maybe even a code that represents when the chip was made. Don't confuse the date code or catalog number with the part number used to identify the device. Manufacturers don't have a universal standard for how they stamp the date code on their integrated circuits, so you may have to do some detective work to pick out the actual part number of the IC.

Packaging is everything

Great things really do come in small packages. Many ICs that can fit in the palm of your hand contain incredibly complex circuitry; for example, an entire AM/FM radio circuit (minus the battery and antenna) fits in an IC package the size of a nickel. The actual circuit is so small that manufacturers have to mount it onto a reasonable-size plastic or ceramic package so humans can use it. During the process of *chip assembly,* leads are attached to the appropriate circuit access points and fed out of the package enclosure so you and others like you can get current flowing to, through, and from the circuit inside.

Many ICs used in hobby electronics projects are assembled into *dual in-line packages (DIPs),* such as the ones in Figure 11-6. DIPs (sometimes called DILs) are rectangular-shaped plastic or ceramic packages with two parallel rows of leads, called *pins,* down either side. DIPs contain between 4 and 64 pins, but the most common sizes are 8-, 14-, and 16-pin.

Meet the logic families

Manufacturers can build digital integrated circuits in many ways. A single gate can be constructed using a resistor and a transistor, or just bipolar transistors, or just MOSFETs (another kind of transistor), or other combinations of components. Certain design approaches make it easier to cram lots of tiny gates together in a chip, while other design approaches result in faster circuits or lower power consumption.

Every digital IC is classified according to the design approach and the processing technology used to build its tiny circuits. These classifications are called *logic families.* There are literally dozens of logic families, but the two most famous families are TTL and CMOS.

TTL, or *transistor-transistor logic,* uses bipolar transistors to construct both gates and amplifiers. Manufacturing TTL ICs is relatively inexpensive, but they generally draw a lot of power and require a specific (5-volt) power supply. Several branches in the TTL family, notably the *Low-Power Schottky* series, draw roughly one-fifth the power of conventional TTL technology. Most TTL ICs use the 74*xx* and 74*xxx* format for part numbers, where *xx* or *xxx* specifies a particular type of logic device. For instance, the 7400 is a quad 2-input NAND gate. The Low-Power Schottky version of this part is coded 74LS00.

CMOS, which stands for *complementary metal-oxide semiconductor,* is one type of technology used to make MOSFETs (metal-oxide semiconductor field-effect transistors). (You can see why this family shortened its name to CMOS!) CMOS chips are a little more expensive than their TTL equivalents, but they draw a lot less power and operate over a wider range of supply voltages (3 to 15 volts). They require special handling because they are very sensitive to static electricity. Some CMOS chips are pin-for-pin equivalents of TTL chips, and are identified by a C in the middle of the part number. For instance, the 74C00 is a CMOS quad 2-input NAND gate with the same pinout as its cousin, the TTL 7400 IC. Chips in the 40xx series, for instance, the 4017 decade counter and 4511 7-segment display driver, are also members of the CMOS family.

Figure 11-6: A popular form of integrated circuit is the dual in-line package (DIP).

DIPs are designed to be *through-hole mounted* onto a printed circuit board (PCB), with the pins extending through holes in the board and soldered on the other side. You can solder DIP pins directly onto a circuit board or use *sockets* designed to hold the chip without bending the pins. You solder the socket connections into your circuit, and then insert the chip into the socket. DIPs also fit nicely into the contact holes of solderless breadboards (which I cover in Chapter 15), making it easy to prototype a circuit.

ICs used in mass-produced products are generally more complex and require a higher number of pins than DIPs can provide, so manufacturers have developed (and continue to develop) clever ways of packaging ICs and connecting them to printed circuit boards. To save space on the board (known as *real estate*), most ICs today are mounted directly to metal connections built onto the PCBs. This is known as *surface-mount technology (SMT),* and many IC packages are specially designed to be used this way. One such surface-mount IC package is the *small-outline integrated circuit (SOIC),* which looks like a shorter, narrower DIP with bent leads (called *gull wing* leads).

SMT packaging has become so widely adopted that it is often difficult to find certain ICs sold in a DIP package. If you want to use a surface-mount IC in a solderless breadboard (because you may not be able to find the DIP variety), look for special DIP adapter modules that convert various surface-mount IC packages to pin-compatible DIP packages you can plug directly into your breadboard. (Enter "DIP adapter" into your favorite Internet search engine to get a list of suppliers of such devices.)

Some ICs are very sensitive to static electricity (which I discuss in Chapter 13), so when you store your ICs, be sure to enclose them in special conductive foam (sold by most electronics suppliers). And before handling an IC, make sure you wear an antistatic wrist strap or discharge yourself by touching a conductive material that is connected to earth ground (such as the grounded metal case of your home computer) so that you don't zap your IC and wonder why it's not working. (Don't count on the metal pipes in your house to provide a conduit for static charge dissipation. Many home plumbing systems use plastic pipes along the way, so the metal pipes you see in your house aren't necessarily electrically connected to the earth.)

Probing IC pinouts

The pins on an IC package provide connections to the tiny integrated circuits inside, but alas, the pins are not labeled on the package so you have to rely on the datasheet for the particular IC to make the proper connections. Among other things, the datasheet provides you with the IC's *pinout,* which describes the function of each pin.

Use a search engine, such as Google or Yahoo!, to find datasheets for most common (and many uncommon) ICs.

To determine which pin is which, you look down on the top of the IC (not up at the little critter's underbelly) for the *clocking mark,* which is usually a small notch in the packaging but might instead be a little dimple or a white or colored stripe. (Some ICs have multiple marks.) By convention, the pins on an IC are numbered counterclockwise, starting with the upper-left pin closest to the clocking mark. So, for example, with the clocking notch orienting the chip at the 12 o'clock position, the pins of a 14-pin IC are numbered 1 through 7 down the left side and 8 through 14 up the right side, as shown in Figure 11-7.

Figure 11-7: IC pin numbering runs counterclockwise from the upper left.

Don't assume that all ICs with the same number of pins have the same *pinouts* (arrangement of external connections — in this case, pins), or even that they use the same pins for power connections. And never — *never!* — make random connections to IC pins, under the misguided notion that you can explore different connections until you get the IC to work. That's a sure-fire way to destroy a poor, defenseless circuit.

Many circuit diagrams (schematics) indicate the connections to integrated circuits by showing an outline of the IC with numbers beside each pin. The numbers correspond to the counterclockwise sequence of the device's pins, as viewed from the top. (Remember, you start with 1 in the upper left and count up as you go counterclockwise around the chip.) You can use these kinds of diagrams to easily wire an IC because you don't need to look up the device in a book or a datasheet. Just make sure that you follow the schematic and that you count the pins properly.

If a schematic lacks pin numbers, you need to find a copy of the pinout diagram. For standard ICs, you can find these diagrams online; for nonstandard ICs, you have to visit the manufacturer's website to get the datasheet.

Relying on IC datasheets

IC datasheets are like owner's manuals, providing detailed information about the insides, outsides, and recommended use of an integrated circuit. They are created by the IC manufacturer and are usually several pages long. Typical information contained in a datasheet includes

- Manufacturer's name
- IC name and part number
- Available packaging formats (for instance, 14-pin DIP) and photos of each format
- Dimensions and pinout diagrams
- Brief functional description
- Minimum/maximum ratings (such as power-supply voltages, currents, power, and temperature)
- Recommended operating conditions
- Input/output waveforms (showing how the chip changes an input signal)

Many datasheets include sample circuit diagrams, illustrating how to use the IC in a complete circuit. You can get lots of guidance and good ideas from IC datasheets. Sometimes it really pays to read the owner's manual!

Manufacturers often publish application notes for their integrated circuits. An *application note* (often called an *app note*) is a multipage document that explains in greater detail than the datasheet how to use the IC in an *application* — a circuit designed for a specific practical task.

Sourcing and sinking current

Because the insides of integrated circuits are hidden from view, it's hard to know exactly how current flows when you connect a load or other circuitry to the IC's output pin or pins. Typically, datasheets will specify how much current an IC output can source or sink. An output is said to *source current* when current flows out of the output pin, and *sink current* when current flows into the output pin.

If you connect a device, say a resistor, between an output pin and the positive terminal of a power supply, and the output goes low (0 volts), current will flow through the resistor into the IC — the IC sinks the current. If you connect a resistor between the output pin and the negative supply (ground), and the output goes high, current will flow out of the IC and through the resistor — the IC sources the current. Refer to the datasheet for the maximum source or sink current (which are usually the same value) of an IC output.

Using Your Logic

In this section, you find out firsthand how to make the right connections to a NAND logic gate IC, and you watch the output change as you fiddle around with various combinations of inputs. Then you find out how to create another type of logic gate, an OR gate, by combining NAND gates in just the right way.

Here are the parts you need to build the two circuits in this section:

- Four 1.5-volt AA batteries
- One four-battery holder (for AA batteries) with battery clip
- One 74C00 or 74HC00 quad 2-input NAND gate IC
- Four 10 kΩ resistors (brown-black-orange)
- Two 470 Ω resistors (yellow-violet-brown)
- Four single-pole, double-throw (SPDT) switches
- Two LEDs (any size, any color)
- One solderless breadboard and assorted jumper wires

The 74C00 and 74HC00 are CMOS chips (see the sidebar "Meet the logic families," earlier in this chapter). You may use a different 2-input NAND, as long as you check the pinout and power supply requirements. For instance, the 4011 is another CMOS 2-input NAND gate, but it has a different pinout from the 74HC00 IC. The 7400 IC is a TTL 2-input NAND chip with the same pinout as the 74HC00 IC, but it requires a fairly steady 5-volt power supply, so you'll need to use a *voltage regulator,* such as the LM7805, along with a couple of capacitors. You supply 9 volts to the LM7805, and it outputs a steady 5 volts that you can use to power the 7400 IC.

Seeing the light at the end of the NAND gate

The circuit in Figure 11-8 uses an LED to indicate the high (1) or low (0) state of the output of a two-input NAND gate.

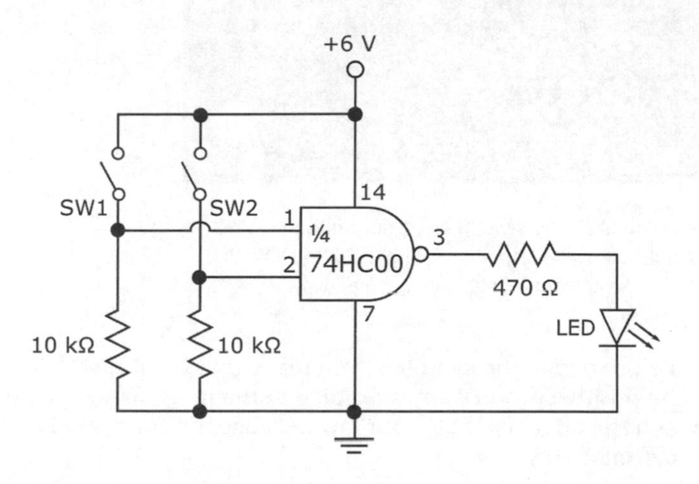

Figure 11-8: Use an LED to show the output of a NAND logic gate.

Set up the circuit using Figure 11-9 as a guide and noting the following important information:

- The 74HC00 IC is a high-speed CMOS chip that is sensitive to static, so be sure to review the precautions outlined in Chapter 13 to avoid damaging the chip, and insert the IC last — but before you apply power — when you build the circuit, as suggested in Chapter 15.
- You use only one side of each SPDT switch in this circuit so that each switch functions as an on/off switch (refer to Chapter 4 for details). You connect the middle terminal and one end terminal to your circuit and leave the other end terminal unconnected.

Remember that the output of a NAND (NOT AND) gate is high whenever either or both inputs are low, and the output of a NAND gate is low only when both inputs are high. High is defined by the positive power supply (6 V) and low is 0 V.

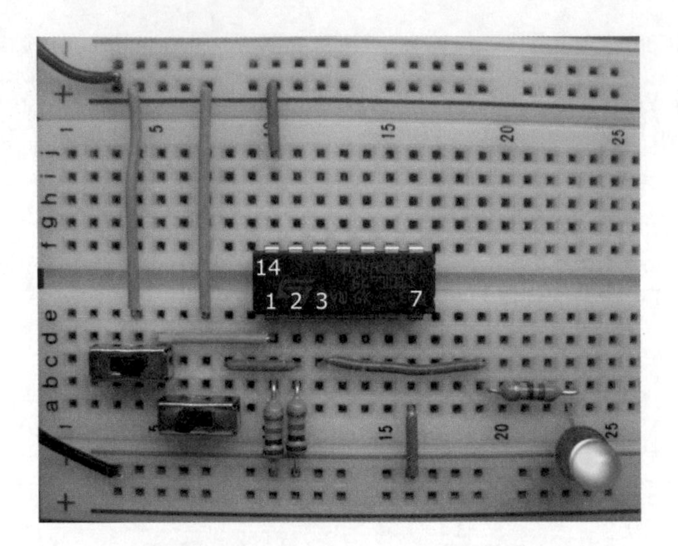

Figure 11-9: With SW1 (top switch) on (slider right) and SW2 (bottom switch) off (slider left), the LED lights up to show that 1 NAND 0 = 1. (Pin number labels have been added.)

When you close one of the switches, you make that input high because you connect the positive power-supply voltage to the input. When you open one of the switches, you make that input low because it's connected through a resistor to ground (0 V).

Test the functionality of the NAND gate by trying all four combinations of open and closed switches, filling in the chart given here (which is essentially a truth table).

Input 1	Input 2	Output (High = LED on; Low = LED off)
Low (SW1 open)	Low (SW2 open)	
Low (SW1 open)	High (SW2 closed)	
High (SW1 closed)	Low (SW2 open)	
High (SW1 closed)	High (SW2 closed)	

Did you see the LED light up when either or both switches were open? Did the LED turn off when both switches were closed? Be sure to tell the truth!

Turning three NAND gates into an OR gate

You can combine several NAND gates to create any other logical function. In the circuit in Figure 11-10, three NAND gates are combined to create an OR gate. The inputs to the OR gate are controlled by switches SW3 and SW4. The output of the OR gate is indicated by the on/off state of the LED.

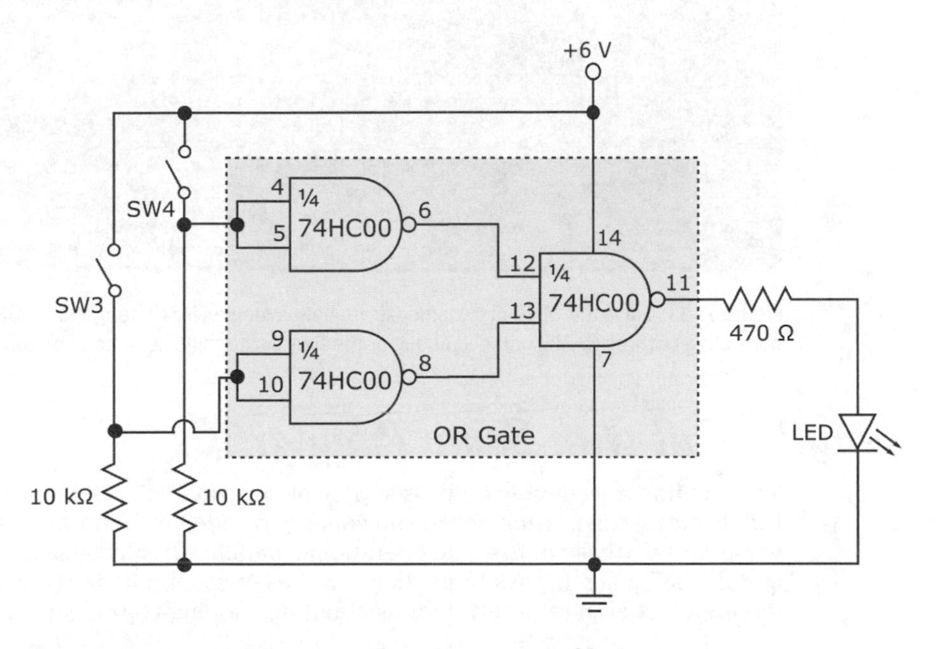

Figure 11-10: Three NAND gates configured to create an OR gate.

Each of the two NAND gates on the left functions as a NOT gate (or inverter). Each NAND gate ties the inputs together so that a low input produces a high output, and a high input produces a low output. The NAND gate on the right produces a high output when either or both of its inputs are low, which happens when either or both switches (SW3 and SW4) are closed. The bottom line is that if either or both switches are closed, the output of the circuit is high. That's an OR gate!

Set up the circuit, being careful to avoid static. You can use the remaining three NAND gates on the 74HC00 IC that you used to build the circuit in Figure 11-9. Figure 11-11 shows the circuits in Figure 11-8 and 11-10 set up together using a single 74HC00. Try not to get your wires crossed!

Test the functionality by opening and closing the switches. The LED should turn on when either or both switches are closed.

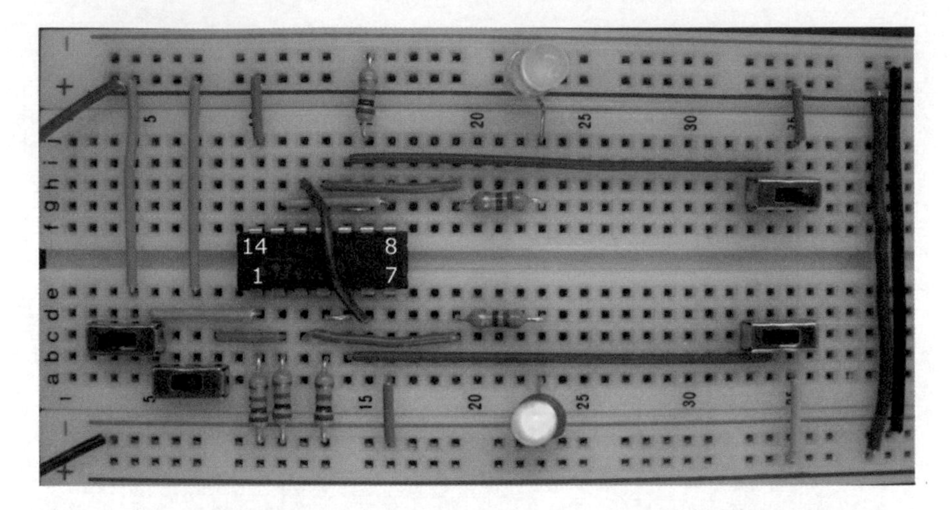

Figure 11-11: Using the 74HC00 to demonstrate both a single NAND gate and an OR gate consisting of three NAND gates. (Labels for the four corner pins have been added.)

Hanging Out with Some Popular ICs

You can find a seemingly endless supply of integrated circuits on the market today, but two in particular are known far and wide for their outstanding versatility and ease of use: the operational amplifier (really a class of ICs) and the 555 timer. It pays to get to know these two circuits fairly well if you intend to get even remotely serious about developing your electronics habit.

In this section, I describe these two popular ICs and one additional IC, the 4017 CMOS decade counter. You encounter the 555 timer IC and the 4017 decade counter IC in projects in Chapter 17, so the upcoming sections provide a quick rundown on how they work.

Operational amplifiers

The most popular type of linear (analog) IC is undoubtedly the *operational amplifier,* nicknamed the *op amp,* which is designed to add muscle to (that is, amplify) a weak signal. An op amp contains several transistors, resistors, and capacitors, and offers more robust performance than a single transistor. For example, an op amp can provide uniform amplification over a much wider range of frequencies *(bandwidth)* than can a single-transistor amplifier.

Most op amps come in 8-pin DIPs (as shown in Figure 11-12), and include two input pins (pin 2, known as the *inverting input,* and pin 3, known as the *non-inverting input*) and one output pin (pin 6). An op amp is one type of *differential amplifier:* The circuitry inside the op amp produces an output signal that is a multiple of the *difference* between the signals applied to the two inputs.

Used a certain way, this setup can help eliminate noise (unwanted voltages) in the input signal by subtracting it out of what's amplified.

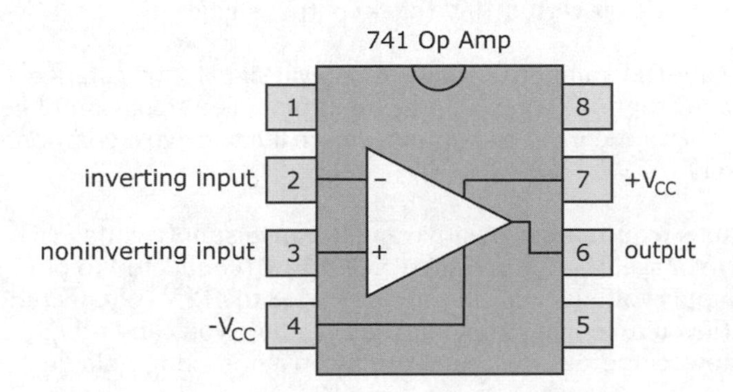

Figure 11-12: Pinout of a standard 8-pin op amp, such as the LM741.

You can configure an op amp to multiply an input signal by a known gain factor that is determined by external resistors. One such configuration, known as an *inverting amplifier,* is shown in Figure 11-13.

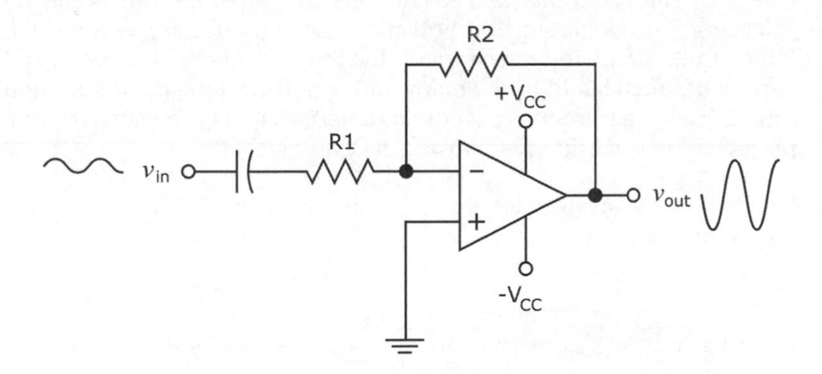

Figure 11-13: An inverting op-amp circuit provides uniform gain over a wide range of frequencies.

The values of the resistors connected to the op amp determine the gain of the inverting amplifier circuit:

$$\text{Gain} = -\frac{R2}{R1}$$

The negative sign indicates that the input signal is flipped, or *inverted,* to produce the output signal. Say, for instance, that the value of *R2* is 10 kΩ and that of *R1* is 1 kΩ. The gain is –10. With a gain of –10, a 1 V input signal (peak value) produces an inverted 10 V (peak) output signal.

To use the inverting amplifier, you just apply a signal (for instance, the output of a microphone) between the input pins. The signal, amplified several times, then appears at the output, where it can drive a component, such as a speaker.

Most op amps require both positive and negative supply voltages. A positive supply voltage, V_{CC}, in the range of 8 to 12 V (connected to pin 7) and a negative supply voltage, $-V_{CC}$, in the range of –8 to –12 V (connected to pin 4) works. (If you're looking for some light reading, you can find application notes on how to operate such dual-supply op amps using a single power supply.)

Gobs of different op amps are available at prices ranging from just a few cents for standard op-amp ICs, such as the LM741 general-purpose op amp, to $10 or more for high-performance op amps.

IC time machine: the 555 timer

One of the most popular and easy-to-use integrated circuits is the versatile 555 timer, introduced in 1971 and still in wide use today, with more than one billion units produced every year. This little workhorse can be used for a variety of functions in both analog and digital circuits, most commonly for timing (ranging from microseconds to hours), and is the cornerstone of many projects you can build (as you see in Chapter 17).

Figure 11-14 illustrates the pin assignments for the 555 timer. Among the pin functions are

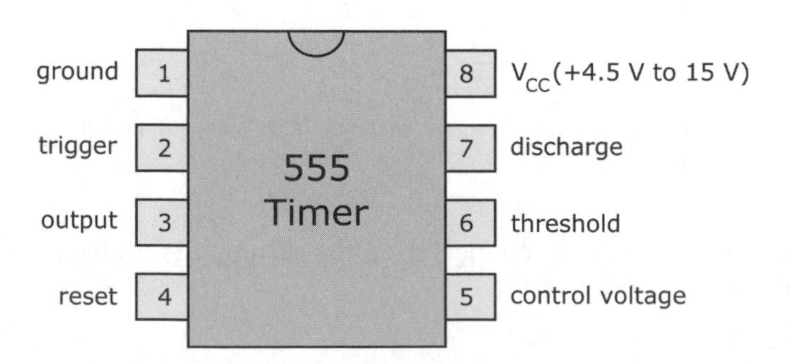

Figure 11-14: 555 timer IC pinout.

- **Trigger input:** When you apply a low voltage to pin 2, you trigger the internal timing circuit to start working. This is known as an *active low* trigger.
- **Output pin:** The output waveform appears on pin 3.
- **Reset:** If you apply a low voltage to pin 4, you reset the timing function, and the output pin (pin 3) goes low. (Some circuits don't use the reset function, and this pin is tied to the positive supply.)
- **Control voltage input:** If you want to override the internal trigger circuit (which normally you don't do), you apply a voltage to pin 5. Otherwise, you connect pin 5 to ground, preferably through a 0.01 μF capacitor.
- **Threshold input:** When the voltage applied to pin 6 reaches a certain level (usually two-thirds the positive power-supply voltage), the timing cycle ends. You connect a resistor between pin 6 and the positive supply. The value of this *timing resistor* influences the length of the timing cycle.
- **Discharge pin:** You connect a capacitor to pin 7. The discharge time of this *timing capacitor* influences the length of the timing intervals.

You can find various models of the 555 timer IC. The 556 timer is a dual version of the 555 timer, packaged in a 14-pin DIP. The two timers inside share the same power supply pins.

By connecting a few resistors, capacitors, and switches to the various pins of the 555 timer, you can get this little gem to perform loads of different functions — and it's remarkably easy to do. You can find detailed, easy-to-read information about its various applications on datasheets. I discuss three popular ways to configure a timing circuit using a 555 here.

Astable multivibrator (oscillator)

The 555 can behave as an *astable multivibrator,* which is just a fancy term to describe a sort of electronic metronome. By connecting components to the chip (as shown in Figure 11-15), you configure the 555 to produce a continuous series of voltage pulses that automatically alternate between low (0 volts) and high (the positive supply voltage, V_{CC}), as shown in Figure 11-16. (The term *astable* refers to the fact that this circuit does not settle down into a stable state, but keeps changing on its own between two different states.) This self-triggering circuit is also known as an *oscillator.*

You can use the 555 astable multivibrator for lots of fun things:

- **Flash lights:** A low-frequency (<10 Hz) pulse train can control the on/off operation of an LED or lamp (see the LED flasher project in Chapter 17).
- **Create an electronic metronome:** Use a low-frequency (<20 Hz) pulse train as the input to a speaker or piezoelectric transducer to generate a periodic clicking sound.

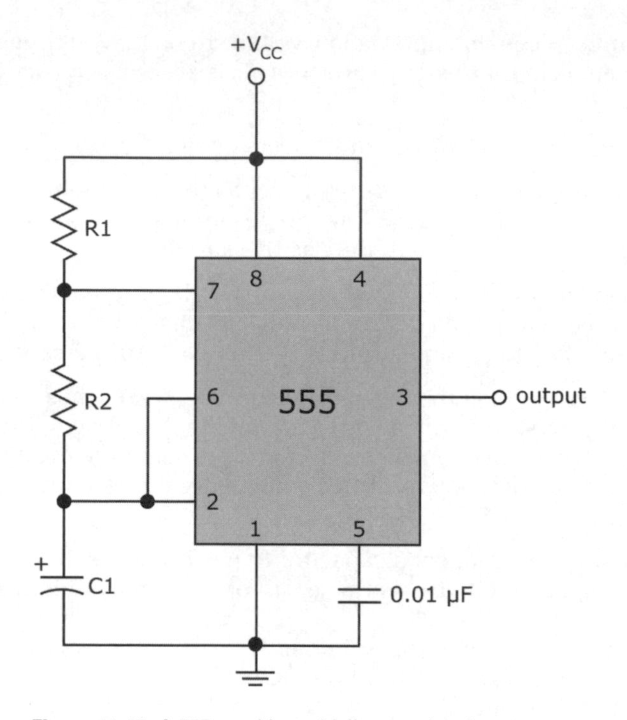

Figure 11-15: A 555 astable multivibrator circuit configuration.

- **Sound an alarm:** By setting the frequency to the audio range (20 Hz to 20 kHz) and feeding the output into a speaker or piezoelectric transducer, you can produce a loud, annoying tone (see the siren and light-sensing alarm projects in Chapter 17).
- **Clock a logic chip:** You can adjust the pulse widths to match the specifications for the signal that clocks the logic inside a chip, such as the 4017 decade counter I describe later in this chapter (see the light chaser and traffic light simulator projects in Chapter 17).

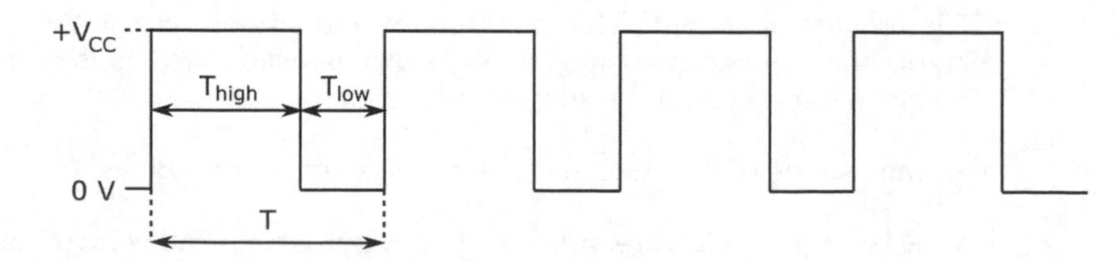

Figure 11-16: Series of voltage pulses from a 555 astable multivibrator circuit. External components control the pulse width.

The frequency *f* (in hertz), which is the number of complete up-and-down cycles per second, of the square wave produced is determined by your choice of three external components, according to this equation:

$$f = \frac{1.44}{(R1 + 2R2) \times C1}$$

If you flip the numerator and dominator in that equation, you get the *time period* (T), which is the length of time (in seconds) of one complete up-and-down pulse:

$$T = 0.693 \times (R1 + 2R2) \times C1$$

You can set up your circuit so that the width of the high part of the pulse is different from the width of the low part of the pulse. To find the width of the high part of the pulse (expressed as T_{high}), use the following equation:

$$T_{high} = 0.693 \times (R1 + R2) \times C1$$

You find the width of the low part of the pulse (expressed as T_{low}) like this:

$$T_{low} = 0.693 \times R2 \times C1$$

If *R2* is much, much bigger than *R1*, the high and low pulse widths will be fairly equal. If *R2*=*R1*, the high portion of the pulse will be twice as wide as the low portion. You get the idea.

You can also use a potentiometer (variable resistor) in series with a small resistor as *R1* or *R2* and adjust its resistance to vary the pulses.

To choose values for *R1, R2,* and *C1,* I suggest you follow these steps:

1. **Choose *C1*.** Decide what frequency range you want to generate and choose an appropriate capacitor. The lower the frequency range, the higher the capacitor you should choose. (Assume that *R1* and *R2* will be somewhere in the 10 kΩ to 1 MΩ range.) For many low-frequency applications, capacitor values of between 0.1 μF and 10 μF work well. For higher-frequency applications, choose a capacitor in the range of 0.01 μF to 0.001 μF.
2. **Choose *R2*.** Decide how wide the low part of the pulse should be, and choose the value of *R2* that will produce that width, given the value of *C1* you've already determined.
3. **Choose *R1*.** Decide how wide the high part of the pulse should be. Using the values of *C1* and *R2* already selected, calculate the value of *R1* that will produce the desired high pulse width.

For instance, one way to produce a pulse that is high for about 3 seconds and low for about 1.5 seconds is to use a 10 μF capacitor for *C1* and 220 kΩ resistors for both *R1* and *R2*. (Plug those values into the equations for T_{high} and T_{low} and see for yourself!)

Monostable multivibrator (one-shot)

By configuring the 555 timer as shown in Figure 11-17, you can use it as a *monostable multivibrator* that generates a single pulse when triggered. This configuration is sometimes called a *one-shot.* Without a trigger, this circuit produces a low (zero) voltage; this is its stable state. When triggered by closing the switch between pin 2 and ground, this circuit generates an output pulse at the level of the supply voltage, V_{CC}. The width of the pulse, T, is determined by the values of *R1* and *C1,* as follows:

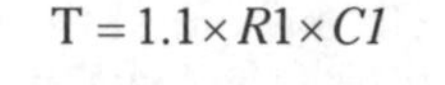

$$T = 1.1 \times R1 \times C1$$

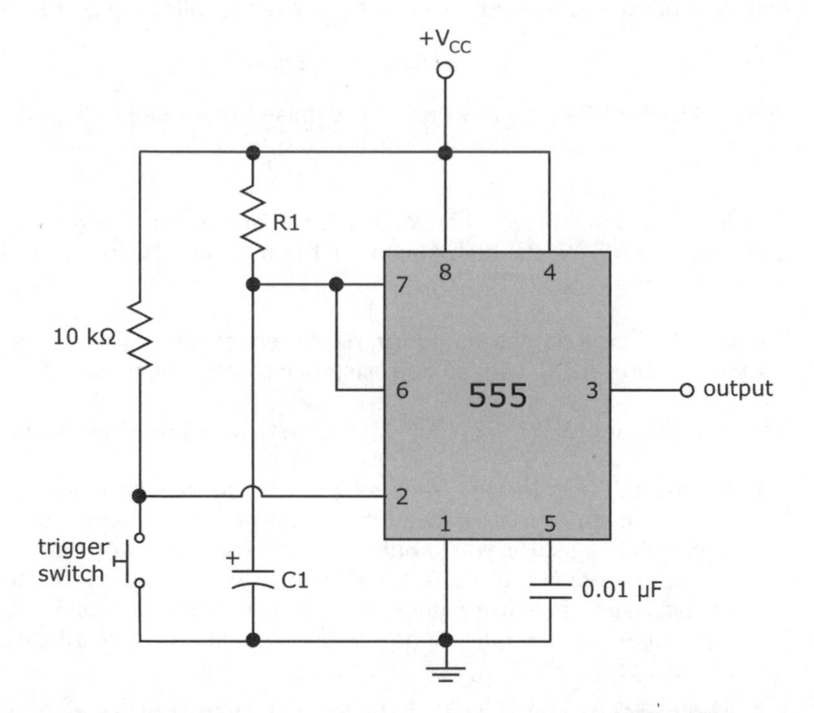

Figure 11-17: Triggered by closing the momentary switch at pin 2, the 555 monostable circuit produces a single pulse whose width is determined by the values of *R1* and *C1*.

Because capacitor values can often vary by as much as 20%, you may need to choose a resistor with a different value than the formula suggests to produce the pulse width you desire.

You can use a one-shot to safely trigger a digital logic device (such as the 4017 CMOS decade counter described later in this chapter). Mechanical switches tend to "bounce" when closed, producing multiple voltage spikes which a digital IC can misinterpret as multiple trigger signals. Instead, if you trigger a one-shot with a mechanical switch, and use the output of the one-shot to trigger the digital IC, you can effectively "de-bounce" the switch.

Bistable multivibrator (flip-flop)

If an astable circuit has no stable state and a monostable circuit has one stable state, what's a bistable circuit? If you guessed that a bistable circuit is a circuit with two stable states, you are correct.

The 555 *bistable multivibrator* shown in Figure 11-18 produces alternating high (V_{CC}) and low (0 V) voltages, switching from one state to the other only when triggered. Such a circuit is commonly known as a *flip-flop.* There's no need to calculate resistor values; activating the trigger switch controls the timing of the pulses generated.

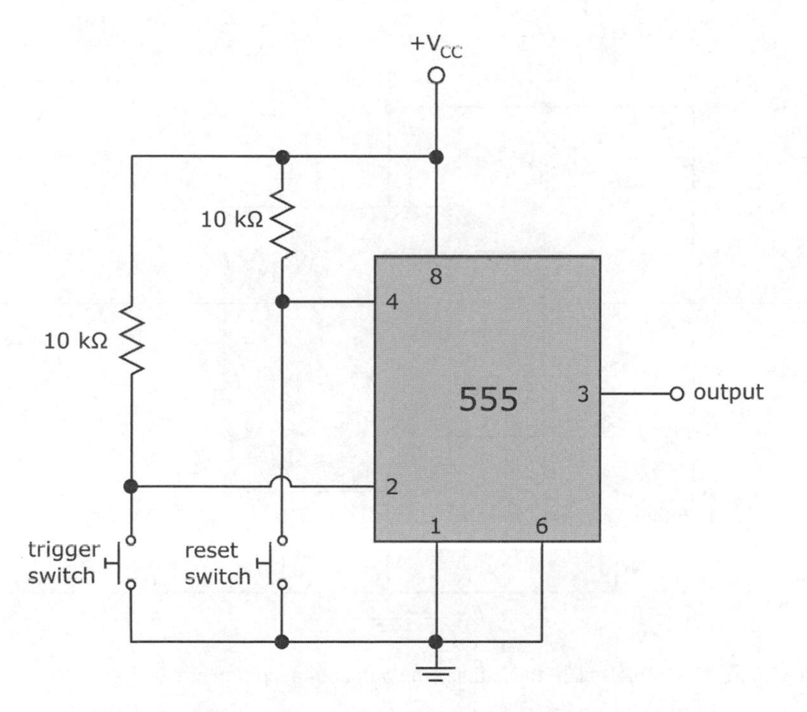

Figure 11-18: The 555 bistable circuit (or flip-flop) produces a high output when triggered by the switch at pin 2, and a low output when reset by the switch at pin 4.

You can witness firsthand how a 555 timer works as a flip-flop by building the circuit shown in Figure 11-19. Here are the parts you need:

- Four 1.5-volt AA batteries
- One four-battery holder (for AA batteries) with battery clip
- One 555 timer IC
- Two 10 kΩ resistors (brown-black-orange)
- One 470 Ω resistor (yellow-violet-brown)
- One LED (any size, any color)
- One single-pole, double-throw (SPDT) switch
- One solderless breadboard
- Assorted jumper wires

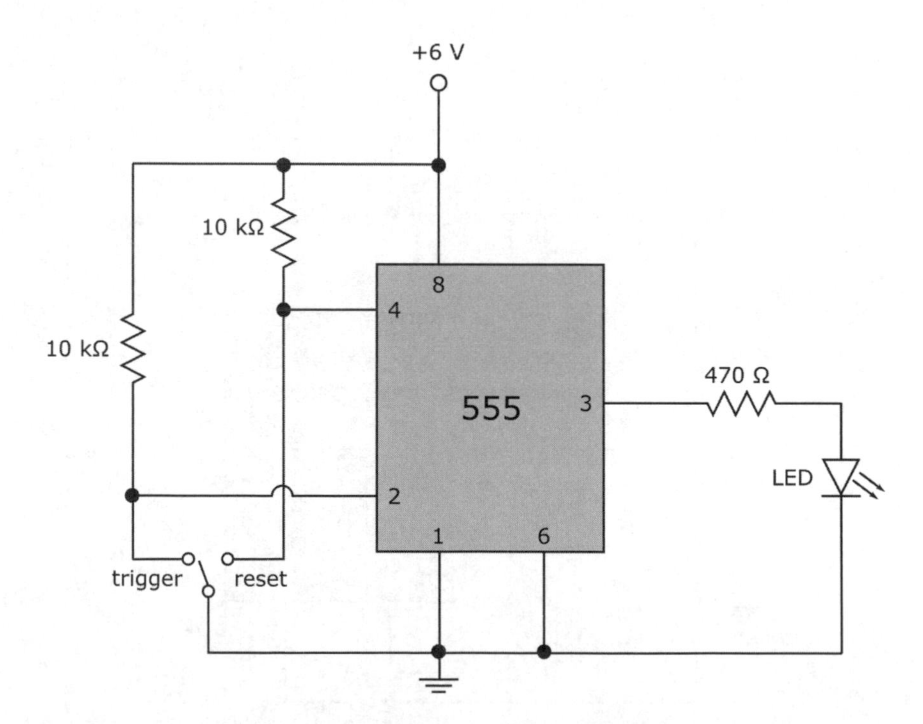

Figure 11-19: Use an LED to visualize the output of a flip-flop circuit.

Set up the circuit, using Figure 11-20 as your guide. Note that one end of the SPDT switch is connected to pin 2 (trigger) of the 555 timer, the other end is connected to pin 4 (reset) of the 555 timer, and the center terminal is connected to ground. Depending on the position of the slider on the SPDT switch, the switch either triggers or resets the 555 timer.

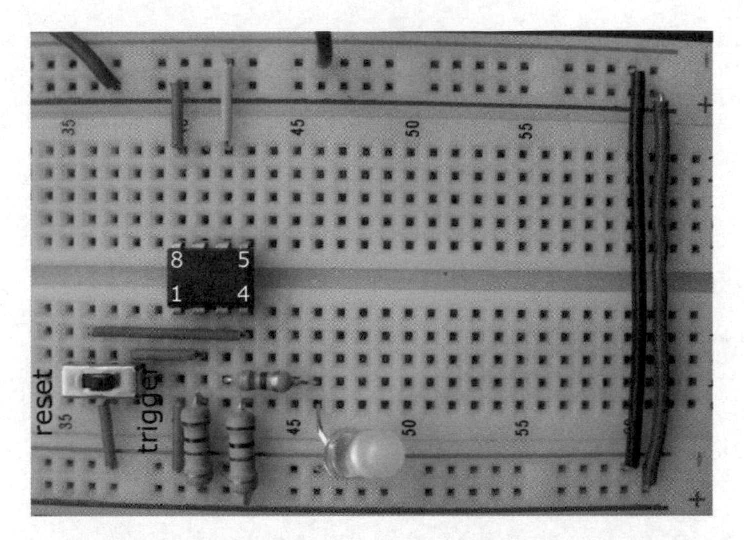

Figure 11-20: The SPDT switch is used to both trigger and reset the 555 timer in this flip-flop circuit. (Pin and switch labels have been added.)

After you set up the circuit, slide the switch to the trigger position. Does the LED light? Does it stay lit as long as you leave the switch in the trigger position? Now slide the switch to the reset position. Does the LED go off and stay off as long as the switch is in the reset position?

Because a flip-flop's output stays high (or low) until triggered (or reset), it can be used to store a data bit. (Remember, a bit is a 0 or a 1, which is a low voltage or a high voltage, respectively.) The registers used to store temporary outputs between stages of logic consist of multiple flip-flops. Flip-flops are also used in certain digital-counter circuits, holding bits in a series of interconnected registers that form an array, the outputs of which make up a bit string representing the count. (Refer to the sidebar, "Storing bits in registers," earlier in this chapter, for more information on registers.)

You can use various types of 555 timer circuits to trigger other 555 timer circuits. For instance, you can use an oscillator to trigger a flip-flop (useful for clocking registers). Or you can use a one-shot to produce a temporary low-volume tone — and when it ends, change the state of a flip-flop, whose output triggers an oscillator that pulses a speaker on and off. Such a circuit might be used in a home alarm system: Upon entering the home, the homeowner (or intruder) has 10 seconds or so to deactivate the system (while hearing a low-volume warning tone) before the siren wakes up the neighbors.

Counting on the 4017 decade counter

The 4017 CMOS decade counter shown in Figure 11-21 is a 16-pin IC that counts from 0 to 9 when triggered. Pins 1–7 and 9–11 go from low to high one at a time when a trigger signal is applied to pin 14. (They *don't* go from low to high in strict counterclockwise order; you have to check the pinout to determine the order.) You can use the count outputs to light LEDs (as in the light chaser and traffic light projects in Chapter 17) or trigger a one-shot that controls another circuit.

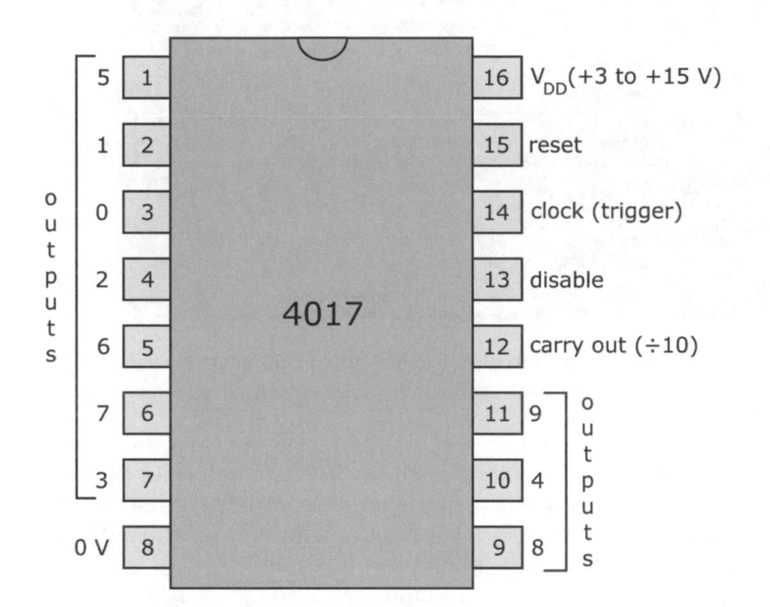

Figure 11-21: Pinout of the 4017 CMOS decade counter.

Counting can take place only when the *disable* pin (pin 13) is low; you can disable counting by applying a high signal to pin 13. You can also force the counter to reset to zero (meaning that the zero-count output, which is pin 3, goes high) by applying a high signal to pin 14.

By piggybacking multiple 4017 ICs, you can count up tens, hundreds, thousands, and so forth. Pin 12 is high when the count is 0 to 4 and low when the count is 5 to 9, so it looks like a trigger signal that changes from low to high at one-tenth the rate of the count. If you feed the output of pin 12 into the trigger input (pin 14) of another decade counter, that second counter will count up tens. By feeding the second counter's pin-12 output into pin 14 of a third counter, you can count up hundreds. With enough 4017 ICs, you may even be able to tally the national debt!

You can also connect two or more of the counter's outputs using diodes to produce a variable timing sequence. To do this, connect each anode (positive side of a diode) to an output pin, and connect all the cathodes (negative sides of the diodes) together and then through a resistor. With this arrangement, when any one of the outputs is high, current will flow through the resistor. For instance, you can simulate the operation of a traffic light by tying outputs 0–4 together and feeding the result (through a resistor) into a red LED, connecting output 5 to a yellow LED, and tying outputs 6–9 together to control a green LED. (See the traffic light project in Chapter 17.)

Microcontrollers

One of the most versatile integrated circuits you can find is the microcontroller. A *microcontroller* is a small, complete computer on a chip. To program it, you place it on a development board that allows the IC to interface with your personal computer. After it's programmed, you mount the microcontroller into a socket on your electronic device (which could be a solderless breadboard). You add a few other components to provide an interface between the microcontroller and LEDs, motors, or switches — and voilà! Your little programmed IC makes things happen (for instance, it can control the motion of a robot). The nice thing about a microcontroller is that you can simply alter a few lines of code (or reprogram it completely) to change what it does; you don't need to swap out wires, resistors, and other components to get this flexible IC to take on a new personality.

Hundreds of microcontrollers are available, but a handful are geared specifically for first-timers. Some, such as the one shown on the PCB in Figure 11-22, can be purchased as part of a complete development package: You get a pre-assembled printed circuit board populated with the microcontroller IC, discrete components, and standard connectors so that you can connect the microcontroller to a circuit during operation or to your computer for programming. Many of these kits also include an *integrated development environment (IDE),* which contains software tools for programming the microcontroller. (See Chapter 18 for more on microcontroller kits.)

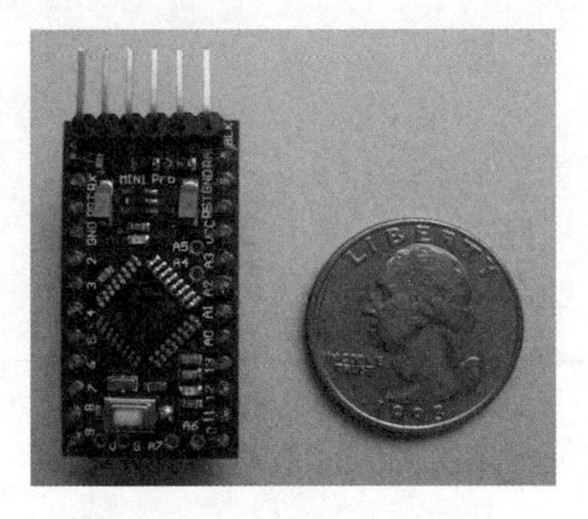

Figure 11-22: The square chip that Mr. Washington is gazing at is a microcontroller that is surface-mounted to a mini PCB populated with additional components and a connector.

Other popular ICs

Among the other common functions provided by ICs are mathematical operations (addition, subtraction, multiplication, and division), *multiplexing* (selecting a single output from among several inputs), and the conversion of signals between analog and digital:

- You use an *analog-to-digital (A/D)* converter to convert a real-world analog signal into a digital signal so you can process it with a computer or other digital electronics system.

- You use a *digital-to-analog (D/A)* converter to convert a processed digital signal back into an analog signal. (For example, you need an analog signal to vibrate the speakers in your home computer system.)

Of course, the *microprocessor* that runs your personal computer (and maybe even your personal life) is also quite popular as ICs go.

There are so many more integrated circuits than I can possibly cover in this book. Really smart circuit designers are always coming up with new ideas and improvements on some old ideas, so there are a lot of choices in the world of integrated circuits.

12

Acquiring Additional Parts

In This Chapter

- Choosing the perfect type of wire
- Powering up with batteries and solar cells
- Controlling connections with switches
- Triggering circuits with sensors
- Turning electricity into light, sound, and motion

Although the individual components and integrated circuits discussed in Chapters 4 through 11 form the A-team when it comes to shaping the flow of electrons in electronic circuits, there are a bunch of other contributing parts that the A-team relies on to help get the job done.

Some of these other parts — such as wires, connectors, and batteries — are essential ingredients in any electronic circuit. After all, you'd be hard-pressed to build an electronic circuit without wires to connect things or a source of power to make things run. As for the other parts I discuss in this chapter, you may use them only now and then for certain circuits. For example, when you need to make some noise, a buzzer sure comes in handy — but you may not want to use one in every circuit you build.

In this chapter, I discuss a mixed bag of components, some of which you should keep in stock (just like toilet paper and toothpaste), while the others can be picked up whenever the spirit moves you.

Making Connections

Making a circuit requires that you connect components to allow electric current to flow between them. The following sections describe wires, cables, and connectors that allow you to do just that.

Choosing wires wisely

Wire that you use in electronics projects is just a long strand of metal, usually made of copper. The wire has only one job: to allow electrons to travel through it. However, you can find a few variations in the types of wire available to you. In the following sections, I give you the lowdown on which type of wire to choose for various situations.

Stranded or solid?

Cut open the cord of any old household lamp (only *after* unplugging the lamp, of course), and you see two or three small bundles of very fine wires, each wrapped in insulation. This is called *stranded* wire. Another type of wire, known as *solid* wire, consists of a single (thicker) wire wrapped in insulation. You can see examples of stranded and solid wires in Figure 12-1.

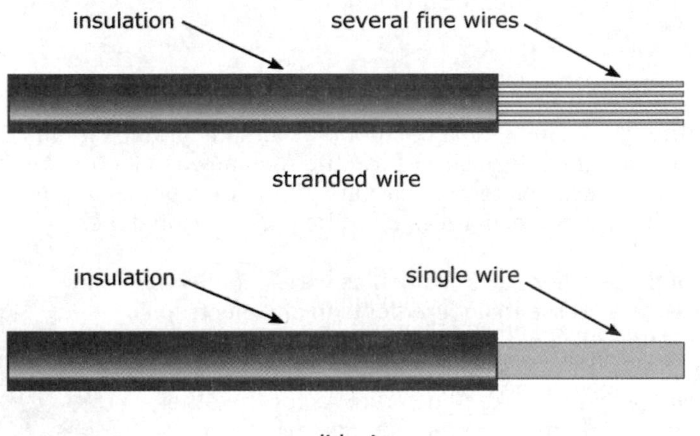

Figure 12-1: Both stranded and solid wire are commonly used in electronics.

Stranded wire is much more flexible that solid wire, and you use it in situations in which the wire will be moved or bent a lot (such as in line cords for lamps and the cables you hook up to your home entertainment system). You use solid wire in places where you don't plan to move the wire around, and to connect components on breadboards (check out Chapter 15 for more on breadboards). It's easy to insert solid wire into holes in the breadboard, but if you try to use a stranded wire, you have to twist the strands to get all of them into the hole, and you may break a strand or two in the process (trust me — it happens), which could short out the circuit.

Sizing up your wire gauge

You refer to the diameter of wire as the *wire gauge.* As luck would have it, the relationship between wire gauge and wire diameter in electronics is essentially backward: The smaller the wire gauge, the larger the wire diameter. You can see common wire gauges in Table 12-1.

Table 12-1 Wires Commonly Used in Electronics Projects

Wire Gauge	*Wire Diameter (inches)*	*Uses*
16	0.051	Heavy-duty electronics applications
18	0.040	Heavy-duty electronics applications
20	0.032	Most electronics projects
22	0.025	Most electronics projects
30	0.01	Connections on small circuit board prototypes or wire-wrap connections

For most electronics projects, including the ones in this book, you use 20- or 22-gauge wire. If you're hooking up a motor to a power supply, you need to use 16- or 18-gauge wire. You may find smaller 28- or 30-gauge wire useful for making connections on small circuit board prototypes. 30-gauge wire was commonly used for connecting components on special circuit boards using a technique known as *wire wrapping*, which is rarely done anymore. (Refer to Chapter 15 for details on circuit construction.)

You sometimes see gauge abbreviated in weird and wonderful ways. For example, you may see 20 gauge abbreviated as 20 ga., #20, or 20 AWG (American Wire Gauge).

If you start working on projects involving higher voltage or current than the ones described in this book, consult the instructions for your project or an authoritative reference to determine the appropriate wire gauge. For example, the *National Electrical Code* lists the required wire gauges for each type of wiring that you use in a house. Make sure that you also have the right skills and sufficient knowledge of safety procedures to work on such a project.

The colorful world of wires

As with the colorful bands that unlock the secrets of resistor values, the colorful insulation around wire can help you keep track of connections in a circuit. When wiring a DC circuit (for example, when you work with a breadboard), using red wire for all connections to positive voltage (+V) and black wire for all connections to negative voltage (–V) or to ground is common practice. For AC circuits, use green wire for ground connections. Yellow or orange wire is often used for input signals, such as the signal from a microphone into a circuit. If you keep lots of different colors of wire handy, you can color-code your component connections so it's easier to tell what's going on in a circuit just by glancing at it (unless, of course, you're colorblind).

Collecting wires into cables or cords

Cables are groups of two or more wires protected by an outer layer of insulation. Line cords that bring AC power from a wall outlet to an electrical device such as a lamp are cables — so are the cords in the mishmash of connections in your home entertainment system. Cables differ from stranded wires because the wires used in cables are separated by insulation.

Plugging in to connectors

If you look at a cable — say, the one that goes from your set-top box to your TV— you see that it has metal or plastic doodads on each end. These doodads are called *plugs,* and they represent one kind of *connector.* There are also metal or plastic receptacles on your set-top box and TV that these cable ends fit into. These *receptacles* (sometimes called *sockets* or *jacks*) are another kind of connector. The various pins and holes in connectors connect the appropriate wire in the cable to the corresponding wire in the device.

Different types of connectors are used for various purposes. Among the connectors you're likely to hook up with in your electronics adventures are these:

- A *terminal* and *terminal block* work together as the simplest type of connector. A terminal block contains sets of screws in pairs. You attach the block to the case or chassis of your project. Then, for each wire you want to connect, you solder (or crimp) a wire to a terminal. Next, you connect each terminal to a screw on the block. When you want to connect two wires, simply choose a pair of screws and connect the terminal on each wire to one of those screws.
- *Plugs* and *receptacles* carry audio and video signals between pieces of equipment that have cables like the ones you see in Figure 12-2. Plugs on each end of the cable connect to receptacles on the equipment being connected. Analog audio cables (see Figure 12-2, left) contain one or two signal wires (which carry the audio signal) and a metal shield surrounding the wires. The metal *shield* protects the signal wires from electrical interference (known as *noise*) by minimizing the introduction of stray current into the wires. Digital multimedia cables, such as the High-Definition Multimedia Interface (HDMI) cable on the right in Figure 12-2, contain multiple shielded wires that carry digital audio and video data.

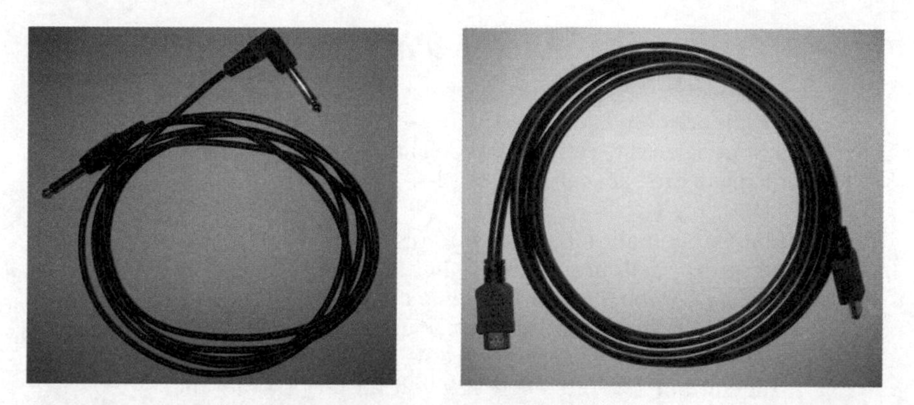

Figure 12-2: An analog audio cable (left) and an HDMI cable (right) facilitate the transmission of information between two pieces of electronic equipment.

- You typically use *pin headers* to bring signals to and from circuit boards, which are thin boards designed to house a permanent circuit. Pin headers come in handy for complex electronics projects that involve multiple circuit boards. Most pin headers consist of one or two rows of metal posts attached to a block of plastic that you mount on the circuit board. You connect the pin header to individual or bundled wires, or to a compatible connector at the end of a *ribbon cable* — a series of insulated wires stuck together side by side to form a flat, flexible cable. The rectangular shape of the connector allows easy routing of signals from each wire in the cable to the correct part of the circuit board. You refer to pin headers by the number of pins (posts) they use, such as a 40-pin header.

Electronics uses a wealth of connectors that you don't have to delve into until you start doing more complex projects. If you want to find out more about the broad array of connectors, you can take a look at some of the catalogs or websites of electronics suppliers listed in Chapter 19. Most devote an entire category of products to connectors.

Powering Up

All the wires and connectors in the world won't do you much good if you don't have a power source. In Chapter 3, I discuss sources of electricity, including AC power from wall outlets and DC power from batteries and solar cells (also known as photovoltaic cells). Here I discuss how to choose a power source and how to feed its power into your circuits.

Turning on the juice with batteries

For most hobby electronics projects, cells or batteries — which are combinations of cells — are the way to go. The symbols used to represent a cell and a battery in a circuit diagram are shown in Figure 12-3.

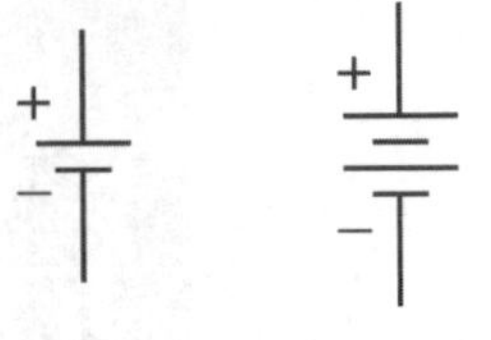

Figure 12-3: Circuit symbols for a cell (left) and a battery (right).

Many schematics use the symbol for a cell to represent a battery. Cells are relatively lightweight and portable, and by combining multiple cells in series, you can create a variety of DC voltage sources. Everyday cells, such as garden-variety AAA-, AA-, C-, and D-cells, all produce about 1.5 volts each. A 9-volt battery (sometimes called a *transistor battery* or PP3 battery) is shaped like a 3-D rectangle and ordinarily contains six 1.5-volt cells. (Some cheap brands may contain only five 1.5-volt cells.) A *lantern battery* (a big boxy thing that can power a flashlight the size of a boom box) produces about 6 volts.

Connecting batteries to circuits

You use a 9-volt (PP3) battery clip (shown in Figure 12-4) to connect an individual 9-volt battery to a circuit. Battery clips snap onto the terminals of the battery (those snaps on the top of the battery are known as a PP3 connector); they contain black and red leads that you connect to your circuit. You strip the insulation from the ends of the black and red wires, and then connect the leads (the bare ends) to your circuit. You can connect the leads to terminals, insert them into holes in a breadboard, or solder them directly to components. I discuss all these techniques in Chapter 15.

Rating the life of the everyday battery

The *amp-hour* or *milliamp-hour rating* for a battery gives you an idea of how much current a battery can conduct for a given length of time. For example, a 9-volt battery usually has about a 500 milliamp-hour rating. Such a battery can power a circuit using 25 milliamps for approximately 20 hours before its voltage begins to drop. (I checked a 9-volt battery that I'd used for a few days, and found that it was producing only 7 volts.) An AA battery that has a 1500 milliamp-hour rating can power a circuit drawing 25 milliamps of current for approximately 60 hours.

Six AA batteries in series, which produce about 9 volts, will last longer than a single 9-volt battery. That's because the six series batteries contain more chemicals than the single battery, and can produce more current over time before becoming depleted. (In Chapter 3, I discuss how batteries are made and why they eventually run out of juice.) If you have a project that uses a lot of current, or you plan to run your circuit all the time, consider using larger C- or D-size batteries, which last longer than smaller batteries or most rechargeable batteries.

See the section "Sorting batteries by what's inside" for more about different types of batteries and how long you can expect them to last.

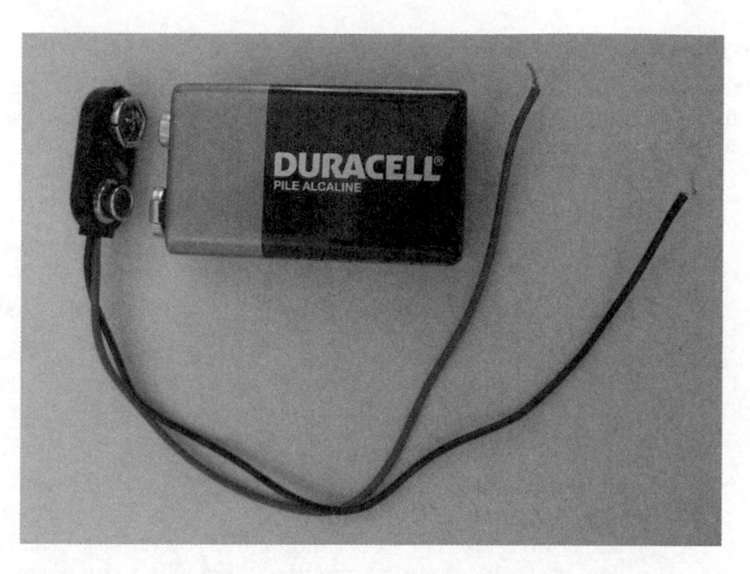

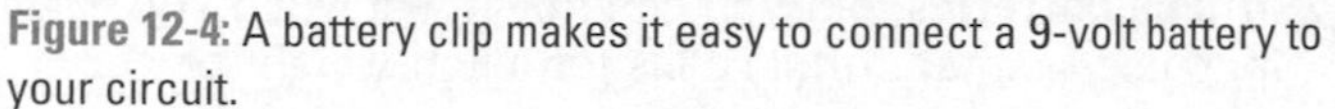
Figure 12-4: A battery clip makes it easy to connect a 9-volt battery to your circuit.

When you connect the positive terminal of one battery to the negative terminal of another battery, the total voltage across this series connection is the sum of the individual battery voltages.

Battery holders make series connections between batteries for you while holding multiple batteries in place. Some battery holders provide red and black leads for access to the total voltage; others, such as the one in Figure 12-5, provide PP3 connector snaps so you can attach a battery clip and access the total voltage across the clip's red and black leads.

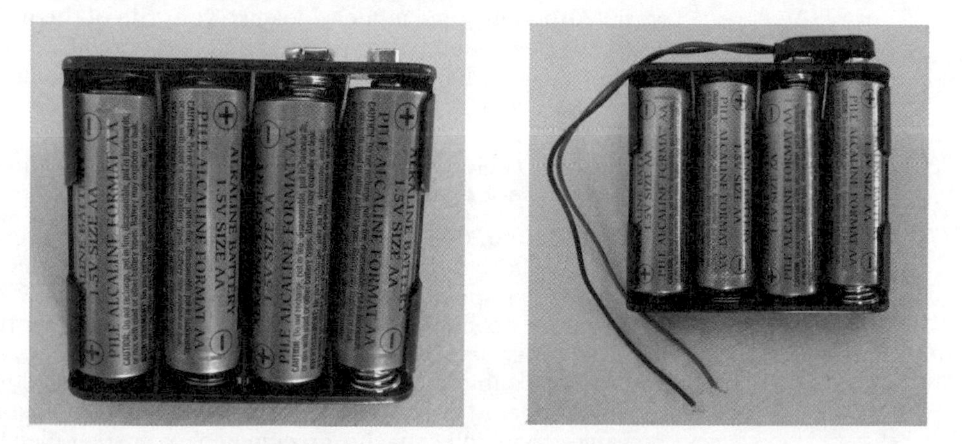

Figure 12-5: Four 1.5 volt batteries in a battery holder produce about 6 volts across the red and black leads.

Sorting batteries by what's inside

Batteries are classified by the chemicals they contain, and the type of chemical determines whether a battery is rechargeable. The following types of batteries are readily available:

- **Disposable (non-rechargeable) batteries:**
  - **Zinc-carbon** batteries come in a variety of sizes (AAA, AA, C, D, and 9 volt, among others) and are at the low end of the battery food chain. They don't cost much, but they also don't last long.
  - **Alkaline** batteries also come in a variety of sizes, and last about three times as long as zinc-carbon batteries. I suggest starting with this type of battery for your projects. If you find yourself replacing them often, you can step up to rechargeable batteries.
  - Lightweight **lithium** batteries generate higher voltages — about 3 volts — than other types and have a higher current capacity than alkaline batteries. They cost more, and you can't recharge most lithium batteries, but you can't beat 'em when your project (for instance, a small robot) calls for a lightweight battery.
- **Rechargeable batteries:**
  - **Nickel-metal hydride (NiMH)** batteries are the most popular type of rechargeable batteries today. They come in a variety of sizes (AAA, AA, C, D, and 9 volt) and generate about 1.2 volts. One drawback to NiMH batteries is that they self-discharge over a few months if not used, but some low self-discharge NiMH battery models are available. I suggest you use NiMH batteries for projects that need rechargeable batteries.
  - **Lithium-ion (Li-ion)** batteries are the newest type of rechargeable battery. Most generate a nominal 3.7 volts and a maximum 4.2 volts, and all require a special charger (an NiMH charger will not work). Although some models look like your run-of-the-mill AA or AAA batteries, Li-ion batteries use a different naming convention (for instance, a 14500 Li-ion battery is the same size as an AA battery). If you choose to use Li-ion batteries, bear in mind the voltage differences between these batteries and conventional cells. (For instance, two Li-ion batteries in series generate 7.4 V while two AA batteries in series generate 3.0 V.)
  - **Nickel-cadmium (NiCd)** batteries, like NiMH batteries, generate about 1.2 volts. NiCds (pronounced "NYE cads") have declined drastically in popularity since their heyday in the mid-1990's, due to their poor capacity, their toxic content (cadmium), improvements in other battery technologies (particularly NiMH), and a flaw NiCds exhibit known as the *memory effect,* which requires you to fully discharge the battery before recharging it, to ensure that it recharges to its full capacity. I recommend you steer clear of NiCd batteries — unless you happen to have some lying around and want to get your money's worth.

Be careful not to mix battery types in the same circuit, and *never* attempt to recharge disposable batteries. These batteries can rupture and leak acid, or even explode. Most disposable batteries contain warnings about the dangers of such misuse right on their labels.

Buying a recharger and a supply of rechargeable batteries can save you a considerable amount of money over time. Just make absolutely sure the battery charger you use is designed for the type of rechargeable battery you select. Check both the chemistry (for instance, Li-ion or NiMH) and the voltage of the battery when selecting a charger.

Be sure to dispose of batteries properly. Batteries containing heavy metals (such as nickel, cadmium, lead, and mercury) can be hazardous to the environment when improperly disposed. For guidelines on proper disposal — which may vary from state to state — check battery manufacturer websites or other websites, such as `www.ehso.com/ehshome/batteries.php`.

Getting power from the sun

If you're building a circuit designed to operate outside — or you just want to use a clean, green source of energy — you may want to purchase one or more solar panels. A *solar panel* consists of an array of solar cells (which are large diodes known as *photodiodes*) that generate current when exposed to a light source, such as the sun. (I discuss diodes in Chapter 9, and photodiodes in the section "Using Your Sensors," later in this chapter.) A panel measuring about 5 x 5 inches may be able to generate 100 milliamps at 5 volts in bright sunlight. If you need 10 amps, you can certainly get it, but you may find the size of the panel problematic — and expensive — for a small or portable project.

Some solar panels contain output leads that you can connect into your circuit, much like the leads from a battery clip or battery holder. Other solar panels have no leads, so you have to solder your own leads to the two terminals.

Here are some criteria to consider to help you determine whether a solar panel is appropriate for your project:

- **Do you plan to have the solar panel in sunlight when you want your circuit to be on, or use the panel to charge a storage battery that can power your project?** If not, look for another power source.
- **Will the solar panel fit on the gadget you're building?** To answer this question, you need to know how much power your gadget will need and the size of the solar panel that can deliver enough power. If the panel is too large for your gadget, either redesign the gadget to use less power or look for another power source.

Using wall power to supply higher DC current or voltage (not recommended)

The AC power supplied by your utility company can cause injury or death if used improperly, so I don't recommend that you power circuits directly off household current. And because the vast majority of hobby electronics projects run on batteries, you may never be tempted to work with AC anyway.

Some projects need more DC current or higher DC voltages than batteries can easily provide. In those cases, you can use an AC adapter, such as the one shown in Figure 12-6, to convert AC to DC and get the higher current or voltage you need. All the working parts are self-contained in the wall transformer, so you aren't exposed to high AC currents.

Under no circumstances should you pry apart the plastic housing for the circuitry inside an AC adapter. Capacitors inside the housing store significant electrical charge. (Refer to Chapter 7 for details on capacitors.)

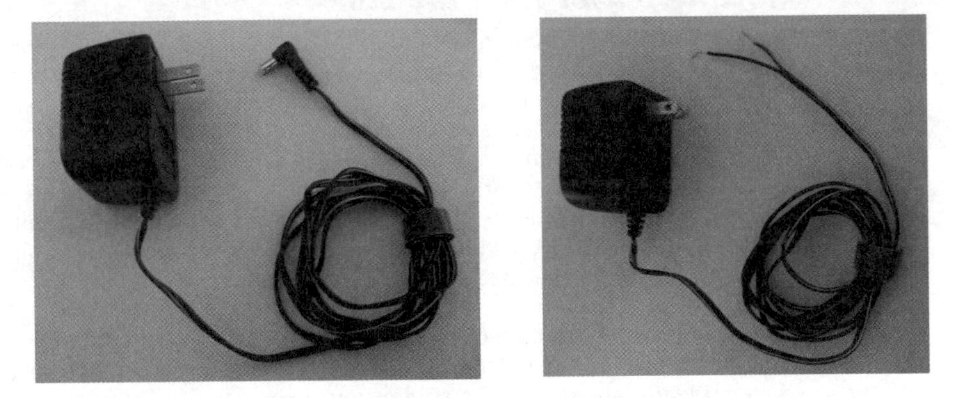

Figure 12-6: This AC adapter converts 120 volts AC to 7.5 volts DC and supplies up to 300 mA of current. I modified the output of this adapter to make it easier to use in DC circuits.

AC adapters supply currents ranging from hundreds of milliamps to a few amps at voltages ranging from 5 volts DC to 20 volts DC. Some provide both a positive DC voltage and a negative DC voltage. Different models use different types of connectors to deliver power. If you purchase an AC adapter, be sure to read the specification sheet (specs) carefully to determine its power rating and how to connect it to your circuit.

Acquiring wall warts

AC adapters are sometimes called wall warts because they stick out of the wall like an ugly wart. You can purchase new or surplus wall warts. (Check out Chapter 19 for some good leads on suppliers.) And, of course, you may already have some old wall warts saved from a discarded cordless phone or other electronic device. If so, verify that it outputs DC (some output AC) and check the voltage and current rating, usually printed on the adapter, to see if it's suitable for your next project. If it is, make sure you know how the connector is wired so you maintain the proper polarity (positive and negative voltage connections) when hooking it up to your circuit. A word of caution: Many wall warts output noisy DC power, rather than the steady DC power you may be expecting, and some newer wall warts are switch-mode (not analog) supplies that may also be noisy or contain voltage spikes.

You may want to prepare your AC adapter for easy use in your circuits by removing its output plug and separating and stripping the two output leads so that you can connect the leads directly to your circuit to supply DC voltage (as shown in Figure 12-6, right). Here are the steps to follow to prepare your adapter:

1. **Make sure the AC adapter is not plugged in to an AC outlet.**
2. **Using your wire cutters (refer to Chapter 13), cut off the output connector.**
3. **Using a utility knife (or your fingernails), separate the two insulated wires along a distance of roughly two inches.**
4. **Using your wire cutters, cut one of the two insulated wires so that it is at least an inch shorter than the other wire.**

   By making one wire shorter that the other, you prevent the two wires from accidentally coming into contact when the AC adapter is powered up, which would short out the adapter, rendering it useless.
5. **Using your wire strippers (refer to Chapter 13), carefully strip the insulation off each of the two wires along a half-inch length.**
6. **Twist the stranded copper wires of each conductor so that there are no loose strands.**
7. **Making sure the two conductors are not touching each other, plug the AC adapter into an AC outlet.**
8. **Set your multimeter to measure DC volts with a range of 20 V or more (refer to Chapter 16), and place one multimeter lead on one AC adapter output wire and the other multimeter lead on the other output wire.**

9. **Observe the voltage reading.**

   If the voltage reading is positive, your positive (usually red) multimeter lead is attached to the positive conductor of the AC adapter. If the voltage reading is negative, your negative (usually black) multimeter lead is attached to the positive conductor of the AC adapter. Label the positive conductor with a marker or by attaching a label.

   Note that the magnitude of the voltage reading you get is likely significantly higher than the nominal voltage specified on the label or data sheet of the AC adapter. This discrepancy is normal and due to the fact that you are using an unregulated power supply and measuring the adapter's output voltage under no-load conditions. *Unregulated power supplies* generate voltages that vary depending on the current drawn by the *load,* that is, the device receiving power. After you connect the adapter's output leads to a circuit, the output voltage will come down. I measured 10.5 V on an AC adapter that was labeled 7.5 VDC (volts DC).

Congratulations! You now have a power supply for your electronics projects that can supply more current that batteries can provide.

Even after you remove your AC adapter from your wall power, you'll still find a DC voltage across the output leads for quite a while because there is a large capacitor inside the housing that is holding its charge. The capacitor will eventually discharge, but it could take hours. To discharge the capacitor quickly, use your insulated needle-nose pliers to grip a 680 Ω resistor, carefully maneuver the pliers to place the resistor's leads across the AC adapter's output leads, and wait about 30 seconds.

Using Your Sensors

When you want to trigger the operation of a circuit as a response to something physical happening (such as a change in temperature), you use electronic components known as *sensors.* Sensors take advantage of the fact that various forms of energy — including light, heat, and motion — can be converted into electrical energy. Sensors are a type of *transducer,* which is an electronic device that converts energy from one form to another. In this section, I describe some of the more common input transducers, or sensors, used in electronic circuits.

The circuit symbols for several types of sensors discussed in this section are shown in Figure 12-7.

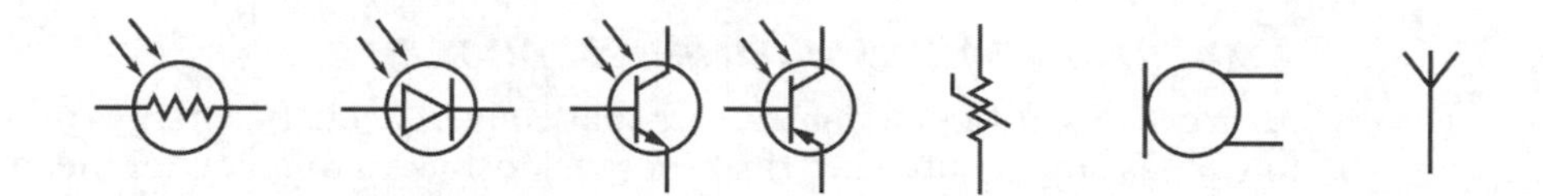

Figure 12-7: Circuit symbols for a variety of sensors.

Seeing the light

Many electronic components behave differently depending on the light they are exposed to. Manufacturers make certain versions of components to exploit this light sensitivity, enclosing them in clear cases so you can use them as sensors in equipment such as burglar alarms, smoke detectors, automatic dusk-to-dawn lighting, and safety devices that stop your electrically controlled garage door from descending when a cat runs under it. You can also use them for communications between your remote control, which sends coded instructions via infrared light using a light-emitting diode (LED, which I discuss in Chapter 9), and your TV or DVD player, which contains a light-sensitive diode or transistor to receive the coded instructions.

Examples of light-sensitive devices used as sensors include the following:

- **Photoresistors (or photocells)** are light-dependent resistors (LDRs) made from semiconductor material. They typically exhibit a high resistance (about 1 MΩ) in darkness and a fairly low resistance (about 100 Ω) in bright light, but you can use a multimeter (as I describe in Chapter 16) to determine the actual resistances exhibited by a specific photoresistor. The typical photoresistor is most sensitive to visible light, especially in the green-yellow spectrum. A photoresistor can be installed with current running either way in your circuits.
- **Photodiodes** are sort of the opposite of light-emitting diodes, which I discuss in Chapter 9. They conduct current or drop voltage only when exposed to sufficient light, usually in the infrared (not visible) range. Like standard diodes, photodiodes contain two leads: The shorter lead is the cathode (negative end) and the longer lead is the anode (positive end).
- Most **phototransistors** are simply bipolar junction transistors (as I discuss in Chapter 10) that are encased in a clear package so that light biases the base-emitter junction. These devices usually contain only two leads (whereas standard transistors contain three leads). That's because you don't need access to the base of the transistor to bias it — light does that job for you. From the outside, phototransistors look just like photodiodes, so you really have to keep track of which is which.

Capturing sound with microphones

Microphones are input transducers that convert acoustic energy (otherwise known as sound) into electrical energy. Most use a thin membrane, or *diaphragm,* that vibrates in response to air pressure changes from sound. The vibrations of the membrane are translated into an AC electrical signal in various ways, depending on the type of microphone:

- In a **condenser microphone,** the vibrating membrane plays the role of one plate of a capacitor, so that variations in sound produce corresponding variations in capacitance. (For more about capacitors, see Chapter 7.)
- In a **dynamic microphone,** the diaphragm is attached to a movable induction coil located inside a permanent magnet. As sound moves the diaphragm, the coil moves inside the magnetic field produced by the magnet, and a current is induced in the coil. (Chapter 8 has the lowdown on this phenomenon, which is known as *electromagnetic induction.*)
- In a **crystal microphone,** a special *piezoelectric crystal* is used to convert sound into electrical energy, taking advantage of the *piezoelectric effect,* in which certain substances produce a voltage when pressure is applied to them.
- In a **fiber-optic microphone,** a laser source directs a light beam toward the surface of a tiny reflective diaphragm. As the diaphragm moves, changes in light reflected off the diaphragm are picked up by a detector, which transforms the differences in light into an electrical signal. Fiber-optic microphones are immune to both electromagnetic interference (EMI) and radio frequency interference (RFI).

Feeling the heat

A *thermistor* is a resistor whose resistance value changes with changes in temperature. Thermistors have two leads and no polarity, so you don't need to worry about which way you insert a thermistor into your circuit.

Following are the two types of thermistors:

- **Negative temperature coefficient (NTC) thermistor:** The resistance of an NTC thermistor decreases with a rise in temperature. This type of thermistor is the more common one.
- **Positive temperature coefficient (PTC) thermistor:** The resistance of an PTC thermistor increases with a rise in temperature.

Using a light sensor to detect motion

Have you ever walked up to someone's dark doorway when suddenly the outdoor lights turn on? Or witnessed a garage door stop descending when a child or wheeled object crosses below it? If so, you've seen motion detectors at work. Motion-detection devices commonly use light sensors to detect either the *presence* of infrared light emitted from a warm object (such as a person or animal) or the *absence* of infrared light when an object interrupts a beam emitted by another part of the device.

Many homes, schools, and stores use *passive infrared (PIR) motion detectors* to turn on lights or detect intruders. PIR motion detectors contain a sensor (which usually consists of two crystals), a lens, and a small electronic circuit. When infrared light hits a crystal, it generates an electric charge. Because warm bodies (such as those of most humans) emit infrared light at different wavelengths than cooler objects (such as a wall), differences in the output of the PIR sensor can be used to detect the presence of a warm body. The electronic circuit interprets differences in the PIR sensor output to determine whether or not a moving, warm object is nearby. (Using two crystals in the PIR sensor enables the sensor to differentiate between events that affect both crystals equally and simultaneously (such as changes in room temperature) and events that affect the crystals differently (such as a warm body moving past first one crystal, and then the other).

Industrial PIR motion detectors use or control 120-volt circuits and are designed to mount on a wall or on top of a floodlight. For hobby projects using a battery pack, you need a compact motion detector that works with about 5 volts. A typical compact motion detector has three leads: ground, positive voltage supply, and detector output. If you supply +5 volts to the detector, the voltage on the output lead reads about 0 volts when no motion is detected or about 5 volts when motion is detected. You can find compact motion detectors through online vendors of security systems, but be sure to buy a *motion detector*, rather than just a *PIR sensor*. The lens included in a motion detector helps the device detect the *motion* of an object rather than just the *presence* of an object.

Suppliers' catalogs typically list the resistance of thermistors as measured at 25 degrees Celsius (77° F). Measure the resistance of the thermistor yourself with a multimeter at a few different temperatures (see Chapter 16 for more about using multimeters). These measurements enable you to *calibrate* the thermistor, or get the exact relationship between temperature and resistance. If you're not sure of a thermistor's type, you can figure that out by identifying whether the value increases or decreases with a rise in temperature.

If you're planning to use the thermistor to trigger an action at a particular temperature, be sure to measure the resistance of the thermistor *at that temperature.*

Other ways to take your temperature

In the section "Feeling the heat," I discuss the temperature sensors called thermistors — but there are several other types of temperature sensors. Here's a brief summary of their characteristics:

- **Semiconductor temperature sensor:** The most common type of temperature sensor, whose output voltage depends on the temperature, contains two transistors (more about those in Chapter 10).
- **Thermocouple:** A thermocouple contains two wires made of different metals (for example, copper and nickel/copper alloy) that are welded or soldered together at one point. These sensors generate a voltage that changes with temperature. The metals it uses determine how the voltage changes with temperature. Thermocouples can measure high temperatures — several hundred degrees or even over a thousand degrees.
- **Infrared temperature sensor:** This sensor measures the infrared light given off by an object. You use it when your sensor must be located at a certain distance from the object you plan to measure. For example, you use this sensor if a corrosive gas surrounds the object. Industrial plants and scientific labs typically use thermocouples and infrared temperature sensors.

More energizing input transducers

Many other types of input transducers are used in electronic circuits. Here are three common examples:

- **Antennas:** An antenna senses electromagnetic waves, and transforms the energy into an electrical signal. (It also functions as an *output transducer,* converting electrical signals into electromagnetic waves.)
- **Pressure or position sensors:** These sensors take advantage of the variable-resistance properties of certain materials when they undergo a deformation. Piezoelectric crystals are one such set of materials.
- **Accelerometers:** One type of accelerometer that detects your smartphone orientation relies on variations in capacitance that result when acceleration forces move one tiny capacitive plate attached to a spring relative to a fixed plate.

Transducers are often categorized by the type of energy conversion they perform, for instance, electroacoustic, electromagnetic, photoelectric, and electromechanical transducers. These amazing devices open up tremendous opportunities for electronic circuits to perform countless useful tasks.

Experiencing the Outcome of Electronics

Sensors, or *input transducers,* take one form of energy and convert it into electrical energy, which is fed into the input of an electronic circuit. *Output transducers* do the opposite: They take the electronic signal at the output of a circuit and convert it into another form of energy — for instance, sound, light, or motion (which is mechanical energy).

You may not realize it, but you're probably familiar with many devices that really are output transducers. Light bulbs, LEDs, motors, speakers, liquid crystal displays (LCDs), and other electronic visual displays all convert electrical energy into some other form of energy. Without these puppies, you might create, shape, and send electrical signals around through wires and components all day long, and never reap the rich rewards of electronics. It's only when you transform the electrical energy into a form of energy you can experience (and use) that you begin to enjoy the fruits of your labor.

The schematic symbols for three output transducers discussed in this section are shown in Figure 12-8.

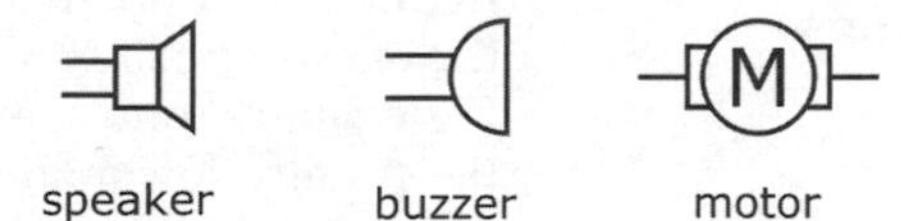

Figure 12-8: Circuit symbols for some popular output transducers.

Speaking of speakers

Speakers convert electrical signals into sound energy. Most speakers consist simply of a permanent magnet, an electromagnet (which is a temporary, electrically controlled magnet), and a vibrating cone. Figure 12-9 shows how the components of a speaker are arranged.

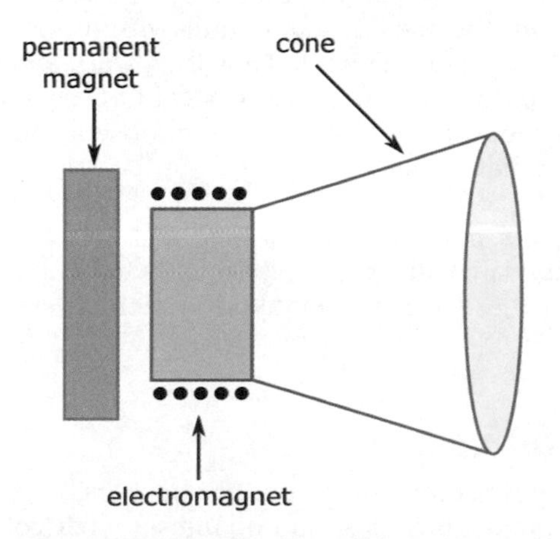

Figure 12-9: The parts of a garden-variety speaker: two magnets and a cone.

The electromagnet, which consists of a coil wrapped around an iron core, is attached to the cone. As electrical current alternates back and forth through the coil, the electromagnet gets pulled toward and then pushed away from the permanent magnet. (Chapter 8 tells you more about the ups and downs of electromagnets.) The motion of the electromagnet causes the cone to vibrate, which creates sound waves.

Most speakers come with two leads that can be used interchangeably. For more serious projects, such as speakers in stereo systems, you must pay attention to the polarity markings on the speakers because of the way they are used in electronic circuits inside the stereo system.

Speakers are rated according to the following criteria:

- **Frequency range:** Speakers can generate sound over different ranges of frequencies, depending on the size and design of the speakers, within the *audible frequency range* (about 20 Hz [hertz] to 20 kHz [kilohertz]). For example, one speaker in a stereo system may generate sound in the bass range (low audible frequency) while another generates sound in a higher range. You need to pay close attention to speaker frequency range only when you're building a high-end audio system.
- **Impedance:** Impedance is a measure of the speaker's resistance to AC current. You can easily find 4 Ω, 8 Ω, 16 Ω, and 32 Ω speakers. It's important to select a speaker that matches the minimum impedance rating of the amplifier you're using to drive the speaker. (You can find that rating in the datasheet for the amplifier on your supplier's website.) If the speaker impedance is too high, you won't get as much volume out of the speaker as you could. If the speaker impedance is too low, you may overheat your amplifier.
- **Power rating:** The power rating tells you how much power *(power = voltage × current)* the speaker can handle without being damaged. Typical power ratings are 0.25 watt, 0.5 watt, 1 watt, and 2 watt. Be sure that you look up the maximum power output of the amplifier driving your speaker (check the datasheet), and choose a speaker with a power rating of at least that value.

For hobby electronics projects, miniature speakers (roughly 2 to 3 inches in diameter) with an input impedance of 8 Ω are often just what you need. Just be careful not to overpower these little noisemakers, which typically handle only 0.25 to 0.5 watt.

Sounding off with buzzers

Like speakers, buzzers generate sound — but unlike speakers, buzzers indiscriminately produce the *same* obnoxious sound no matter what voltage you apply (within reason). With speakers, "Mozart in" creates "Mozart out"; with buzzers, "Mozart in" creates nothin' but noise.

One type of buzzer, a *piezoelectric buzzer,* contains a diaphragm attached to a piezoelectric crystal. When a voltage is applied to the crystal, the crystal expands or contracts (the piezoelectric effect); this effect, in turn, makes the diaphragm vibrate, generating sound waves. (Note that piezoelectric buzzers work in pretty much the opposite way a crystal microphone works, as described earlier in this chapter.)

Buzzers have two leads and come in a variety of packages. Figure 12-10 shows a typical buzzer. To connect the leads the correct way, remember that the red lead connects to a positive DC voltage.

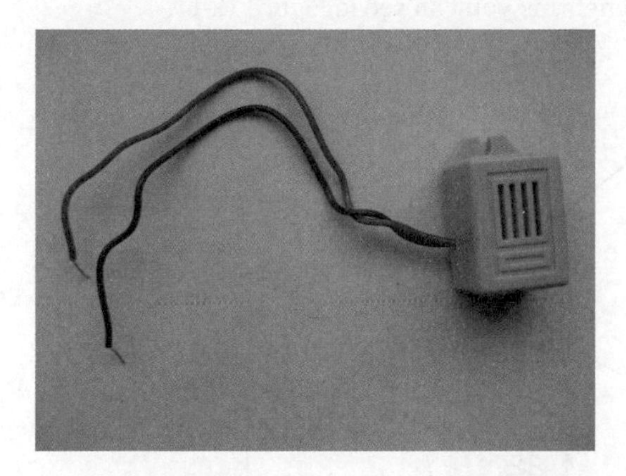

Figure 12-10: This noisy little buzzer is simple to operate.

When shopping for a buzzer, you should consider three specifications:

- **The frequency of sound it emits:** Most buzzers give off sound at one frequency, somewhere in the range of 2 kHz to 4 kHz.
- **The operating voltage and voltage range:** Make sure you get a buzzer that works with the DC voltage that your project supplies.
- **The level of sound it produces in unit of decibels (dB):** The higher the decibel rating, the louder (and more obnoxious) the sound emitted. Higher DC voltage provides a higher sound level.

Be careful that the sound doesn't get so loud that it damages your hearing. You can suffer permanent hearing loss if exposed to sound at 90 dB or higher for a sustained interval — but you won't feel pain until sound reaches at least 125 dB.

Creating good vibrations with DC motors

Have you ever wondered what causes your smartphone to vibrate? No, it's not Mexican jumping beans: These devices normally use a *DC motor,* which changes electrical energy (such as the energy stored in a battery) into motion. That motion may involve turning the wheels of a robot that you built or shaking your smartphone. In fact, you can use a DC motor in any project in which you need motion.

Electromagnets make up an important part of DC motors because these motors consist of, essentially, an electromagnet on an axle rotating between two permanent magnets, as you can see in Figure 12-11.

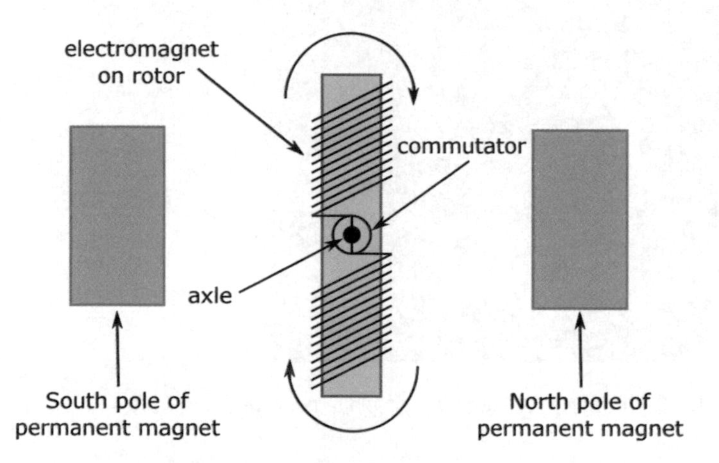

Figure 12-11: How the parts of a simple DC motor fit together.

The positive and negative terminals of the battery connect so that each end of the electromagnet has the same polarity as the permanent magnet next to it. Like poles of magnets repel each other. This repelling action moves the electromagnet and causes the axle to spin. As the axle spins, the positive and negative connections to the electromagnet swap places, so the magnets continue to push the axle around.

A simple mechanism — consisting of a *commutator* (a segmented wheel with each segment connected to a different end of the electromagnet) and brushes that touch the commutator — causes the connections to change. The commutator turns with the axle and the brushes are stationary, with one brush connected to the positive battery terminal and the other brush to the negative battery terminal. As the axle — and (therefore) the commutator — rotates, the segment in contact with each brush changes. This, in turn, changes which end of the electromagnet is connected to negative or positive voltage.

If you want to get a feel for the mechanism inside a DC motor, buy a cheap one for a few dollars and tear it apart.

The axle in a DC motor rotates a few thousand times per minute — a bit fast for most applications. Suppliers sell DC motors with something called a *gear head* premounted; this device reduces the speed of the output shaft to under a hundred revolutions per minute (rpm). This technique is similar to the way that changing gears in your car changes the speed of the car.

Suppliers' catalogs typically list several specifications for the motors they carry. When you shop for electric motors, consider these two key characteristics:

- **Speed:** The speed (in rpm) that you need depends on your project. For example, when turning the wheels of a model car, you may aim for 60 rpm, with the motor rotating the wheels once per second.
- **Operating voltage:** The operating voltage is given as a range. Hobby electronics projects typically use a motor that works in the 4.5 V to 12 V range. Also note the manufacturer's nominal voltage and stated rpm for the motor. The motor runs at this rpm when you supply the nominal voltage. If you supply less than the nominal voltage, the motor runs slower than the stated rpm. If you supply more, it may run faster but it'll probably burn out.

DC motors have two wires (or terminals that you solder wires to), one each for the positive and negative supply voltage. You run the motor by simply supplying a DC voltage that generates the speed that you want, and you switch off the voltage when you want the motor to stop. For many DC motors, changing the polarity of the supply voltage changes the direction of the axle rotation.

You can use a more efficient method of controlling the speed of the motor called *pulse-width modulation.* This method turns voltage on and off in quick pulses. The longer the on intervals, the faster the motor goes. If you're building a kit for something motor controlled (such as a robot), this type of speed control should be included with the electronics for the kit.

If you're attaching things such as wheels, fan blades, and so on to the motor shaft, make sure that you've attached the component *securely* before you apply power to the motor. If you don't, the item may spin off and hit you, or someone near and dear to you, in the face.

Part III

Getting Serious about Electronics

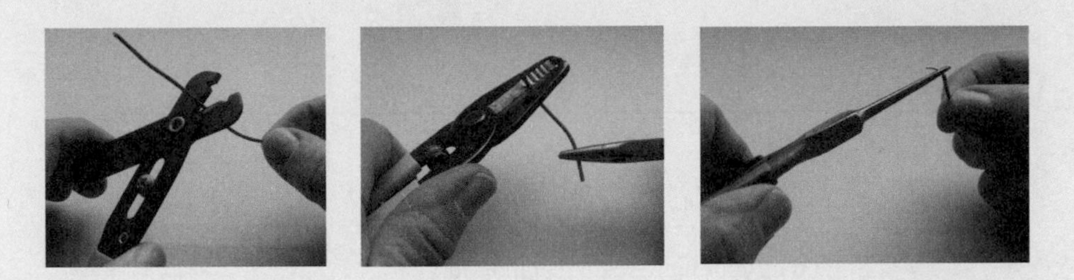

Check out www.dummies.com/extras/electronics to learn what an oscilloscope is and how it can help you visualize what's going on in your circuits.

In this part . . .

- Setting up your mad-scientist laboratory
- Learning to read schematics
- Mastering the art of soldering
- Building plug-and-play circuits on a solderless breadboard
- Creating permanent circuits
- Exploring circuit operation using a multimeter
- Building cool projects that control lights, sound an alarm, make music, and more

13

Preparing Your Lab and Ensuring Your Safety

In This Chapter

- Designing a workspace that works for you
- Stockpiling tools and other supplies
- Creating a starter kit of electronics components
- Realizing that Ohm's Law applies to humans too
- Avoiding electrocution
- Keeping your components from turning into lumps of coal

Finding out how resistors, diodes, transistors, and other electronic components work is great, but creating real projects that make things buzz, beep, and go bump in the night is where the real fun is! To get the most out of your journey into the world of electronics, you'd be wise to spend some time getting properly prepared.

In this chapter, I give you guidelines for setting up a little electronics laboratory in your own home. I outline the tools and supplies you need to accomplish circuit-building jobs, and I give you a shopping list of electronic components to purchase so you can build a bunch of different projects.

Because building circuits isn't for the faint of heart (even small currents can affect your heart), I run you through the safety information you need to know to remain a healthy hobbyist.

A word to the wise: It doesn't take much electrical current to seriously hurt or even kill you. Even the most seasoned professionals take appropriate precautions to stay safe. I strongly suggest (insist, even) that you thoroughly read the safety information provided (hey, I took great pains putting it together) and, before you start each project, review the safety checklist at the end of this chapter. Promise?

Picking a Place to Practice Electronics

Where you put your workshop is just as important as the projects you make and the tools you use. Just as in real estate, the guiding words for electronics work are location, location, location. By staking out just the right spot in your house or apartment, you'll be better organized and enjoy your experiments much more. Nothing is worse than working with a messy workbench in dim lighting while breathing stale air.

The top ingredients for a great lab

The prime ingredients for the well set-up electronics laboratory are the following:

- A comfortable place to work, with a table and chair
- Good lighting
- Ample electrical outlets, with at least 15-amp service
- Tools and toolboxes on nearby shelves or racks
- A comfortable, dry climate
- A solid, flat work surface
- Peace and quiet

The ideal workspace isn't disturbed if you have to leave it for hours or days. Also, the worktable should be off-limits or inaccessible to your children. Curious kids and electronics don't mix!

The garage is an ideal setting because it gives you the freedom to work with solder and other messy materials without worrying about soiling the carpet or nearby furniture. You don't need much space; about 3 by 4 feet ought to do it. If you can't clear that much space in your garage (or you don't have a garage), you can use a room in the house, but try to designate a corner or a section of the room for your electronics work. When working in a carpeted room, you can prevent static electricity by spreading a protective cover, such as an antistatic mat, over the floor. I discuss this in detail later in this chapter.

If your work area must be exposed to other family members, find ways to make the area off-limits to others with less knowledge about electronics safety (which I cover later in this chapter), especially young children. Keep your projects, tools, and supplies out of reach or behind locked doors. And be sure to keep integrated circuits and other sharp parts off the floor — they're painful when stepped on!

No matter where you set up shop, consider the climate. Extremes in heat, cold, or humidity can have a profound effect on your electronics circuits. If you find a work area chilly, warm, or damp, take steps to control the climate in that area, or don't use that area for electronics work. You may need to add insulation, an air conditioner, or a dehumidifier to control the temperature and humidity of your work area. Locate your workbench away from open doors and windows that can allow moisture and extreme temperatures in. And for safety reasons, never — repeat, *never* — work in an area where the floor is wet or even slightly damp.

Workbench basics

The types of projects that you do determine the size of the workbench you need, but for most applications, a table or other flat surface spanning about 2 by 3 feet will suffice. You may even have a small desk, table, or drafting table that you can use for your electronics bench.

You can make your own workbench easily by using an old door as a table surface. If you don't have an old door lying around the house, pick up an inexpensive hollow-core door or a sturdier solid-core door at your local home improvement store. Build legs using 30-inch lengths of 2-by-4 lumber and attach the legs using joist hangers. As an alternative, you can use 3/4-inch plywood or particle board to fashion your work surface.

If you prefer, forgo the 2-by-4 legs and make a simple workbench using a door and two sawhorses. This way, you can take your workbench apart and store it in a corner when you're not using it. Use bungee cords to secure the door to the sawhorses, to prevent accidentally flipping the top of your workbench off the sawhorses.

Remember, as you work on projects, you crouch over your workbench for hours at a time. You can skimp and buy or build an inexpensive worktable, but if you don't already own a good chair, put one on the top of your shopping list. Be sure to adjust the seat for the height of the worktable. A poor-fitting chair can cause backaches and fatigue.

Acquiring Tools and Supplies

Every hobby has its special assortment of tools and supplies, and electronics is no exception. From the lowly screwdriver to the high-speed drill, you'll enjoy playing with electronics much more if you have the right tools and an assortment of supplies, organized and stored so that you can put your hands on them when you need them without cluttering your work area.

This section tells you exactly what tools and supplies you need to have to complete basic-to-intermediate electronics projects.

If you have a permanent place in your house to work on electronics, you can hang some of the hand tools mentioned in this section on the wall or a pegboard. Reserve this special treatment for the tools you use the most. You can stash other small tools and some supplies in a small toolbox, which you can keep on your workbench. A plastic fishing tackle box with lots of small compartments and one large section can help you keep your things organized.

Amassing a multimeter

One of the most important tools you'll need is a *multimeter,* which you use to measure AC and DC voltages, resistance, and current when you want to explore what's going on in a circuit. Most multimeters you find today are of the digital variety (see Figure 13-1), which just means they use numeric displays, like a digital clock or watch. (You can use them to explore analog as well as digital circuits.) An older-style analog multimeter uses a needle to point to a set of graduated scales.

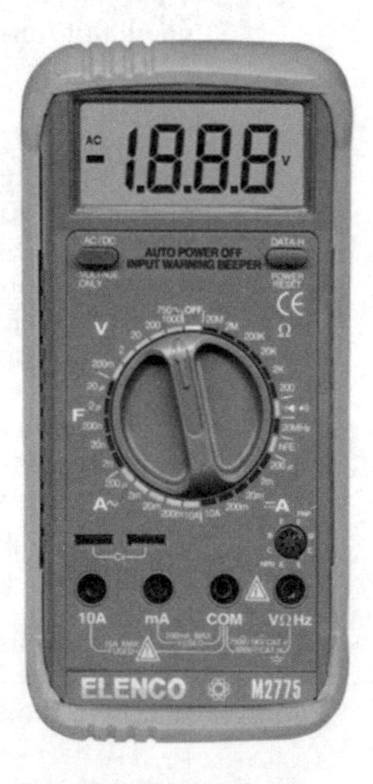

Figure 13-1: Multimeters measure voltage, resistance, and current.

Each multimeter comes with a pair of test leads: one black (for the ground connection) and one red (for the positive connection). On small pocket units, the test leads are permanently attached to the meters, whereas on larger models, you can unplug the leads. Each test lead has a cone-shaped metal tip used for probing circuits. You can also purchase test clips that slip over the tips, making testing much easier because you can attach these clips onto wires or component leads.

Prices for new multimeters range from $10 to over $100. The higher-priced meters include additional features, such as built-in testing capabilities for capacitors, diodes, and transistors. Think of a multimeter as a set of eyes into your circuits, and consider purchasing the best model you can afford. That way, as your projects grow more complex, you still get a magnificent view of what's going on inside.

I give you the lowdown on how to use a multimeter in Chapter 16.

Stockpiling soldering equipment

Soldering (pronounced "SOD-er-ing") is the method you use to make semipermanent connections between components as you build a circuit. Instead of using glue to hold things together, you use small globs of molten metal called *solder* applied by a device called a *soldering iron.* The metal provides a conductive physical joint, known as a *solder joint,* between the wires and component leads of your circuit.

You'll be glad to know that you need only a few simple tools for soldering. You can purchase a basic, no-frills soldering setup for under $10, but the better soldering tools cost a bit more. At a minimum, you will need the following basic items for soldering:

- **Soldering iron:** A *soldering iron,* also called a *soldering pencil,* is a wand-like tool that consists of an insulating handle, a heating element, and a polished metal tip. (See Figure 13-2.) Choose a soldering iron that is rated at 25–30 watts, sports a replaceable tip, and has a three-prong plug so that it will be grounded. Some models allow you to use different size tips for different types of projects, and some include variable controls that allow you to change the wattage. (Both are nice but not absolutely necessary.)

Figure 13-2: Some soldering iron models are temperature-adjustable and come with their own stands.

- **Soldering stand:** The stand holds the soldering iron and keeps the very hot tip from coming into contact with anything on your work surface. Some soldering irons come with stands. (Usually, these combos are known as *soldering stations.*) The stand should have a weighted base; if not, clamp it to your worktable so it doesn't tip over. A stand is a must-have — unless you want to burn your project, your desk, or yourself!

- **Solder:** *Solder* is a soft metal that is heated by a soldering iron, and then allowed to cool, forming a conductive joint. Standard solder used for electronics is *60/40 rosin core,* which contains roughly 60 percent tin and 40 percent lead and has a core of rosin flux. (Avoid solder formulated for plumbing, which corrodes electronic parts and circuit boards.) The wax-like *flux* helps to clean the metals you're joining, and it improves the molten solder's capability to flow around and adhere to the components and wire, ensuring a good solder joint. Solder is sold in spools, and I recommend diameters of 0.031 inch (22 gauge) or 0.062 inch (16 gauge) for hobby electronics projects.

The lead content in 60/40 rosin core solder may pose a health hazard if you don't handle it carefully. Be sure to keep your hands away from your mouth and eyes whenever you've been touching this solder. Above all, don't use your teeth to hold a piece of solder while your hands are busy.

I recommend that you also get these additional soldering tools and accessories:

- **Wetted sponge:** You use this to wipe off excess solder and flux from the hot tip of the soldering iron. Some soldering stands include a small sponge and a built-in space to hold it, but a clean household sponge also works fine.
- **Solder removal tools:** A *solder sucker,* also known as a *desoldering pump,* is a spring-loaded vacuum you can use to remove a solder joint or excess solder in your circuit. To use it, melt the solder that you want to remove, quickly position the solder sucker over the molten blob, and activate it to suck up the solder. Alternatively, you can use a *solder* (or *desoldering*) *wick* or *braid,* which is a flat, woven copper wire that you place over unwanted solder and apply heat to. When the solder reaches its melting point, it adheres to the copper wire, which you then remove and dispose of.
- **Tip cleaner paste:** This gives your soldering tip a good cleaning.
- **Rosin flux remover:** Available in a bottle or spray can, use the remover after soldering to clean any remaining flux and prevent it from oxidizing (or rusting, in unscientific terms) your circuit, which can weaken the metal joint.
- **Extra soldering tips:** For most electronics work, a small (3/64-inch through 7/64-inch radius) conical or chiseled tip, or one simply described as a fine tip, works well. You can find larger or smaller tips used for different types of projects. Be sure to purchase the correct tip for your make and model of soldering iron. Replace the tip when it shows signs of corrosion, pitting, or peeling plating; a worn tip doesn't pass as much heat.

In Chapter 15, I explain in detail how to use a soldering iron.

Hoarding hand tools

Hand tools are the mainstay of any toolbox. These tools tighten screws, snip off wires, bend little pieces of metal, and do all those other mundane tasks. Make sure you have the following tools available at your workbench:

- **Wire cutter:** You can find general-purpose wire cutters at hardware and home improvement stores, but it pays to invest $5 or so in a *flush,* or *diagonal cutter,* shown in Figure 13-3, for making cuts in tight places, such as above a solder joint.

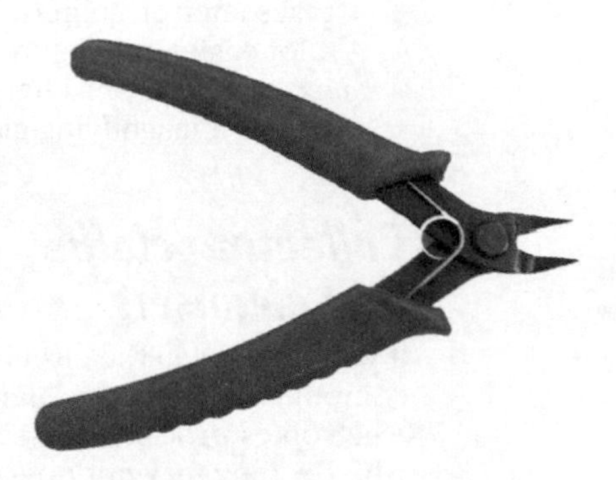

Figure 13-3: Diagonal cutters trim wire ends flush to the surface.

- **Wire stripper:** You often need to expose a half-inch or so of bare wire so you can solder a connection or insert the wire into the holes of a solderless breadboard (which I discuss next). A good wire stripper contains notches allowing you to neatly and easily strip just the plastic insulation from wires of various sizes (known as gauges, as described in Chapter 12), without nicking the copper wire inside. You can also find a combination wire cutter and stripper, but you may have to perform your own gauge control.
- **Needle-nose pliers (two sets):** These pliers help you bend wires, insert leads into breadboard holes, and hold parts in place. Get two pairs: a mini (5-inch long) set for intricate work and a standard-size set to use when you need to apply a wee bit more pressure.
- **Precision screwdrivers:** Make sure you have both straight and *Phillips head* (cross-shaped tip) screwdrivers that are small enough for your electronics needs. Use the right size for the job to avoid damaging the head of the screw. To make it easier to work with small screws, use a magnetized screwdriver or place a small amount of rubber holdup putty into the head of the screw before inserting the screwdriver tip. Works wonders.
- **Magnifying glass:** A 3X (or more) magnifying glass can help you check solder joints and read teeny tiny part numbers.

- **Third hand:** No, this isn't a body part from your buddy. It's a tool that clamps onto your worktable and has adjustable clips that hold small parts while you're working. A third hand makes tasks such as soldering a heck of a lot easier. See Figure 13-4 for an example of a third hand that also sports a magnifying glass.

Figure 13-4: These helping hands combine alligator clips with a magnifying glass.

Collecting cloths and cleansers

If you don't keep the circuitry, components, and other parts of your electronics projects as clean as a whistle, they may not operate as advertised. It's especially important to start with a clean slate if you're soldering parts together or to a circuit board. Dirt makes for bad solder joints, and bad solder joints make for faulty circuits.

Here's a list of items that can help you keep your projects spick-and-span:

- **Soft cloth or gauze bandage:** Keep your stuff dust-free by using a soft cloth or sterilized lint-free bandage. Don't use household dusting sprays because some generate static charges that can damage electronics.
- **Compressed air:** A shot of compressed air, available in cans, can remove dust from delicate electronic innards. But keep it locked away when you're not using it; if misused as an inhalant, compressed air can cause death.
- **Water-based household cleaner:** Lightly spray to remove stubborn dirt and excess grease from tools, work surfaces, and the exterior surfaces of your projects. Don't use them around powered circuits, or you may short something out.
- **Electronics cleaner/degreaser:** Use only a cleaner/degreaser specifically made for use on electronic components.
- **Artist brushes:** Get both a small brush and a wide brush to dust away dirt, but avoid cheap brushes that shed bristles. A dry, clean toothbrush works well, too.
- **Photographic bulb brush:** Available at any photo shop, a bulb brush combines the whisking action of a soft brush with the cleaning action of a strong puff of air.

- **Contact cleaner:** Available in a spray can, contact cleaner enables you to clean electrical contacts. Spray it onto a brush, and then whisk the brush against the contacts to give them a good cleaning.
- **Cotton swabs:** Soak up excess oil, lubricant, and cleaner with these swabs.
- **Cuticle sticks and nail files:** Scrape junk off circuit boards and electrical contacts, and then give yourself a manicure!
- **Pink pencil eraser:** Great for rubbing electrical contacts clean, especially contacts that have been contaminated by the acid from a leaky battery. Must be pink; other erasers can leave a hard-to-remove residue. Avoid rubbing the eraser against a circuit board because it may create static electricity.

Loading up on lubricants

Motors and other mechanical parts used in electronics projects require a certain amount of grease or oil to operate, and you need to re-lubricate them periodically. Two types of lubricants are commonly used in electronics projects — and there's one type of lubricant you should avoid using with electronics projects.

Avoid using a spray-on synthetic lubricant (such as WD-40 and LPS) with your electronics projects. Because you can't control the width of the spray, you're bound to get some on parts that shouldn't be lubed. Also, some synthetic lubricants are nonconductive, and their fine mist can get in the way, interrupting electrical contacts.

The okey-dokey lubricants are

- **Light machine oil:** Use this type of oil for parts that spin. Avoid using oil with antirust ingredients that may react with plastic parts, causing them to melt. A syringe oiler with a long, thin spout is ideal for hard-to-reach places.
- **Synthetic grease:** Use lithium grease or another synthetic grease for parts that mesh or slide.

You can find light machine oil and synthetic grease at electronics supply houses as well as many music, sewing machine, hobby, and hardware stores.

Don't apply a lubricant unless you know for sure that a mechanical part needs it. Certain self-lubricating plastics used for mechanical components can break down when exposed to petroleum-based lubricants. If you're fixing a CD player or other piece of electronic equipment, check with the manufacturer for instructions regarding use of lubrication.

Stocking up on sticky stuff

Many electronics projects require that you use an adhesive of some type. For example, you may need to secure a small printed circuit board to the inside of a pocket-sized project box. Depending on the application, you can use one or more of the following adhesives:

- **White household glue** is best used for projects that involve wood or other porous materials. Allow 20–30 minutes for the glue to dry, and about 12 hours to cure.
- **Epoxy cement** creates strong, moisture-resistant bonds and can be used for any material. Allow 5–30 minutes for the epoxy to set, and 12 hours for it to cure.
- **Cyanoacrylate (CA) glue, or super glue,** bonds almost anything (including fingers, so use caution), almost instantly. Use ordinary CA glue when bonding smooth and perfectly matching parts; use the heavier-bodied gap-filling CA glue if the parts don't mate 100 percent.
- **Double-sided foam tape** is a quick way to secure circuit boards to enclosures or to make sure that loosely fitting components remain in place.
- **A hot-melt glue gun** allows you to glue things with a drying time of only about 30 seconds. The waterproof, gap-sealing glue comes in a stick that you slide into a slot in the gun, which heats the glue to about 250°F–350°F — hot enough to hurt you but not hot enough to melt solder.

Other tools and supplies

I highly recommend that you acquire three other items before you begin any electronics work:

- **Safety glasses:** Stylish plastic safety glasses never go out of fashion. They are a must-have to protect your eyes from flying bits of wire, sputtering solder, exploding electronics parts, and many other small objects. If you wear prescription glasses, place safety glasses over them to ensure complete protection all around your eyes.
- **Antistatic wrist strap:** This inexpensive strap prevents electrostatic discharge from damaging sensitive electronic components. I discuss this device later in this chapter.
- **First-aid kit and guide:** Burns (or worse) can happen when working with electronic circuits. Keeping a first-aid kit at your workbench is a good idea. Make sure you include guidelines for applying first aid.

The time will come when you'll want to enclose an electronics project in a container with wires or knobs sticking out. For instance, say you build a holiday light display with a controllable blink rate. You may want to place the main circuit in a box, cut a hole through the front of the box, and insert a *potentiometer* (variable resistor) through the hole so you (or someone else) can control how fast the lights blink. Or you may want to build a circuit that detects intruders opening your refrigerator. You could disguise the circuit as a breadbox and place it next to the fridge. In any case, you'll need some additional tools and supplies to enclose your project.

Here's a list of supplies and associated tools you may need to box up your project:

- **Ready-made box:** You can find simple unfinished wooden boxes at craft stores, and ABS plastic boxes at most electronics suppliers. Or you can make your own box out of plywood or PVC plastic, using contact cement or another adhesive to keep it together.
- **Wire clips:** Adhesive-backed plastic clips hold wires in place along the insides of your box.
- **Cable ties:** Use cable ties to attach wires to non-flat surfaces, such as a wooden dowel.
- **Electric drill:** A drill with a 3/8-inch *chuck* (the opening in the drill where you insert the drill bit) comes in handy for making holes in your box for knobs and switches. You can also use it to attach wheels or other external parts to your box.
- **Hand saws:** You can use a hack saw to cut wood or plastic to make your box, and a coping saw to cut broad openings in the box.

Stocking Up on Parts and Components

Okay, so you have your workbench set up, complete with screwdrivers, pliers, and hand saws, you've donned your antistatic wrist strap and safety glasses (along with your everyday clothes, please!), and you have your soldering iron plugged in and ready to go. So what's missing? Oh yeah, circuit components!

When you shop for circuit components, you usually don't go out and purchase only the parts listed for a particular circuit diagram, or *schematic.* You purchase an assortment of parts so you can build several different projects without having to run out for parts each time you try something new. Think of this like gathering ingredients for cooking and baking. You keep many basic ingredients, such as flour, sugar, oil, rice, and spices, on hand all the time, and you purchase enough other ingredients to enable you to cook the sorts of things you like for a week or two. Well, the same is true when stocking up on electronics parts and components.

In this section, I tell you what parts and how many you should keep on hand to build some basic electronics projects.

Solderless breadboards

A *solderless breadboard* is similar, in a way, to a LEGO table: It's a surface on which you can build temporary circuits simply by plugging components into holes arranged in rows and columns across the board. You can easily take one circuit apart and build another different circuit on the same surface.

The holes in a solderless breadboard aren't just ordinary holes; they are *contact holes* with copper lines running underneath so that components plugged into two or more holes within a particular row are connected below the surface of the breadboard. You plug in your *discrete components* (resistors, capacitors, diodes, and transistors) and *integrated circuits (ICs)* in just the right way, and — *voilà* — you have a connected circuit without soldering. When you're tired of the circuit, you can simply remove the parts and build something else using the same breadboard.

Figure 13-5 shows a small solderless breadboard with a battery-powered circuit connected. The breadboard in the figure has sections of rows and columns connected in a certain way underneath the board. I discuss just how the various contact holes are connected in Chapter 15, where I also discuss how to build circuits using breadboards. For now, just know that different sizes of breadboards with different numbers of contact holes are available.

Figure 13-5: You can build a circuit on a small solderless breadboard in just minutes.

A typical small breadboard has 400 contact holes, and it's useful for building smaller circuits with no more than two ICs (plus other discrete components). A typical larger breadboard contains 830 contacts, and you can use it to build somewhat more complex circuits. You can also link multiple breadboards simply by connecting one or more wires between contact holes on one board and contact holes on the other board.

I recommend that you purchase at least two solderless breadboards, with at least one of them a larger (830-contact) breadboard. Also, buy some adhesive-backed Velcro strips to help hold the breadboards in place on your work surface.

You commonly use solderless breadboards to test your circuit design ideas or explore circuits as you're learning how things work. If you create and test a circuit using a breadboard and you want to use the circuit on a long-term basis, you can re-create it on a soldered or printed circuit board (PCB). A PCB is a kind of breadboard, but instead of contact holes, it has ordinary holes with copper pads surrounding each hole and lines of metal connecting the holes within each row. You make connections by soldering component leads to the copper pads, ensuring that the components you're connecting are located in the same row. In this book, I focus exclusively on circuit construction using solderless breadboards.

Circuit-building starter kit

You need an assortment of discrete electronic components (those with two or three individual leads), a few ICs, several batteries, and lots of wire to connect things. Some components, such as resistors and capacitors, come in packages of ten or more pieces. You'll be happy to know that these components are inexpensive (cheap, even); it'll cost you one or two weeks' worth of lattes to stock up.

You may want to refresh your memory on what these components are and how they work by referring to other chapters in this book. Resistors and potentiometers are covered in Chapter 5, capacitors in Chapter 7, diodes (including LEDs) in Chapter 9, and transistors in Chapter 10. Integrated circuits are covered in Chapter 11, and batteries and wires in Chapter 12.

Here are the discrete components I recommend that you start with:

- **Fixed resistors (1/4-watt or 1/2-watt carbon film):** 10–20 (1 or 2 packages) of each of these resistances: 1 kΩ, 10 kΩ, 100 kΩ, 1 MΩ, 2.2 kΩ, 22 kΩ, 220 kΩ, 33 kΩ, 470 Ω, 4.7 kΩ, 47 kΩ, 470 kΩ.
- **Potentiometers:** Two each of 10 kΩ, 50 kΩ, 100 kΩ, 1 MΩ.

- **Capacitors:** 10 each (1 package) of 0.01 μF and 0.1 μF nonpolarized (polyester or ceramic disc); 10 each (1 package) of 1 μF, 10 μF, 100 μF electrolytic; 3–5 each of 220 μF and 470 μF electrolytic.
- **Diodes:** 10 each of 1N4001 (or any 1N400x) rectifier diode and 1N4148 small-signal diode, 1 4.3-volt Zener diode (or other Zener breakdown voltage between 3 and 7 volts).
- **LEDs (light-emitting diodes):** 10 each (1 package) of red, yellow, and green 5mm diffused LEDs.
- **Transistors:** 3–5 general-purpose, low-power bipolar transistors (such as the 2N3904 NPN or the 2N3906 PNP) and 3–5 medium-power bipolar transistors (such as the NTE123A NPN or NTE159M PNP).

I suggest that you obtain a few of these popular ICs:

- **555 timer IC:** Get 3–5 of these. You'll use 'em!
- **Op-amp ICs:** Get one or two op amps, such as the LM741 general-purpose amplifier.
- **4017 CMOS decade counter IC:** Get two or three of these. You need two if you want to make a tens counter, too, as I discuss in Chapter 11, and it's wise to have an extra one on hand if you think you might accidentally zap one with electrostatic discharge.

Don't forget these essential power and wire components:

- **Batteries:** Pick up an assortment of 9 V batteries as well as some 1.5 V batteries. (Size depends on how long you think you're going to run your circuit.)
- **Battery clips and holders:** These devices connect to batteries and provide wire leads to make it easy for you to connect battery power to your circuit. Get 3–5 clips for the size batteries you plan to use.
- **Wire:** Ample 20–22 gauge solid wire. You can buy a 100-foot roll in any one of a variety of colors for about $7. You cut it to various lengths and strip the insulation off each end to connect components. You can solder each end to a component lead, or insert each end into contact holes on your solderless breadboard. Some electronics suppliers sell kits containing dozens of precut, prestripped *jumper wires* of various lengths and colors, ideal for use in solderless breadboards. A kit with 140–350 jumper wires may cost you $8–$12, but it can save you the time (and trouble) of cutting and stripping your own wire. (Plus, you get rainbow colors!)

You can use a jumper wire as a sort of on/off switch in your circuit, connecting or disconnecting power or components. Just place one end of the jumper wire in your solderless breadboard and alternatively place and remove the other end to make or break the connection.

Adding up the extras

Lots of other parts and components that can enrich your circuits are out there. I recommend you get a few of the ones listed here:

- **Alligator clips:** So-named because they look like the jaws of a fierce gator, these insulated clips can help you connect test equipment to component leads, and they can double as heat sinks! Get a bunch (10 or so).
- **Speakers:** You gotta build a circuit that makes noise, so purchase one or two miniature 8-ohm speakers. (Chapter 12 discusses speakers.)
- **Switches:** Purchase 5–10 single-pole, double-throw (SPDT) switches with 0.1" spacing between terminals for use in a solderless breadboard. These SPDT switches can double as on/off switches in your circuits. You may also want to buy a few pushbutton (momentary on) switches. If you think you might enclose one or more projects in a box and you'd like a robust front-panel on/off control, pick up a couple of SPST (single-pole, single-throw) switches, such as an SPST mini rocker switch. For a little more money, you can get a mini rocker switch with a built-in LED that lights up when the switch is in the on position. (Chapter 4 provides details on switches.)

Organizing all your parts

Keeping all these parts and components organized is essential — unless you're the type who enjoys sorting through junk drawers looking for some tiny, yet important, item. An easy way to get it together is to run over to your local big-box discount store and purchase one or more sets of clear-plastic drawer organizers. Be sure to spray the plastic boxes with an antistatic ESD (electrostatic discharge) spray, which you can find on Amazon.com. Then label each drawer for a particular component (or group of components, such as LEDs, 10–99 Ω resistors, and so forth). You'll know in a glance where everything is and be able to see when your stock is getting low.

Protecting You and Your Electronics

You probably know that Benjamin Franklin "discovered" electricity in 1752 by flying a kite during a lightning storm. Actually, Franklin already knew about electricity and was well aware of its potential power — and potential danger. As Franklin carried out his experiment, he was careful to insulate himself from the conductive materials attached to the kite (the key and a metal wire) and to stay dry by taking cover in a barn. Had he not, we might be looking at someone else's face on the $100 bill!

Respect for the power of electricity is necessary when working with electronics. In this section, you take a look at keeping yourself — and your electronic projects — safe. This is one section you should read from start to finish, even if you already have some experience in electronics.

As you read this section, remember that you can describe electrical current as being one of the following:

- **Direct current (DC):** The electrons flow one way through a wire or circuit.
- **Alternating current (AC):** The electrons flow one way, and then another, in a continuing cycle.

Refer to Chapter 1 for more about these two types of electrical current.

Understanding that electricity can really hurt

By far, the single most dangerous aspect of working with electronics is the possibility of electrocution. Electrical shock results when the body reacts to an electrical current — this reaction can include an intense contraction of muscles (namely, the heart) and extremely high heat at the point of contact between your skin and the electrical current. The heat leads to burns that can cause death or disfigurement. Even small currents can disrupt your heartbeat.

The degree to which electrical shock can harm you depends on a lot of factors, including your age, your general health, the voltage, and the current. If you're well over 50 or in poor health, you probably won't stand up to injury as well as if you're 14 and as healthy as an Olympic athlete. But no matter how young and healthy you may be, voltage and current can pack a wallop, so it's important that you understand how much they can harm you.

The two most dangerous electrical paths through the human body are hand-to-hand and left hand to either foot. If electrical current passes from one hand to the other, on its way it passes through the heart. If current passes from the left hand to either foot, it passes through the heart as well as several major organs.

Seeing yourself as a giant resistor

Your body exhibits some resistance to electrical current, mostly due to the poor conductive qualities of dry skin. The amount of resistance can vary tremendously, depending on body chemistry, level of moisture in the skin, the total path across which resistance is measured, and other factors. You'll see figures ranging anywhere from 50,000 ohms to 1,000,000 ohms of resistance for an average human being. (I discuss what resistance is and how it's measured in Chapter 5.)

If your skin is moist (say you have sweaty hands), you're wearing a metal ring, or you're standing in a puddle, you can bet you've lowered your resistance. Industry figures indicate that such activity can result in resistances as low as 100–300 Ω from one hand to the other or from one hand to one foot. That's not a whole lot of resistance.

To make matters worse, if you're handling high AC voltages (which you shouldn't be), your skin's resistance — wet or dry — won't help you at all. When you're in contact with a metal, your body and the metal form a capacitor: The tissue underneath your skin is one plate, the metal is the other plate, and your skin is the dielectric. (See Chapter 7 for the lowdown on capacitors.) If that metal wire you're holding is carrying an AC current, the capacitor that is your body acts like a short circuit, allowing current to bypass your skin's resistance. Voltage shocks of more than 240 volts will burn right through your skin, leaving deep third-degree burns at the entry points.

Knowing how voltage and current can harm you

You've seen the signs: WARNING! HIGH VOLTAGE. So you might think that voltage is what causes harm to the human body, but it's actually current that inflicts the damage. So why the warning signs? That's because the higher the voltage, the more current can flow for an equal amount of resistance. And because your body is like a giant resistor, you should shy away from high voltages.

So how much current does it take to hurt the average human being? Not much. Table 13-1 summarizes some estimates of just how much — or how little — DC and 60-Hz (hertz) AC current it takes to affect the human body. Remember that a milliamp (mA) is one one-thousandth of an amp (or 0.001 A). Please note that these are *estimates* (no one has performed experiments on real humans to derive these figures), and that each person is affected differently depending on age, body chemistry, health status, and other factors.

Table 13-1 Effects of Current on Average Human Body

Effect	*DC current*	*60-Hz AC current*
Slight tingling sensation	0.6–1.0 mA	0.3–0.4 mA
Noticeable sensation	3.5–5.2 mA	0.7–1.1 mA
Pain felt, but muscle control maintained	41–62 mA	6–9 mA
Pain felt, and unable to let go of wires	51–76 mA	10–16 mA
Difficulty breathing (paralysis of chest muscles)	60–90 mA	15–23 mA
Heart fibrillation (within 3 seconds)	500 mA	65–100 mA

As Table 13-1 shows, the average human body is four to six times more sensitive to AC current than to DC current. Whereas a DC current of 15 mA isn't all that dangerous, 15 mA of alternating current has the potential to cause death.

So what does all this mean to you as you pursue your electronics hobby? You probably know enough to stay away from high voltages, but what about getting up close and personal with low voltages? Well, even low voltages can be dangerous — depending on your resistance.

Remember that Ohm's Law (which I cover in Chapter 6) states that voltage is the product of current and resistance:

$$voltage = current \times resistance$$

$$V = I \times R$$

Let's say your hands are dry and you aren't wearing a metal ring or standing in a puddle, and your hand-to-hand resistance is about 50,000 ohms. (Keep in mind that your resistance under these dry, ringless conditions may actually be lower.) You can calculate an estimate (repeat: *estimate*) of the voltage levels that might hurt you by multiplying your resistance by the different current levels in Table 13-1. For instance, if you don't want to feel even the slightest tingling sensation in your fingers, you need to avoid coming into contact with wires carrying DC voltages of 30 V (that's $0.6\text{ mA} \times 50{,}000\ \Omega$). To avoid involuntary muscle contractions (grabbing the wires), you need to keep AC current below 10 mA, so avoid close proximity to 500 volts AC (VAC) or more.

Now, if you're not so careful, and you wear a ring on your finger while tinkering around with electronics, or you step in a little puddle of water created by a dog or small child, you may accidentally lower your resistance to a dangerous level. If your resistance is 5,000 ohms — and it may be even lower — you'll notice a sensation if you handle just 17.5 VDC (because $0.0035\text{ A} \times 5{,}000\ \Omega = 17.5\text{ V}$), and you'll lose muscle control and have difficulty breathing if you handle 120 VAC line power (because $\frac{120\text{ V}}{5{,}000\ \Omega} = 0.024\text{ A} = 24\text{ mA}$).

Household electrical systems in the US and Canada operate at about 120 VAC. This significantly high voltage can, and does, kill. You must exercise *extreme caution* if you ever work with 120 VAC line power.

Until you become experienced in the ways of electronics, you're best off avoiding circuits that run directly off household current. Stick with circuits that run off standard-size batteries, or those small plug-in wall transformers. (You can read about these DC sources in Chapter 12.) Unless you do something silly, like licking the terminal of a 9 V battery (and, yes, that will deliver a shock!), you're fairly safe with these voltages and currents.

Working with AC-powered circuits

Although I strongly recommend that you avoid working with circuits that run directly off household current, I realize you can't always do this. Here are some tips designed to help you avoid electrocution if you choose to work with AC power:

- **Use a self-contained power supply.** If your project requires an AC power supply (which converts the AC to lower-voltage DC), using a self-contained power supply, such as a plug-in wall transformer, is much safer than using a homemade power supply. A *wall transformer* is a little black box with plug prongs, such as the one you use to charge your cellphone.
- **Keep your AC away from your DC.** Physically separating the AC and DC portions of your circuit can help prevent a bad shock if a wire comes loose.
- **Keep AC circuits covered.** A little sheet of plastic works wonders.
- **Use the proper fuse.** Don't use a fuse with a too-high rating and never bypass the fuse on any device.
- **Double- and triple-check your work before applying power.** Ask someone who knows about circuits to inspect your handiwork before you switch the circuit on for the first time. If you decide to test it further, first remove the power by unplugging the power cord from the wall.
- **When troubleshooting a live circuit, keep one hand in your pocket at all times.** By using just one hand to manipulate the testing apparatus, you avoid the situation where one hand touches ground and the other a live circuit, allowing AC to flow through your heart.
- **Take care when enclosing your project.** Use a metal enclosure only if the enclosure is fully grounded. You need to use a three-prong electrical plug and wire for this. Be sure to firmly attach the green wire (which is always connected to earth ground) to the metal of the enclosure. If you can't guarantee a fully grounded metal enclosure, use a plastic enclosure. The plastic helps insulate you from any loose wires or accidental electrocution. For projects that aren't fully grounded, use only an isolated power supply, such as a wall transformer.
- **Secure all wiring inside your project.** Use a strain relief or a cable mount to secure the AC line cord to your project enclosure so you don't expose a live wire. A *strain relief* (available at hardware stores and electronics suppliers) clamps around the wire and prevents you from tugging the wire out of the enclosure.
- **Periodically inspect AC circuits.** Look for worn, broken, or loose wires and components, and promptly make any necessary repairs — with the power off!
- **Err on the side of caution.** Take a lesson from Mr. Murphy, and assume that if something can go wrong, it will. Keep your work environment free of all liquids, pets, and small children. Post a first-aid chart nearby. Don't work when you're tired or distracted. Be serious and focused while you're working around electricity.

One final word: If you simply must work with AC voltages, *don't do it alone.* Make sure you have a buddy — preferably someone *not* named in your will — nearby who is willing and able to dial 911 when you're lying on the ground unconscious. Seriously.

The main danger of household current is the effect it can have on your heart muscle. It takes only 65–100 mA to send your heart into fibrillation, which means the muscles contract in an uncontrolled, uncoordinated fashion — and the heart isn't pumping blood. At much lower levels (10–16 mA), AC current can cause severe muscle contractions, so what might start out as a loose grip on a high-voltage wire (just to move it a little bit, or something like that) ends up as a powerful, unyielding grip. Trust me: You won't be able to let go. A stronger grip means a lower resistance (you're just making it easier for electrons to travel through your hand and into your body), and a lower resistance means a higher (often fatal) current. (Situations like this really do happen. The body acts like a variable resistor, with its resistance decreasing sharply as the hands tighten around the wire.)

The potential dangers of DC currents are not to be ignored either. Burns are the most common form of injury caused by high DC current. Remember that voltage doesn't have to come from a power plant to be dangerous. It pays to respect even a 9 V transistor battery: If you short its terminals, the battery may overheat and can even explode. Battery explosions often send tiny battery pieces flying out at high velocities, burning skin or injuring eyes. Many people have been burned by placing a battery in a pocket along with coins, keys, or other metallic objects. When the battery terminals are shorted, the battery heats up quickly.

Maximizing your resistance — and your safety

When working with electronics, it pays to maximize your resistance just in case you come into contact with an exposed wire. Make sure any tools you pick up are insulated, so that you add more resistance between you and any voltages you may encounter.

Take simple precautions to ensure that your work area starts out dry and stays dry. For example, don't place a glass of water or cup of coffee too close to your work area; if you accidentally knock it over, you may lower your own resistance or short out circuit components.

Keeping a first-aid chart handy

Even if you're the safest person on earth, it's still a good idea to get one of those emergency first-aid charts that include information about what to do in case of electrical shock. You can find these charts on the Internet; try a search for *first aid wall chart.* You can also find them in school and industrial supply catalogs.

Helping someone who has been electrocuted may require cardio-pulmonary resuscitation (CPR). Be sure that you're properly trained before you administer CPR on anyone. Check out `www.redcross.org` to get more information about CPR training.

Soldering safely

The soldering iron you use to join components in an electronics project operates at temperatures in excess of 700 degrees Fahrenheit. (You can read up on soldering in Chapter 15.) That's about the same temperature as an electric stove burner set at high heat. You can imagine how much that hurts if you touch it.

When using a soldering iron, keep the following safety tips in mind:

- **Solder only in a well-ventilated area.** Soldering produces mildly caustic and toxic fumes that can irritate your eyes and throat.
- **Wear safety glasses when soldering.** Solder has been known to sputter.
- **Always place your soldering iron in a stand designed for the job.** Never place the hot soldering iron directly on a table or workbench. You can easily start a fire or burn your hands that way.
- **Be sure that the electrical cord doesn't snag on the table or any other object.** Otherwise, the hot soldering iron can get yanked out of its stand and fall to the ground. Or worse, right into your lap!
- **Use the appropriate soldering setting.** If your soldering iron has an adjustable temperature control, dial the recommended setting for the kind of solder that you're using. Too much heat can spoil a good circuit.
- **Never solder a *live circuit* (a circuit to which you've applied voltage).** You may damage the circuit or the soldering iron — and you may receive a nasty shock.
- **Never grab a tumbling soldering iron.** Just let it fall, and buy a new one if the iron is damaged.
- **Consider using silver solder.** If you're concerned about health issues — or tend to stick your fingers in your mouth or rub your eyes a lot — you may want to avoid solders that contain lead. Instead, use silver solder specifically designed for use on electronic equipment. (Never use acid-flux solder in electronics; it wrecks your circuits.)
- **Unplug your soldering iron when you're finished.**

Avoiding static like the plague

One type of everyday electricity that can be dangerous to people and electronic components is static electricity. It's called *static* because it's a form of current that remains trapped in some insulating body, even after you remove the power source. Static electricity hangs around until it dissipates in some way. Most static dissipates slowly over time, but in some cases, it gets released all at once. Lightning is one of the most common forms of static electricity.

Safety checklist

After you've read all the safety warnings in this chapter, you may want to review this simple checklist of *minimal* safety requirements before you get started on an electronics project. Better yet, you can make a copy of this checklist, laminate it, and post it at your workbench as a reminder of the simple steps that can ensure your safety — and the well-being of your electronics projects.

Workspace check:

- Ample ventilation
- Dry working surface, dry floor
- No liquids, pets, or small children within a 10-foot range
- Dangerous tools and materials locked up
- First-aid chart within view
- Phone (and caring buddy) nearby
- Grounded soldering iron with weighted stand

Personal check:

- Safety glasses
- Antistatic wrist band (attached to you and to earth ground)
- No rings, wristwatches, or loose jewelry
- Cotton or wool clothing
- Dry hands (or use gloves)
- Alert and well-rested

If you drag your feet across a carpeted floor, your body takes on a static charge. If you then touch a metal object, such as a doorknob or a metal sink, the static quickly discharges from your body, and you feel a slight shock. This is known as *electrostatic discharge (ESD),* and can run as high as 50,000 V. The resulting current is small — in the μA range — because of the high resistance of the air that the charges arc through as they leave your fingertips, and it doesn't last very long. So static shocks of the doorknob variety generally don't inflict bodily injury — but they can easily destroy sensitive electronic components.

On the other hand, static shocks from certain electronic components can be harmful. The *capacitor,* an electronic component that stores energy in an electric field, is designed to hold a static charge. Most capacitors in electronic circuits store a very minute amount of charge for extremely short periods of time, but some capacitors, such as those used in bulky power supplies, can store near-lethal doses for several minutes — or even hours.

Use caution when working around capacitors that can store a lot of charge so that you don't get an unwanted shock.

Being sensitive to static discharge

The ESD that results from dragging your feet across the carpet or combing your hair on a dry day may be several thousand volts — or higher. Although you probably just experience an annoying tickle (and maybe a bad hair day), your electronic components may not be so lucky. Transistors and integrated circuits that are made using metal-oxide semiconductor (MOS) technology are particularly sensitive to ESD, regardless of the amount of current.

A MOS device contains a thin layer of insulating glass that can easily be zapped away by 50 V of discharge or less. If you, your clothes, and your tools aren't free of static discharge, that MOS field-effect transistor (MOSFET) or complementary MOS (CMOS) IC you planned to use will be nothing more than a useless lump. Because bipolar transistors are constructed differently, they are less susceptible to ESD damage. Other components — resistors, capacitors, inductors, transformers, and diodes — don't seem to be bothered by ESD.

I recommend that you develop static-safe work habits for all the components you handle, whether they're overly sensitive or not.

Minimizing static electricity

You can bet that most of the electronic projects you want to build contain at least some components that are susceptible to damage from electrostatic discharge. You can take these steps to prevent exposing your projects to the dangers of ESD:

- **Use an antistatic wrist strap.** Pictured in Figure 13-6, an antistatic wrist strap grounds you and prevents static build-up. It's one of the most effective means of eliminating ESD, and it's inexpensive (less than $10). To use one, roll up your shirt sleeves; remove all rings, watches, bracelets, and other metals; and wrap the strap around your wrist tightly. Then securely attach the clip from the wrist strap to a proper earth ground connection, which can be the bare (unpainted) surface of your computer case — with the computer plugged in — or simply the ground receptacle of a properly installed wall outlet. Be sure to review the instruction sheet that comes with the strap.

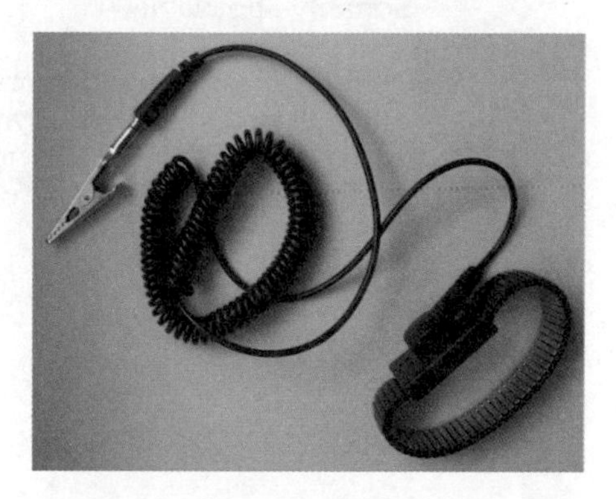

Figure 13-6: An antistatic wrist strap reduces or eliminates the risk of electrostatic discharge.

- **Wear low-static clothing.** Whenever possible, wear natural fabrics, such as cotton or wool. Avoid polyester and acetate clothing because these fabrics have a tendency to develop a whole lot of static.

- **Use an antistatic mat.** Available in both tabletop and floor varieties, an antistatic mat looks like a sponge, but it's really conductive foam. It can reduce or eliminate the build-up of static electricity on your table and your body.

Usually, wearing cotton clothing and using an antistatic wrist strap is sufficient for preventing ESD damage.

Grounding your tools

The tools you use when building electronics projects can also build up static electricity — a lot of it. If your soldering iron operates from AC current, ground it to defend against ESD. There's a double benefit here: A grounded soldering iron not only helps prevent damage from ESD but also lessens the chance of a bad shock if you accidentally touch a live wire with the iron.

Cheapo soldering irons use only two-prong plugs and don't have ground connections. Some soldering irons that have three-prong plugs still pose an ESD threat because their tips aren't grounded, even if their bodies are. Because you can't find a really safe-and-sure means of attaching a grounding wire to a low-end soldering iron, your best bet is to fork over some money for a new, well-grounded soldering iron. The popular Weller WES51 is ESD safe and reasonably affordable.

As long as you ground yourself by using an antistatic wrist strap, you generally don't need to ground your other metal tools, such as screwdrivers and wire cutters. Any static generated by these tools is dissipated through your body and into the antistatic wrist strap.

Chapter 14

Interpreting Schematics

In This Chapter

- Understanding the role of schematics
- Getting to know the most common symbols
- Using (and not abusing!) component polarity
- Diving into some specialized components
- Having fun with schematics from around the world

Imagine driving cross-country without a road map. Chances are, you'd get lost along the way and end up driving in circles. Road maps exist to help you find your way. You can use the equivalent of road maps for building electronic circuits as well. They're called *schematic diagrams,* and they show you how all the parts of the circuits are connected. Schematics show these connections with symbols that represent electronic parts and lines that show how you attach the parts.

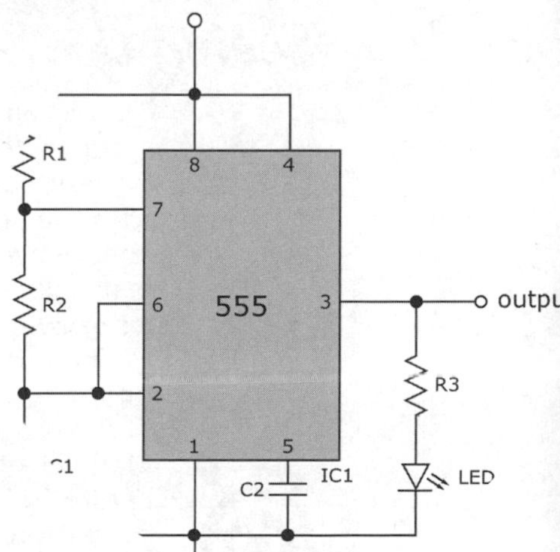

Although not all electronics circuits you encounter are described in the form of a schematic, many are. If you're serious at all about studying electronics, (sooner or later) you need to understand how to read a schematic. Not to worry! The language of schematics isn't all that hard to learn. Most schematic diagrams use only a handful of symbols for components, such as resistors, capacitors, and transistors.

This chapter tells you all you really need to know to read almost any schematic diagram you come across.

What's a Schematic and Why Should I Care?

A *schematic* is a circuit diagram that shows all the components of a circuit, including power supplies and their connections. When you're reading a schematic, the most important things to focus on are the *connections* because the positioning of components in a schematic diagram does not necessarily correspond to the physical layout of components in a constructed circuit. (In fact, for complex circuits, it's unlikely that the physical circuit layout reflects the positioning shown in the schematic. Complex circuits often require separate *layout* diagrams, sometimes known as *artwork*.)

Schematics use symbols to represent resistors, transistors, and other circuit components, and lines to show connections between components. By reading the symbols and following the interconnections, you can build the circuit shown in the schematic. Schematics can also help you understand how a circuit operates, which comes in handy when you're testing or repairing the circuit.

Discovering how to read a schematic is a little like learning a foreign language. On the whole, you'll find that most schematics follow fairly standard conventions. However, just as many languages have different dialects, the language of schematics is far from universal. Schematics can vary depending on the age of the diagram, its country of origin, the whim of the circuit designer, and many other factors.

This book uses conventions commonly accepted in North America. But to help you deal with the variations that you may encounter, I include some other conventions, such as those commonly used in Europe.

Seeing the Big Picture

There's an unwritten rule in electronics about how to orient certain parts of a circuit schematic — especially when drawing diagrams of complex circuits. Batteries and other power supplies are almost always oriented vertically, with the positive terminal on top. In complex schematics, power supplies are split between two symbols (as you will see later), but the positive terminal is usually shown at the top of the schematic (sometimes extending across a horizontal line, or *rail*) and the negative terminal appears at the bottom (sometimes along a rail). Inputs are commonly shown on the left and outputs on the right.

Many electronic systems, such as the radio-receiver system shown in Figure 14-1, are represented in schematics by several stages of circuitry — even though the system really consists of one large, complex circuit. The schematic for such a system shows the sub-circuits for each stage in a left-to-right progression (for instance, the tuner sub-circuit on the left, the detector in the middle, and the amplifier on the right), with the output of the first stage feeding into the input of the second stage, and so forth. Organizing schematics in this way helps make complex circuits more understandable.

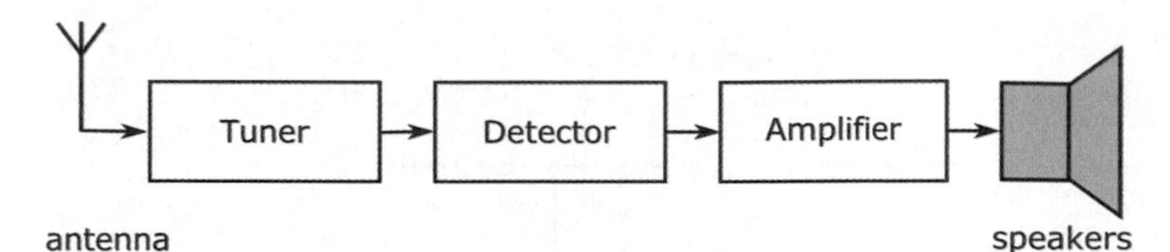

Figure 14-1: Block diagram representing a radio-receiver system.

It's all about your connections

In all schematics, simple or complex, components are arranged as neatly as possible, and connections in a circuit are drawn as lines, with any bends shown as 90° angles. (No squiggles or arcs allowed!) It's absolutely critical to understand what all the lines in a schematic really mean — and their meaning is not always obvious.

The more complex the schematic, the more likely it is that some lines will crisscross each other (due to the 2-D nature of schematic drawings). You need to know when crossed lines represent an actual wire-it-together connection and when they don't. Ideally, a schematic will clearly distinguish connecting and nonconnecting wires like this:

- A break or a loop (think of it as a bridge) in one of the two lines at the intersection indicates wires that should *not* be connected.
- A dot at the intersection of two lines indicates that the wires *should* be connected.

You can see some common variations in Figure 14-2.

This method of showing connections isn't universal, so you have to figure out which wires connect and which don't by checking the drawing style used in the schematic. If you see an intersection of two lines without a dot to positively identify a real connection, you simply cannot be sure whether the wires should be connected or not. It's best to consult with the person who drew the schematic to determine how to interpret the crisscrossing lines.

To physically implement the connections shown in a schematic, you typically use insulated wires or thin traces of copper on a circuit board. Most schematics don't make a distinction about how you connect the components; that connection is wholly dependent on how you choose to build the circuit. The schematic's representation of the wiring merely shows you the connections that must be made between components.

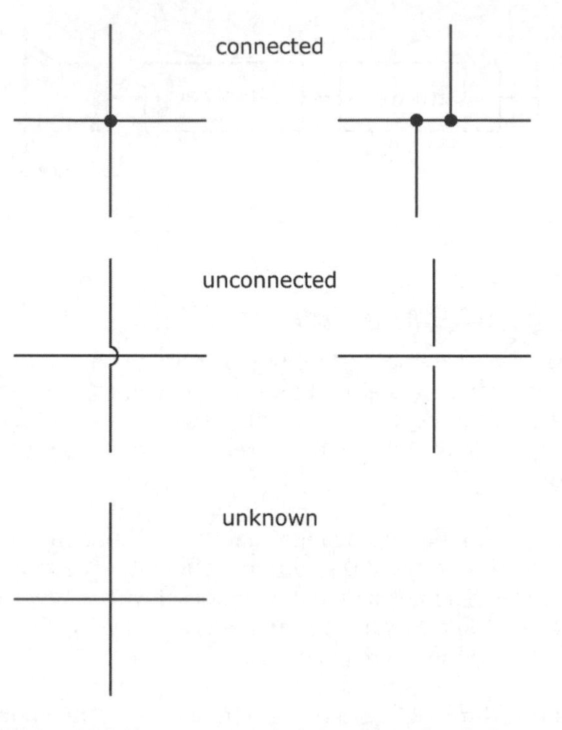

Figure 14-2: You may encounter a number of variations in how a schematic shows connections and nonconnections.

Looking at a simple battery circuit

Figure 14-3 shows a simple DC circuit with a 1.5 V battery connected to a resistor labeled *R1*. The positive side of the battery (labeled +) is connected to the lead on one side of the resistor; the negative side of the battery is connected to the lead on the other side of the resistor. With these connections made, current flows from the positive terminal of the battery through the resistor, and back to the negative terminal of the battery.

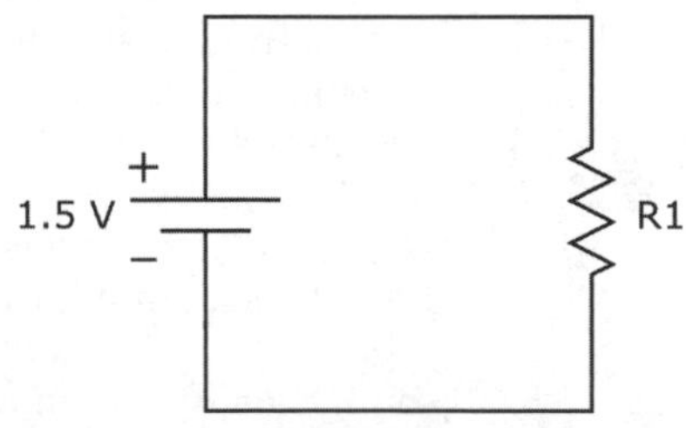

Figure 14-3: A simple schematic shows the connections between a battery and a resistor.

In schematics, *current* is assumed to be *conventional* current, which is described as the flow of positive charges, traveling in a direction opposite to that of real electron flow. (For more about conventional current and electron flow, see Chapter 3.)

Recognizing Symbols of Power

Power for a circuit can come from an alternating current (AC) source, such as the 120 VAC (volts AC) outlet in your house or office (line power), or a direct current (DC) source, such as a battery or the low-voltage output from a wall transformer. DC supplies can be positive or negative with respect to the zero-volt reference (known as *common ground,* or simply *common*) in a circuit. Figure 14-4 shows various symbols used to represent power and ground connections. These symbols are discussed more fully in the next two subsections.

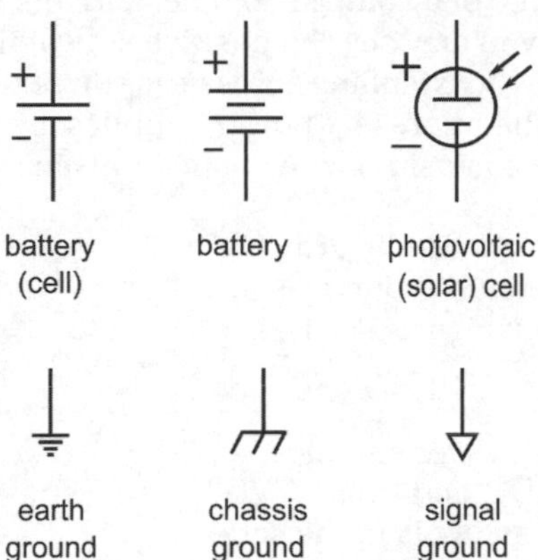

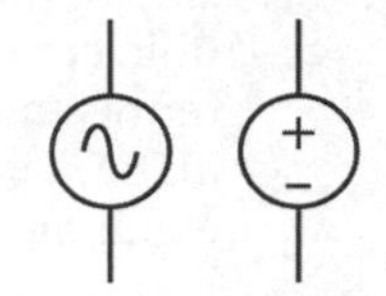

Figure 14-4: Symbols for power and ground.

Figuring out the various power connections in a complex schematic is sometimes a task unto itself. This section aims to clear things up a bit. As you read through this section, refer to Figure 14-4 to see the symbols as they're discussed.

Showing where the power is

DC power supplies are shown in one of two ways:

- **Battery or solar-cell symbol:** Each of the battery symbols in Figure 14-4 represents a DC source with two leads. Technically, the battery symbol that includes two parallel lines (the first symbol for battery) represents a single electrochemical *cell;* the symbol with multiple pairs of lines (the second symbol) represents a *battery* (which consists of multiple cells).

  Many schematics (including those in this book) use the symbol for a cell to represent a battery.

  Each symbol includes a positive terminal (indicated by the larger horizontal line) and a negative terminal. The polarity symbols (+ and –) and nominal voltage are usually shown next to the symbol. The negative terminal is often assumed to be at 0 volts, unless clearly distinguished as different from the zero-voltage reference (known as *common ground* and detailed later in this chapter). Conventional current flows out of the positive terminal and into the negative terminal when the battery is connected in to a complete circuit.

- **Split DC power and ground symbols:** To simplify schematics, a DC power supply is often shown using two separate symbols. These symbols are a small circle at the end of a line representing one side of the supply, with or without a specific voltage label, and the symbol for ground (vertical line with three horizontal lines at the bottom) representing the other side of the supply, with a value of 0 volts. In complex circuits with multiple connections to power, you may see the positive side of the supply represented by a rail labeled +V extending across the top of the schematic. These split symbols that represent power supplies are used to eliminate a lot of (otherwise confusing) wire connections in a schematic.

The circuit shown in Figure 14-3 can also be drawn using separate symbols for power and ground, as in Figure 14-5. Note that the circuit in Figure 14-5 is, in fact, a complete circuit.

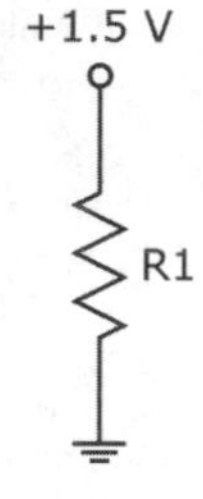

Figure 14-5: A simpler way to show the connections between a battery and a resistor.

Many DC circuits use multiple DC power supplies, such as +5 VDC (volts DC), +12 VDC, and even –5 VDC or –2 VDC, so the voltage-source symbols in the schematics are usually labeled with the nominal voltage. If a schematic doesn't specify a voltage, you're often (but not always!) dealing with 5 VDC. And remember: Unless otherwise specifically noted, the voltage in a schematic is almost always DC, *not* AC.

Some circuits (for instance, some op-amp circuits, which are discussed in Chapter 11) require both positive and negative DC power supplies. You'll often see the positive supply represented by an open circle labeled +V and the negative supply represented by an open circle labeled –V. If the voltages are not specified, they may be +5 VDC and –5 VDC. Figure 14-6 shows how these power supply connection points are really implemented.

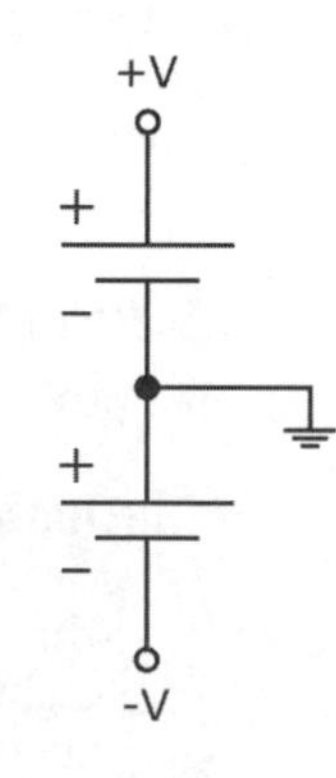

Figure 14-6: Some circuits require positive and negative power supplies.

An AC power supply is usually represented by a circle with two leads, either with or without a waveform shape and polarity indicators:

- **Circle containing waveform:** A squiggly line or other shape inside an open circle represents one cycle of the alternating voltage produced by the power supply. Usually, the source is a sine wave, but it could be a square wave, a triangle wave, or something else.

- **Circle with polarity:** Some schematics include one or both polarity indicators inside or outside the open circle. Polarity indicators are just for reference purposes, so you can relate the direction of current flow to the direction of voltage swings.

Power for a circuit can come from an AC source, such as the 120 VAC outlet in your house or office (such circuits are called *line-powered*). You typically use an internal power supply to *step down* (or lower) the 120 VAC and convert it to DC. This lower-voltage DC power is then delivered to the components in your circuit. If you're looking at a schematic for a DVD player or some other gadget getting its power from a wall outlet, that schematic probably shows both AC and DC power.

Marking your ground

Ready for some electronics schematic double-talk? When it comes to labeling ground connections in schematics, it's common practice to use the symbol for *earth ground* (which is a real connection to the earth) to represent the *common ground* (the reference point for zero volts) in a circuit. (Chapter 3 details these two types of grounds.) More often than not, the "ground" points in low-voltage circuits are not connected to earth ground; instead, they're tied to each other — hence the term *common ground* (or simply *common*). Any voltages labeled at specific points in a circuit are assumed to be relative to this common ground. (Remember, voltage is really a differential measurement between two points in a circuit.)

So what symbol should *really* be used for ground points that are not truly connected to the earth? It's the symbol labeled *chassis ground.* Common ground is sometimes called chassis ground because in older equipment, the metal chassis of the device (hi-fi, television, or whatever) served as the common ground connection. Using a metal chassis for a ground connection is not as common today, but the term is still often used.

You may also see the symbol for *signal ground* used to represent a zero-volt reference point for signals (information-carrying waveforms) carried by two wires. One wire is connected to this reference point and the other wire carries a varying voltage representing the signal. Again, in many schematics, the symbol for earth ground is used instead.

In this book, I use only the schematic symbol for earth ground because most schematics you see these days use that symbol.

As you can see in Figure 14-7, a schematic may show the ground connections in a number of ways:

- **No ground symbol:** The schematic can show two power wires connected to the circuit. In a battery-powered circuit, common ground is assumed to be the negative terminal of the battery.

- **Single ground symbol:** The schematic shows all the ground connections connected to a single point. It doesn't often show the power source or sources (for instance, the battery), but you should assume that ground connects to the positive or negative DC power sources (refer to Figure 14-6).
- **Multiple ground symbols:** In more complex schematics, it's usually easier to draw the circuit with several ground points. In the working circuit, all these ground points connect together.

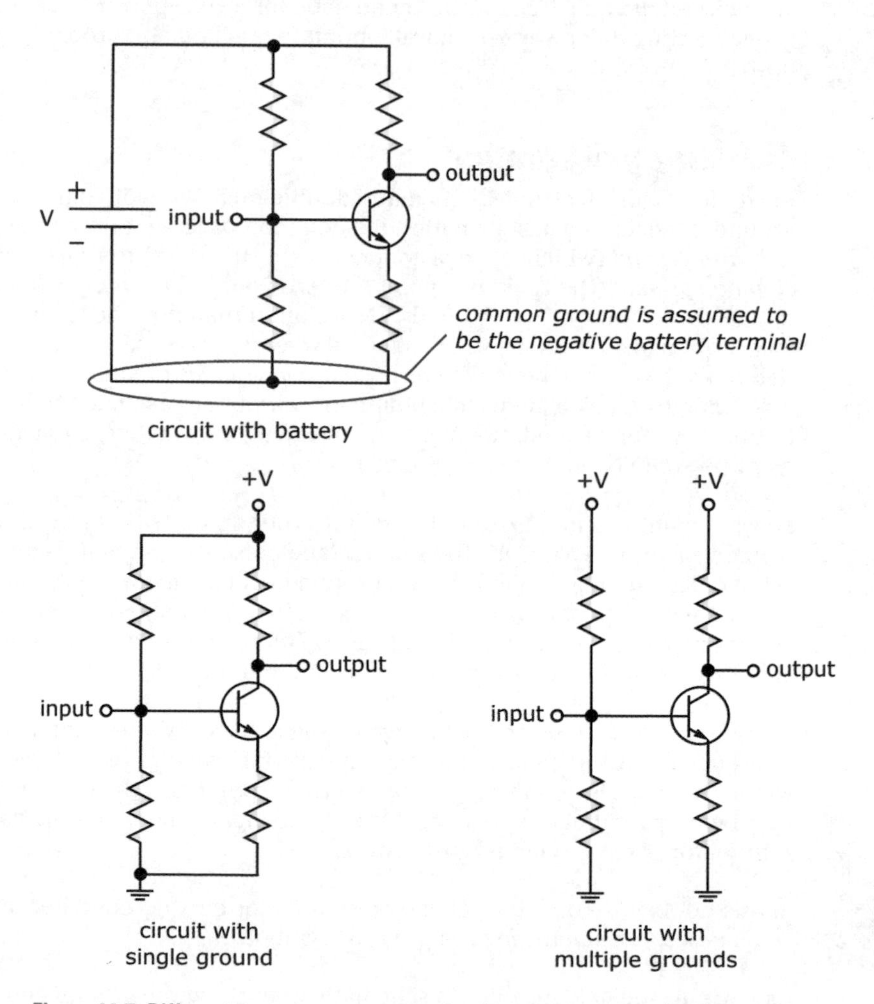

Figure 14-7: Different ways to represent a circuit's common ground connections.

Labeling Circuit Components

You can find hundreds of symbols for electronic components out there, because there are hundreds of component types to depict. Fortunately, you probably encounter only a small number of these symbols in schematics for hobby electronics projects.

Along with the circuit symbol for a particular electronic component, you may see additional information to help uniquely identify the part:

- **Reference ID:** An identifier, such as R1 or Q3. The convention is to use one or more letters to represent the type of component, and a numerical suffix (sometimes subscripted) to distinguish one particular component from others of the same type. The most common type designators are R for resistor, C for capacitor, D for diode, L for inductor, T for transformer, Q for transistor, and U or IC for integrated circuit.
- **Part number:** Used if the component is standard (as with a transistor or an integrated circuit) or you have a manufacturer's customer product part. For example, a part number may be something like 2N2222 (that's a commonly used transistor) or 555 (a type of IC used in timing applications).
- **Value:** Component values are sometimes shown for passive parts, such as resistors and capacitors, that don't go by conventional part numbers. For example, when indicating a resistor, the value (in ohms) could be marked beside the resistor symbol or the reference ID. Most often, you'll see just the value without a label for the unit of measurement (ohms, microfarads, and so on). Normally resistor values are assumed to be in ohms, and capacitor values are assumed to be in microfarads.
- **Additional information:** A schematic may include additional specifics about one or more components, such as the wattage for a resistor when it isn't your typical 1/4- or 1/8-watt value. If you see *10W* next to a resistor value, you know you need a power resistor.

Many schematics show only the reference ID and the circuit symbol for each component, and then include a separate *parts list* to provide the details of part numbers, values, and other information. The parts list maps the reference ID to the specific information about each component.

Analog electronic components

Analog components control the flow of continuous (analog) electrical signals. Table 14-1 shows the circuit symbols used for basic analog electronic components. The third column in the table provides the chapter reference in this book where you can find detailed information about the functionality of each component.

Reference ID primer

Components are often identified in a schematic using an alphabetic type designator, such as C for capacitor, followed by a numerical identifier (1, 2, 3, and so on) to distinguish multiple components of the same type. Together, these identifiers form a *reference ID* that uniquely identifies a specific capacitor or other component. If that value isn't printed beside the component symbol, don't worry; you can find the reference ID in a parts list to indicate the precise value of the component to use. The following type designators are among those most commonly used:

C	Capacitor
D	Diode
IC (or U)	Integrated circuit
L	Inductor
LED	Light-emitting diode
Q	Transistor
R	Resistor
RLY	Relay
T	Transformer
XTAL	Crystal

Table 14-1 Symbols for Analog Components

Component	*Symbol*	*Chapter*
Resistor		Chapter 5
Variable resistor (potentiometer)	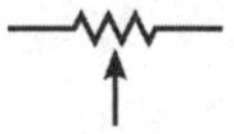	Chapter 5
Photoresistor (photocell)	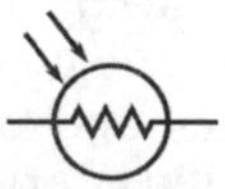	Chapter 12
Capacitor	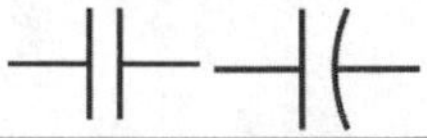	Chapter 7
Polarized capacitor	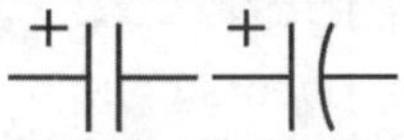 + +	Chapter 7
Variable capacitor	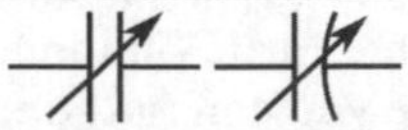	Chapter 7
Inductor		Chapter 8

Component	*Symbol*	*Chapter*
Air core transformer	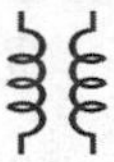	Chapter 8
Solid core transformer	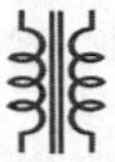	Chapter 8
Crystal		Chapter 8
NPN (bipolar) transistor		Chapter 10
PNP (bipolar) transistor	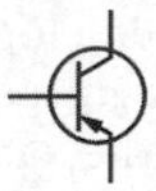	Chapter 10
N-channel MOSFET		Chapter 10
P-channel MOSFET		Chapter 10
Phototransistor (NPN)		Chapter 12
Phototransistor (PNP)		Chapter 12
Standard diode	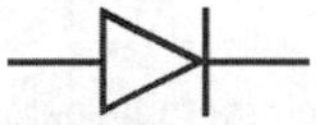	Chapter 9
Zener diode		Chapter 9
Light-emitting diode (LED)	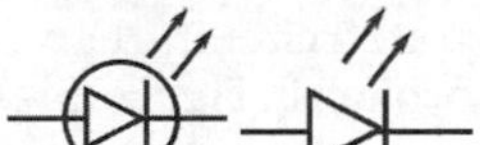	Chapter 9
Photodiode		Chapter 12
Operational amplifier (op amp)	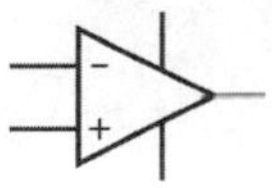 − +	Chapter 11

The circuit symbol for an op amp represents the interconnection of dozens of individual components in a nearly complete circuit (power is external to the op amp). Schematics always use a single symbol to represent the entire circuit, which is packaged as an integrated circuit (IC). The circuit symbol for an op amp is commonly used to represent many other amplifiers, such as an LM386 audio power amplifier.

Digital logic and IC components

Digital electronic components, such as logic gates, manipulate digital signals that consist of just two possible voltage levels (high or low). Inside each digital component is a nearly complete circuit (power is external), consisting of individual transistors or other analog components. Circuit symbols for digital components represent the interconnection of several individual components that make up the circuit's logic. You can build the logic from scratch or obtain it in the form of an integrated circuit. Logic ICs usually contain several gates (not necessarily all of the same type) sharing a single power connection.

Figure 14-8 shows the circuit symbols for individual digital logic gates. You can find detailed information about the functionality of each logic gate in Chapter 11.

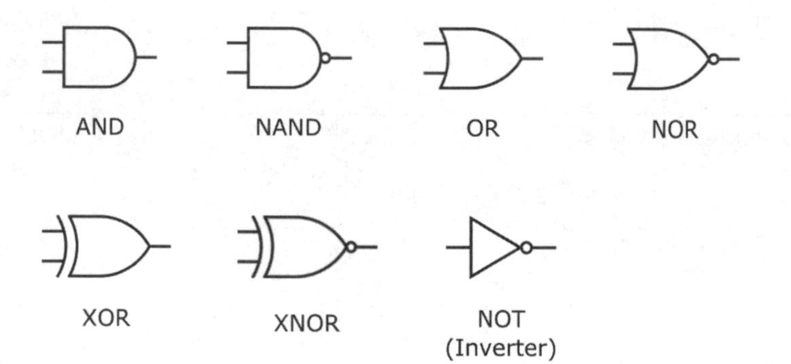

Figure 14-8: Symbols for logic gates.

Some schematics show individual logic gates; others show connections to the full integrated circuit, represented by a rectangle. You can see an example of each approach in Figure 14-9.

The 74HC00 IC shown in Figure 14-9 is a CMOS quad 2-input NAND gate. In the top circuit diagram, each NAND gate is labeled *1/4 74HC00* because it is one of four NAND gates in the IC. (This type of gate labeling is common in digital circuit schematics.) Note that the fourth NAND gate is not used in this particular circuit (which is why pins 11, 12, and 13 are not used). Whether the schematic uses individual gates or an entire IC package, it usually notes the external power connections. If it doesn't, you have to look up the pinout of the device on the IC datasheet to determine how to connect power. (For more about pinouts and datasheets, see Chapter 11.)

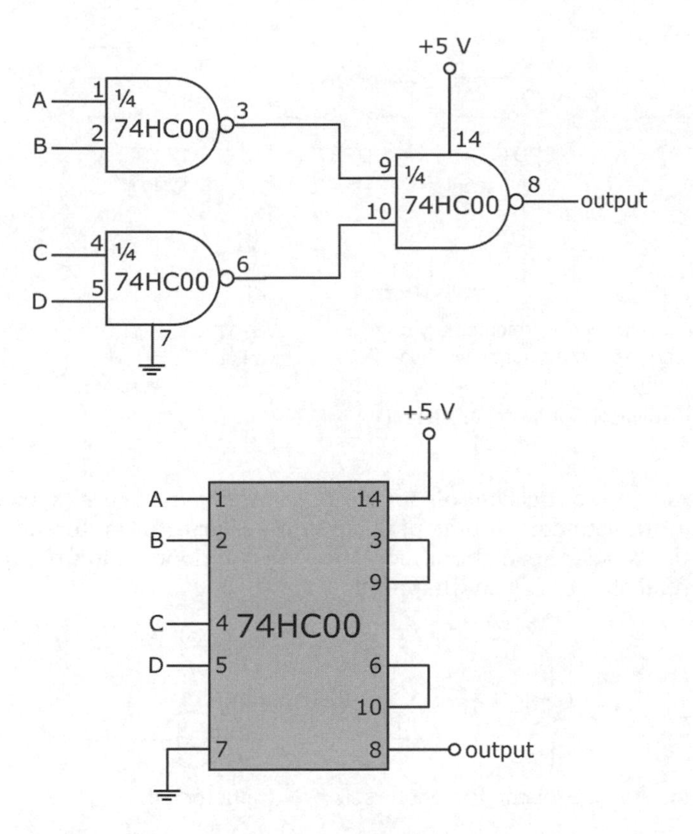

Figure 14-9: Two different schematic representations of the same circuit.

You'll find many more digital ICs other than those containing just logic gates. You'll also find linear (analog) ICs that contain analog circuits and mixed-signal ICs that contain a combination of analog and digital circuits. Most ICs — except for op amps — are shown the same way in schematics: as a rectangle, labeled with a reference ID (such as IC1) or the part number (such as 74CH00), with numbered pin connections. The function of the IC is usually determined by looking up the part number, but the occasional schematic may include a functional label, such as *one shot*.

Miscellaneous components

Figure 14-10 shows the symbols for switches and relays. Refer to Chapter 4 for detailed information on each of these components.

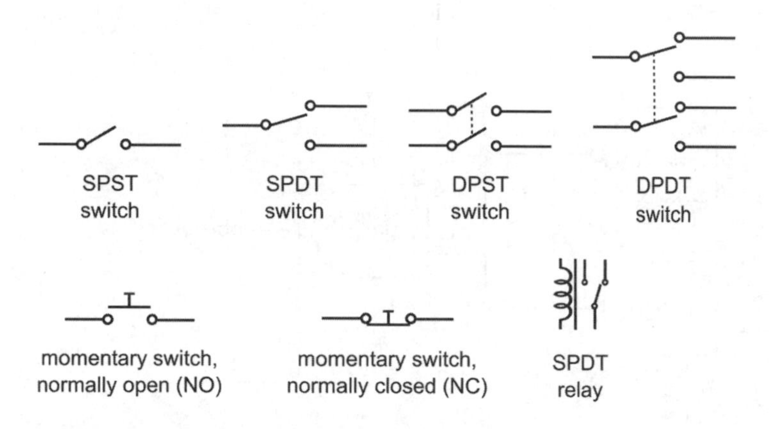

Figure 14-10: Symbols for switches and relays.

Figure 14-11 shows the symbols for various input transducers (sensors) and output transducers. (Some of these symbols are cross-referenced in Table 14-1.) You can read about most of these components in Chapter 12, and you can read about LEDs in Chapter 9.

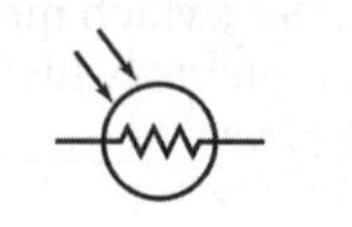
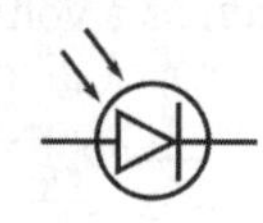

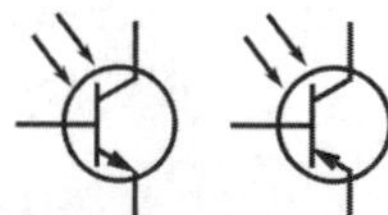

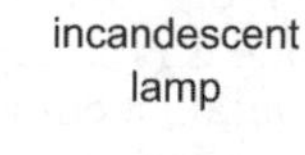

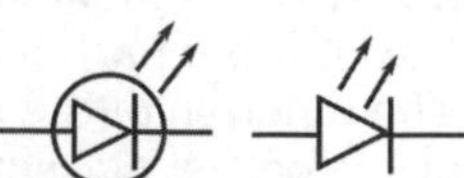
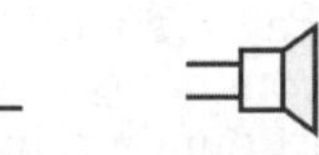
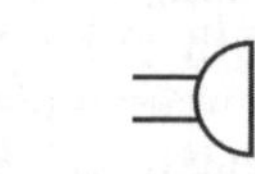

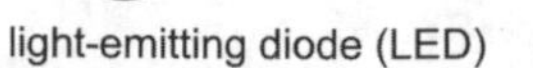

Figure 14-11: Symbols for input and output transducers.

Some circuits accept inputs from and send outputs to other circuits or devices. Schematics often show what looks like a loose wire leading into or out of the circuit. Usually it's labeled something like *signal input,* or *input from doodad #1,* or *output* so you know you're supposed to connect something up to it. (You connect one wire of the signal to this input point and the other to signal ground.) Other schematics may show a symbol for a specific connector, such as a *plug* and *jack* pair, which connect an output signal from one device to the input of another device. (Chapter 12 provides a closer look.)

Figure 14-12 shows a few of the ways input and output connections to other circuits are shown in schematics. Symbols for input/output connections can vary greatly among schematics. The symbols used in this book are among the most common. Although the exact style of the symbol may vary from one schematic to the next, the idea is the same: The symbol is telling you to make a connection to something external to the circuit.

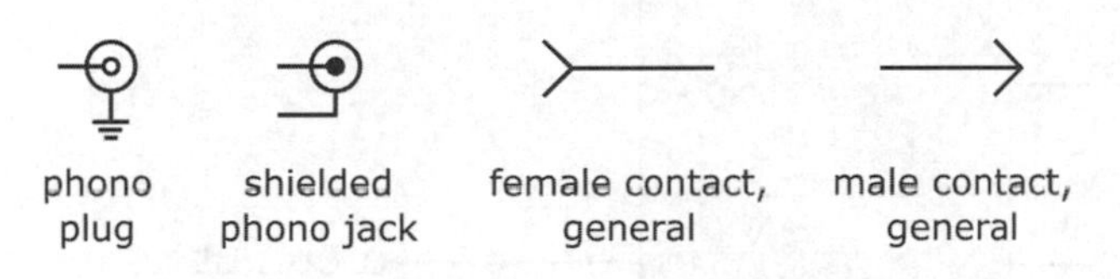

Figure 14-12: Symbols for connections to other circuits.

Knowing Where to Take Measurements

You may run across a schematic or two that include symbols for test instruments, such as a voltmeter (which measures voltage), an ammeter (which measures current), or an ohmmeter (which measures resistance). (As Chapter 16 explains, a multipurpose multimeter can function as any of these meters — and more.) You may see these symbols in schematics on educational websites or in documents designed for educational purposes. They point out exactly where to place your meter's leads to take the measurement properly. (The abbreviation *TP,* for test point, is often used to indicate where to take a measurement.)

When you see one of the symbols shown in Figure 14-13 in a schematic, remember that it represents a test instrument — not some newfangled "vulcanistor" or other electronic component you've never heard of before.

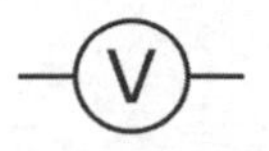

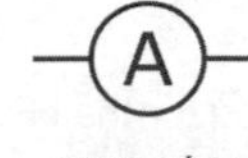

Figure 14-13: Symbols for common test instruments.

Exploring a Schematic

Now that you're familiar with the ABCs of schematics, it's time to put it all together and walk through each part of a simple schematic. The schematic shown in Figure 14-14 shows the LED flasher circuit used in Chapter 17. This circuit controls the on/off blinking of an LED, with the blinking rate controlled by turning the knob of a potentiometer (variable resistor).

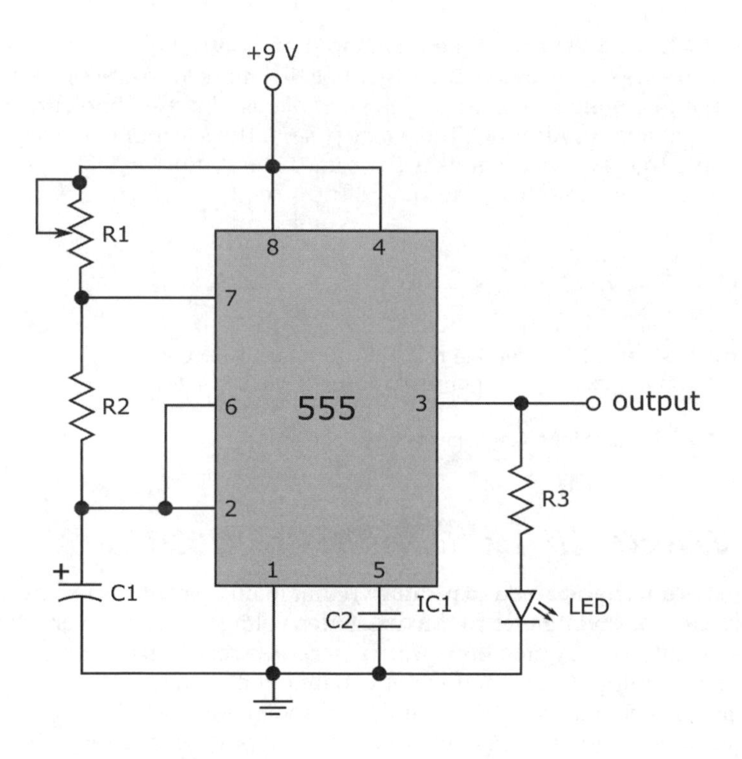

Parts List:

IC1: LM555 timer IC
R1: 1 MΩ potentiometer
R2: 47 kΩ resistor
R3: 330 Ω resistor
C1: 1 μF tantalum capacitor
C2: 0.1 μF disc capacitor
LED: light-emitting diode

Figure 14-14: The schematic and parts list used for the LED flasher project in Chapter 17.

Here's what this schematic is saying:

- At the heart of the schematic is ***IC1,* an 8-pin 555 timer IC,** with all eight pins connecting to parts of the circuit. Pins 2 and 6 are connected together.
- The circuit is powered by a 9-volt power supply, which can be a **9 V battery.**
  - The positive terminal of the power supply is connected to pins 4 and 8 of *IC1,* and to one fixed lead and the variable contact (wiper) lead of *R1,* which is a variable resistor (potentiometer).
  - The negative terminal of the power supply (shown as the common ground connection) is connected to pin 1 of *IC1,* to the negative side of capacitor *C1,* to capacitor *C2,* and to the cathode (negative side) of the LED.
- ***R1* is a potentiometer** with one fixed lead connected to pin 7 of *IC1* and to resistor *R2,* and both the other fixed lead and the wiper lead connected to the positive battery terminal (and to pins 4 and 8 of *IC1*).
- ***R2* is a fixed resistor** with one lead connected to pin 7 of *IC1* and to one fixed lead of *R1,* and the other lead connected to pins 2 and 6 of *IC1* and the positive side of capacitor *C1.*
- ***C1* is a polarized capacitor.** Its positive side is connected to *R2* and to pins 2 and 6 of *IC1,* and its negative side is connected to the negative battery terminal (as well as to pin 1 of *IC1,* capacitor *C2,* and the cathode of the LED).
- ***C2* is a nonpolarized capacitor** connected on one side to pin 5 of *IC1* and on the other side to the negative battery terminal (as well as the negative side of capacitor *C1,* pin 1 of *IC1,* and the cathode of the LED).
- The anode (positive side) of the **LED** is connected to resistor *R3* and the cathode of the LED is connected to the negative battery terminal (as well as to the negative side of capacitor C1, capacitor C2, and pin 1 of IC1).
- ***R3* is a fixed resistor** connected between pin 3 of *IC1* and the anode of the LED.
- Finally, the **output** shown at pin 3 of *IC1* can be used as a signal source (input) for another stage of circuitry.

Each item in the walk-through list just given focuses on one circuit component and its connections. Although the list does mention the same connections multiple times, that's consistent with good practice; it pays to check and double-check your circuit connections by making sure *each lead or pin of each individual component* is connected correctly. (Ever hear the rule of thumb "measure twice, cut once"? Well, the same principle applies here.) You can't be too careful when you're connecting electronic components.

Alternative Schematic Drawing Styles

The schematic symbols in this chapter belong to the drawing style used in North America (particularly in the United States) and in Japan. Some countries — notably European nations as well as Australia — use somewhat different schematic symbols. If you're using a schematic for a circuit not designed in the United States or Japan, you need to do a wee bit o' schematic translation to understand all the components.

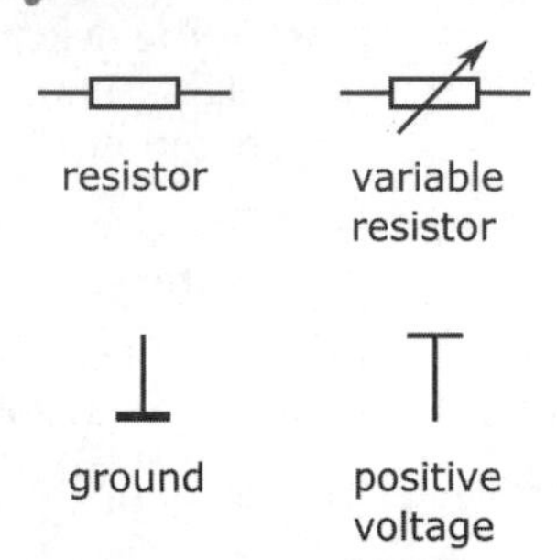

Figure 14-15: Schematic symbols used for circuits designed in Europe.

Figure 14-15 shows a sampling of schematic symbols commonly used in the United Kingdom and other European nations. Note the obvious differences in the resistor symbols, both fixed and variable.

This style organizes its symbols differently from the American style. In the United States, you express resistor values over 1,000 Ω in the form of 6.8k or 10.2k, with the lowercase *k* following the value. The European schematic style eliminates the decimal point. Typical of schematics you'd find in the United Kingdom are resistor values expressed in the form 6k8 or 10k2. This style substitutes the lowercase *k* (which stands for *kilohms,* or thousands of ohms) for the decimal point.

You may encounter a few other variations in schematic drawing styles, but all are fairly self-explanatory and the differences are not substantial. After you learn how to use one style of drawing, the others come relatively easily.

Chapter 15

Building Circuits

In This Chapter

- Probing the depths of a solderless breadboard
- Creating a no-fuss, no-muss circuit with a solderless breadboard
- Soldering — safely — like the pros
- Owning up to and fixing soldering mistakes (like the pros)
- Solidifying your circuit relationship with a solder breadboard or a perfboard
- Achieving circuit-building nirvana with a custom printed circuit board

You've carefully set up your workbench, strategically positioned your shiny new toys — oops! I meant *tools* — to impress your friends, and shopped around for the best deals on resistors and other components. Now you're ready to get down to business and build some light-flashing, noisemaking circuits. So how do you transform an unassuming two-dimensional circuit diagram into a real, live, *working* (maybe even moving) electronic circuit?

In this chapter, I show you various ways to connect electronic components together into circuits that push electrons around at your command. First, I describe how to make flexible, temporary circuits using plug-and-play solderless breadboards, which provide the ideal platform for testing and tweaking your designs. Next, I give you the lowdown on how to safely fuse components together using a molten substance called solder (what fun!). Finally, I outline your options for creating permanent circuits using an assortment of today's most popular circuit boards.

So arm yourself with screwdrivers, needle-nose pliers, and a soldering iron, and don your safety glasses and antistatic wristband: You are about to enter the electronics construction zone!

Taking a Look at Solderless Breadboards

Solderless breadboards, also called *prototyping boards* or *circuit breadboards,* make it a snap to build (and dismantle) temporary circuits (see Figure 15-1). These reusable rectangular plastic boards contain several hundred square sockets, or *contact holes,* into which you plug your components (for instance, resistors, capacitors, diodes, transistors, and integrated circuits). Groups of contact holes are electrically connected by flexible metal strips running underneath the surface. You poke a wire or lead into a hole and it makes contact with the underlying metal. By plugging in components in just the right way and running wires from your breadboard to your power supply, you can build a working circuit without permanently bonding components together.

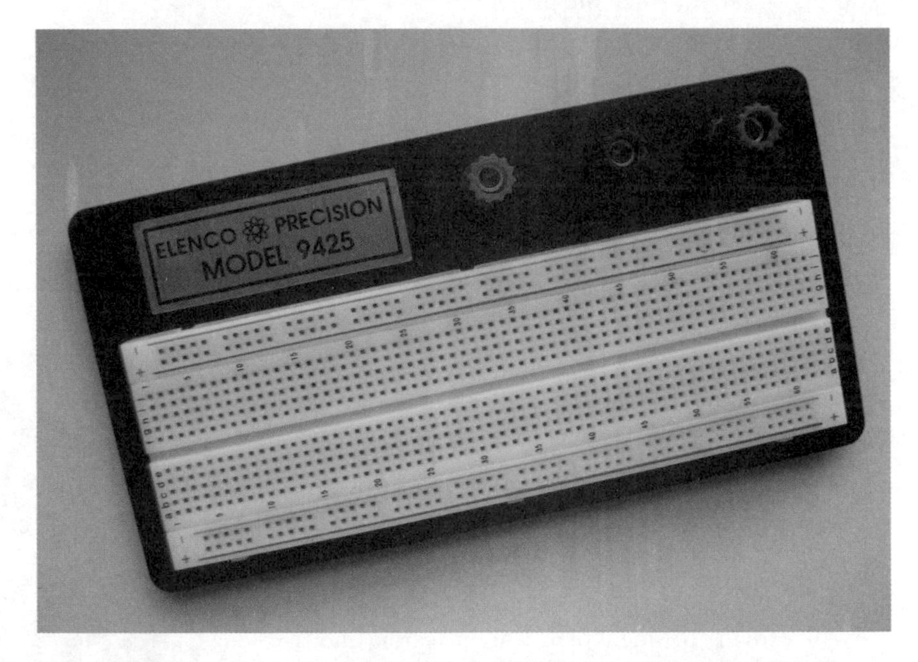

Figure 15-1: Solderless breadboards come in a variety of sizes. This one has 830 holes and includes four power rails and three binding posts for connecting external power.

I highly recommend that you use a solderless breadboard (or two) when you first build a circuit. That way, you can test the circuit to make sure it works properly and make any necessary adjustments. Often you can improve the performance of a circuit just by tweaking a few component values. You can easily make such changes by simply removing one component and inserting another on the board — without having to unsolder and resolder. (For the details of soldering, flip to the "Soldering 101" section, later in this chapter.)

When you're sure your circuit works the way you want it to, you create a permanent circuit on other types of boards (as described in the section "Creating a Permanent Circuit," later in this chapter).

Solderless breadboards are designed for low-voltage DC circuits. Never use a breadboard for 120 VAC house current. Excessive current or voltage can melt the plastic or cause arcing between contacts — ruining the breadboard and possibly exposing you to dangerous currents.

Exploring a solderless breadboard

The photo in Figure 15-2 shows a solderless breadboard with yellow lines added to help you visualize the underlying connections between contact holes. Most breadboards contains three notable features:

- **Terminal strips:** In the center of the breadboard, holes are linked horizontally in blocks of five called *terminal strips,* as indicated by the yellow lines on the photo in Figure 15-2. For instance, the holes in row 1, columns a, b, c, d, and e are electrically connected to each other, as are the holes in row 1, columns f, g, h, i, and j. There are no vertical connections within these two center sections, so, for instance, hole 1a is not electrically connected to hole 2a.
- **Center gap:** The ditch between columns e and f on the breadboard in Figure 15-2 provides an electrical gap between the two center sections of the board. *It's important to remember that there are no internal connections across this center gap.* Many integrated circuits (ICs) are packaged in *dual inline packages (DIPs)* so that you can straddle them across the center gap and instantly set up independent sets of connections for each of its pins.
- **Power rails:** Along the left and right of most breadboards are four columns (two on each side) of contact holes that are linked vertically. These columns, labeled + and –, are known as *bus strips* or *power rails* because they are typically used to distribute power along the length of the board. If you connect power (or ground) to just one hole in each of these columns, you can access that power (or ground) by running a wire from any other hole in those columns to a hole in a terminal strip.

  You can't tell exactly how many holes in each power rail are electrically connected just by looking at the board. On some boards, such as the one shown in Figure 15-2, all contacts in each of the power rails are electrically connected, but other boards have a break in the connections halfway down each column. If there is a break, you can connect a wire between neighboring contacts to bridge the two separate sets of connections in each column.

There's nothing special about the power rails — except their labels and color coding — so you are free to connect anything you want to these columns. If you're smart, however, you'll take advantage of the labels to help keep track of your power connections. I recommend that you connect the positive power supply to the positive power rail on one side of the board, and use the negative power rail on the other side of the board for your ground connection.

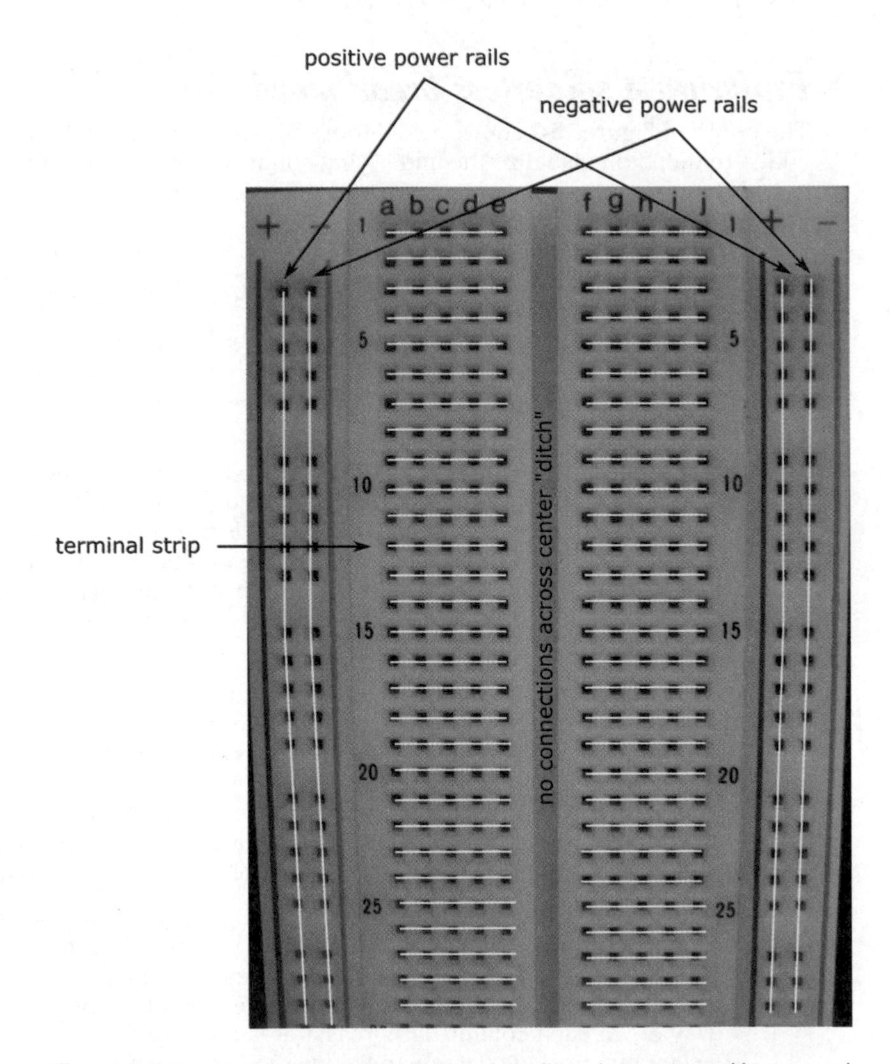

Figure 15-2: The contact holes in a solderless breadboard are arranged in rows and columns that are electrically connected in small groups underneath the surface.

You can use a multimeter to check whether two points within a row — or between rows — are electrically connected. Stick a jumper wire in each hole, and then touch one multimeter probe to one wire and the other probe to the other wire. If you get a low ohm reading, the two points are connected. If you get an infinite ohm reading, they aren't connected. (See Chapter 16 for more about testing things with your multimeter.)

Holes are spaced 1/10 of an inch apart (0.1 inch), a size just right for DIP ICs, most transistors, and discrete components, such as capacitors and resistors. You just plug DIP ICs, resistors, capacitors, transistors, and 20- or 22-gauge solid wire into the proper contact holes to create your circuit. Typically, you use the two center sections of the board to make connections between components; use the left and right sections of the board to connect power.

Breadboard manufacturers make contact strips from a springy metal coated with a plating. The plating prevents the contacts from oxidizing, and the springiness of the metal allows you to use wire component leads of different diameters without seriously deforming the contacts. Note, however, that you can damage the contacts if you attempt to use wire larger than 20 gauge or use components with very thick leads. If the wire is too thick to go into the hole, don't try to force it. If you do force it, you can loosen the fit of the contact, and your breadboard may not work the way you want it to.

When you aren't using it, keep your breadboard in a resealable plastic bag. Why? To keep out the dust. Dirty contacts make for poor electrical connections. Although you can use a spray-on electrical cleaner to remove dust and other contaminants, you make things easier on yourself by keeping the breadboard clean in the first place.

Sizing up solderless breadboard varieties

Solderless breadboards come in many sizes. Smaller breadboards (with 400 to 550 holes) accommodate designs with up to three or four ICs plus a small handful of other discrete components. Larger boards, such as the 830-hole board shown in Figures 15-1 and 15-2, provide more flexibility and accommodate five or more ICs. If you're into elaborate design work, purchase extra-large breadboards with anywhere from 1,660 to more than 3,200 contact holes. These boards can handle one to three dozen ICs plus other discrete components.

Don't overdo it when buying a solderless breadboard. You don't need a breadboard the size of Wyoming if you're making small- to medium-sized circuits, such as the ones I show you in Chapter 17. And if you get into the middle of designing a circuit and find that you need a little more breadboard real estate, you can always make connections between two breadboards. Some solderless breadboards have interlocking ridges so you can put several together to make a larger breadboard.

Building Circuits with Solderless Breadboards

Essentially, breadboarding consists of sticking components into the board, connecting power to the board, and making connections with wire. But there's a right way and a wrong way to do these things. This section gives you the lowdown on what type of wire to use, efficient breadboarding techniques, and how to give your board a neat, logical design.

Preparing your parts and tools

Before you start randomly sticking things into your breadboard, you should make sure that you have everything you need. Check the parts list — the list of electronic ingredients you need to build your circuit — and set aside the components you need. Gather essential tools, such as needle-nose pliers, wire cutters, and a wire stripper (see Figure 15-3).

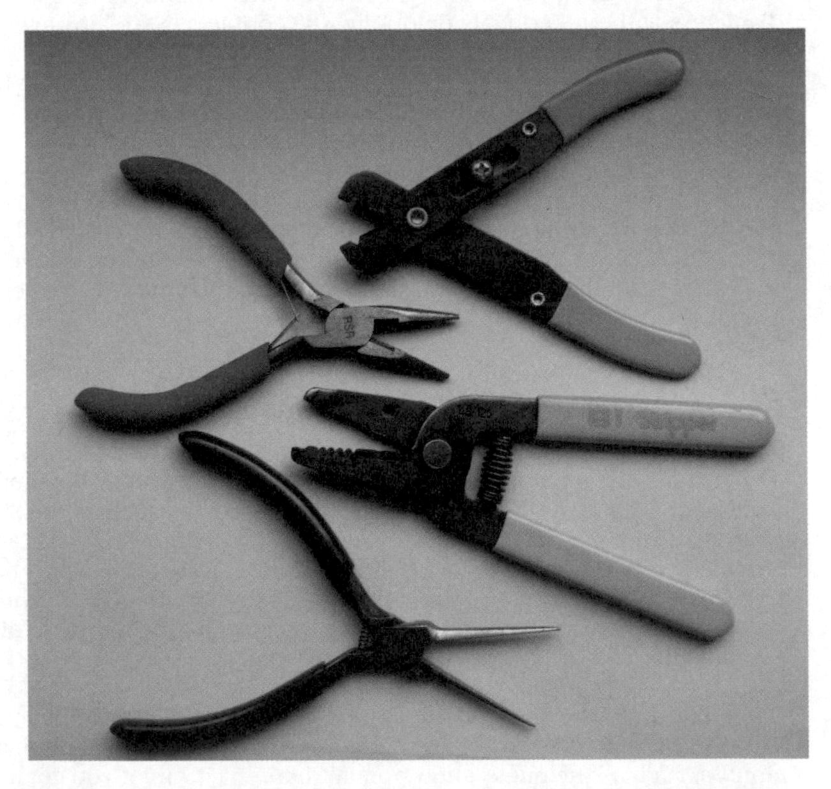

Figure 15-3: Tools of the trade (top to bottom): wire cutter/stripper, mini needle-nose pliers, wire stripper, needle-nose pliers.

Make sure all your component leads are suitable for inserting into breadboard holes. Clip long leads, if possible, so the components will lie flat and snug against the board. (Don't worry if you can't reuse them for another circuit — they're cheap enough.) Some components, such as potentiometers, may not have leads, so you'll need to solder single-core wires to their terminals (see the section "Soldering 101," later in this chapter for how to do this). Familiarize yourself with the polarity of parts, which leads are what on transistors, potentiometers, and ICs. And finally, get interconnect wires ready, as described in the next section.

Saving time with prestripped wires

Many of the connections between components on your breadboard are made by the breadboard itself, underneath the surface, but when you just can't make a direct connection via the board, you use interconnect wires (sometimes called *jumper wires*). You use solid (not stranded) 20- to 22-gauge insulated wire to connect components on your breadboard. Thicker or thinner wire doesn't work well in breadboard: too thick, and the wire won't go into the holes; too thin, and the electrical contact will be poor.

WARNING!

Do not use stranded wire in a breadboard. The individual strands can break off, lodging inside the metal contacts of the breadboard.

TIP

While you're buying your breadboard, purchase a set of prestripped jumper wires, as I suggest in Chapter 13. (Don't get cheap now; this purchase is worth it.) Prestripped wires come in a variety of lengths and are already stripped (obviously) and bent, ready for you to use in breadboards. For instance, one popular assortment contains 10 each of 14 wire lengths, ranging from 0.1 inch to 5 inches (see Figure 15-4). A set of 140-350 prestripped wires may cost you $6–$15, but you can bet the price is well worth the time you'll save. The alternative is to buy a bunch of wire, cut segments of various lengths, and painstakingly strip about 1/3-inch of the insulation from each end.

Even if you purchase a large assortment of prestripped wires, there may come a time when you have to make an interconnect wire or two of your own. Start with 20- or 22-gauge wire (or a longer prestripped section that you want to cut into smaller sections), and cut it to the desired length. If you have a wire stripper with a gauge-selection dial, set the dial for the gauge of wire you're using. Other wire strippers may have several cutting notches labeled for various gauges. Using one of these gauge-specific devices instead of a generic wire stripper prevents you from nicking the wire when you're stripping the insulation. Nicks weaken the wire, and a weak wire can get stuck inside a breadboard hole and ruin your whole day.

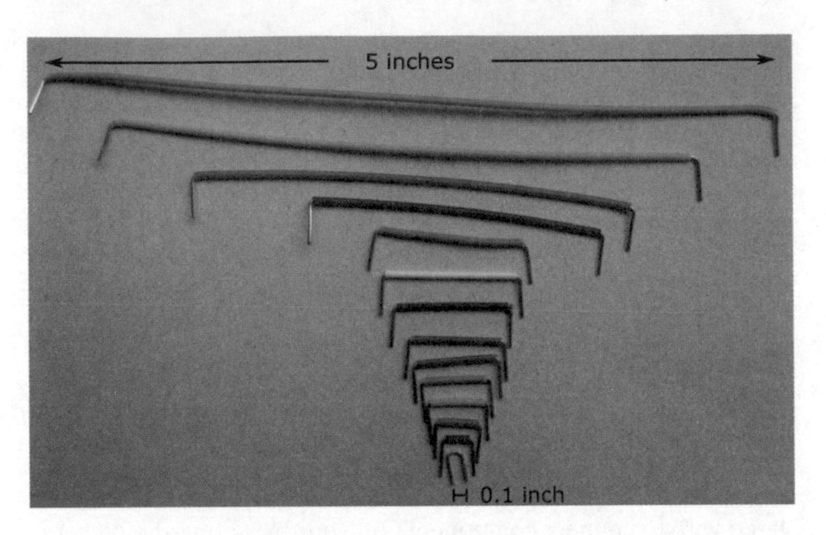

Figure 15-4: Prestripped jumper wires can save you time and agony.

To make your own breadboard wire, follow these steps (see Figure 15-5):

1. **Cut the wire to the length you need, using a wire-cutting tool.**
2. **Strip off about 1/4- to 1/3-inch of insulation from each end.**

   If you use a gauge-specific tool, insert one end of the wire into the stripping tool, hold the other end with a pair of needle-nose pliers, and draw the wire through the stripping tool. If you use a generic wire stripper, you provide the gauge control by how much you squeeze the tool around the wire: too much, and you nick the wire; too little, and you don't cut through all the insulation.
3. **Bend the exposed ends of wire at a right (90°) angle.**

   Use needle-nose pliers for this task.

Figure 15-5: Cutting, stripping, and bending an interconnect wire.

Laying out your circuit

Your parts and tools are ready and the schematic (the circuit diagram) is in hand. Now you want to build a circuit on the breadboard. But where should you start? What's the best way to connect everything?

Welcome to the world of circuit layout — figuring out where everything should go on the board so it all fits together and is neat, tidy, and error-free. Don't expect your circuit layout to look exactly like your schematic — doing so is not only difficult but usually impractical. You can, however, orient your key circuit elements so your circuit is easier to understand and debug.

When you're building a circuit on a breadboard, concentrate on the *connections between components,* rather than on the position of components in your schematic.

Here are some guidelines for building your breadboard circuit:

- **Orient your breadboard so that the power rails are along the top and bottom of the board, as shown in Figure 15-1.**
- **Use one of the top power rails (preferably the one labeled +) for the positive power supply, and one of the bottom rails (preferably the one labeled –) for ground (and the negative power supply, if there is one).** These rails give you plenty of interconnected sockets so you can easily connect components to power and ground.
- **Orient circuit inputs on the left side of the board and outputs on the right side.** Plan your component layout to minimize the number of jumper wires. The more wiring you have to insert, the more crowded and confusing the board becomes.
- **Place ICs first, straddling the center gap.** Allow at least three — and preferably ten — columns of holes (that is, terminal strips) between each IC. You can use a chip inserter/extractor tool to implant and remove ICs to reduce the chances of damaging the IC while handling it.

  If you're working with CMOS chips, be sure that you ground the tool to eliminate stray static electricity. (Refer to Chapter 13 for details.)
- **Work your way around each IC, starting from pin 1, inserting the components that connect to each pin. Then insert any additional components to complete the circuit.** Use needle-nose pliers to bend leads and wires to a 90-degree angle and to insert them into the sockets, keeping leads and wires as close to the board as possible to prevent them from getting knocked loose.
- **If your circuit requires common connection points in addition to power, and you don't have enough points in one column of holes, use longer pieces of wire to bring the connection out to another part of the board where you have more space.** You can make the common connection point one or two columns between a couple of ICs, for instance

Figure 15-6 shows the breadboard set up for a simple resistor-LED (light-emitting diode) circuit before and after power is applied.

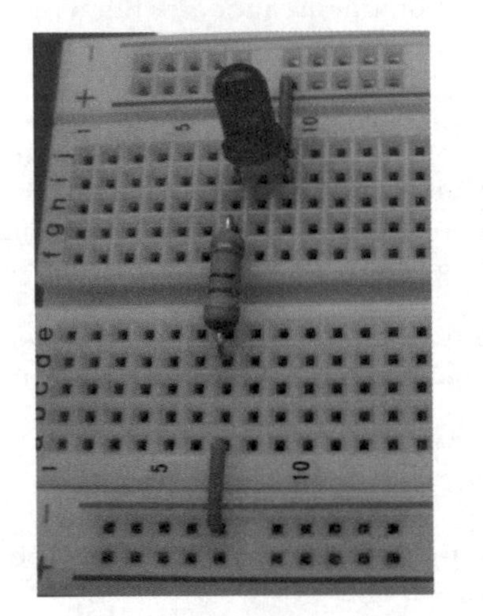

Figure 15-6: Strip and bend the ends of jumper wire, and clip the leads of components so they fit neatly on the breadboard.

My breadboard circuit doesn't work right!

As you work with solderless breadboards, you may encounter the fairly common problem of *stray capacitance,* which is unwanted capacitance (stored electrical energy) in a circuit. All circuits have an inherent capacitance that can't be avoided, but when lots of wires are going every which way, the capacitance can unexpectedly increase. At a certain point (and it differs from one circuit to the next), this stray capacitance can cause the circuit to misbehave.

Because solderless breadboards contain strips of metal and require somewhat longer component leads, they tend to introduce a fair amount of stray capacitance into unsuspecting circuits. As a result, solderless breadboards have a tendency to change the characteristics of some components — most notably capacitors and inductors. These variations can change the way a circuit behaves. Keep this fact in mind if you're working with RF (radio frequency) circuits, such as radio receivers and transmitters, with digital circuits that use signals that change at a very fast rate (on the order of a couple of million Hertz), and with more sensitive timing circuits that rely on exact component values.

If you're building a radio or other circuit that stray capacitance can affect, you may have to forego the step of first building the circuit on a solderless breadboard. Instead, you may have to go straight to a perfboard (described in the section "Creating a Permanent Circuit," later in this chapter).

Don't worry about urban sprawl on your breadboard. You do better to place components a little farther apart than to jam them too close together. Keeping a lot of distance between ICs and components also helps you to tweak and refine the circuit. You can more easily add parts without disturbing the existing ones.

Messy wiring makes it difficult to debug a circuit, and a tangle of wires greatly increases the chance of mistakes. Wires pull out when you don't want them to, or the circuit can malfunction. To avoid chaos, carefully plan and construct your breadboard circuits. The extra effort can save you lots of time and frustration down the road.

Avoiding damaged circuits

You need to know just a few other things to keep your breadboard and circuits in good working order:

- **If your circuit uses one or more CMOS chips, insert the CMOS chips last.** If you need to, use a dummy TTL IC to make sure that you wired everything properly. TTL chips aren't nearly as sensitive to static as the CMOS variety. Be sure to provide connections for the positive and negative power supply — and to connect all inputs (tie the inputs that you aren't using to the positive or negative supply rail). When you're ready to test the circuit, remove the dummy chip and replace it with the CMOS IC.
- **Never expose a breadboard to heat because you can permanently damage the plastic.** ICs and other components that become very hot (because of a short circuit or excess current, for example) may melt the plastic underneath them. Touch all the components while you have the circuit under power to check for overheating.

- **Never use a solderless breadboard to carry 120 VAC house current.** Current may arc across the contacts, damaging the board and posing a danger.
- **If a small piece of a lead or wire becomes lodged in a socket, use needle-nose pliers to gently pull it out of the hole — with the power switched off.**
- **You won't always be able to finish and test a circuit in one sitting. If you have to put your breadboard circuit aside for a while, put it out of the reach of children, animals, and the overly curious.**

Soldering 101

Soldering (pronounced "SOD-er-ing") is the method you use to make conductive connections between components or wire or both. You use a device called a *soldering iron* to melt a soft metal called *solder* in a way that the solder flows around the two metal leads you are joining. When the soldering iron is removed, the solder cools and forms a conductive physical joint, known as a *solder joint,* between the wires or component leads.

Should you care about soldering when you plan to use solderless breadboards for your circuit construction projects? The answer is yes. Almost every electronics project involves a certain amount of soldering. For instance, you may purchase components (such as potentiometers, switches, and microphones) that don't come with leads — in which case, you have to solder two or more wires to their terminals to create leads so you can connect them to your breadboard.

Of course, you use this technique extensively when you build permanent circuits on solder breadboards, perfboards, or printed circuit boards (as described in the section. "Creating a Permanent Circuit," later in this chapter).

Preparing to solder

To solder, you need a soldering iron (25–30 watts), a spool of 16- or 22-gauge standard solder (60/40 rosin-core), a secure soldering stand, and a small sponge. Make sure you secure your soldering iron in its stand and position it in a safe place on your workbench, where it's unlikely to be knocked over.

I recommend you take a look at Chapter 12 for detailed information on how to choose soldering equipment for your electronics projects, including a discussion of the use of 60/40 rosin-core solder — which contains lead — versus lead-free solder.

Gather up a few other items, such as safety glasses (to protect your eyes from sputtering solder), an alligator clip (which doubles as a heat sink for temperature-sensitive components), an antistatic wristband (described in Chapter 13), isopropyl alcohol, a piece of paper, a pencil, and some sticky tape. Place each part that you need to solder on the paper, securing it with sticky tape. Write a label, such as R1, on the paper next to each part so it matches the label on your schematic. Put on your safety glasses and antistatic wristband, and make sure your work area is properly ventilated.

Wet the sponge, squeezing out excess water. Turn on the soldering iron, wait a minute or so for it to heat up (to about 700°F), and then wet the tip of the soldering iron by *briefly* touching it to the sponge. If the tip is new, *tin* it before soldering to help prevent solder from sticking to the tip. (Sticky solder can form an ugly globule, which can wreak havoc if it falls off into your circuit.) You tin the tip by applying a small amount of molten solder to it. Then you wipe off any excess solder on the sponge.

Periodically tin your soldering iron's tip to keep it clean. You can also purchase soldering tip cleaners if dirt becomes caked on and you just can't get it off during regular tip retinning.

Soldering for success

Successful soldering requires that you follow some simple steps and get a lot of practice. It's important to remember that timing is critical when it comes to the art of soldering. As you read through the soldering procedure steps, pay close attention to words like *immediately* and *a few seconds* — and interpret them literally. Here are the steps for soldering a joint:

1. **Clean the metal surfaces to be soldered.**

   Wipe leads, wire ends, or etched circuit board surfaces (described later in this chapter) with isopropyl alcohol so that the solder adheres better. Let surfaces dry thoroughly before soldering: You don't want them to catch on fire!

2. **Secure the items being joined.**

   You can use a *third hand* clamp (as described in Chapter 13) or a vise and an alligator clip to hold a discrete component steady as you solder a wire to it, or use needle-nose pliers to hold a component in place over a circuit board. For leaded components, such as a resistor, bending the leads slightly will help hold the component in place.

3. **Position the soldering iron.**

   Holding the iron like a pen, position the tip at a 30- to 45-degree angle to the work surface, as shown in Figure 15-7.

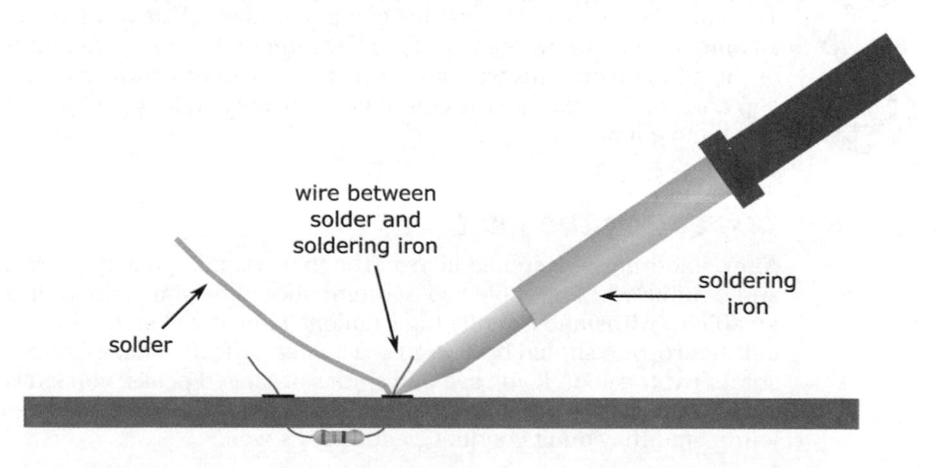

Figure 15-7: Holding the iron at an angle, first, apply heat to the metal parts you are soldering, and then feed the cold solder into the joint.

4. **Apply the tip to the joint you're working on. (*Do not* use any solder at this point in the process.)**

   Make sure that you touch the tip of the soldering iron to both parts you are trying to join (for instance, a resistor lead and a copper pad on the back of a printed circuit board). You want to heat both, so do *not* apply heat directly to the solder. Allow the metal a few seconds to heat up.

5. **Feed the cold solder to the heated metal area.**

   The solder will melt and flow around the joint within a couple of seconds.

6. **Immediately remove the solder, and then remove the iron.**

   As you remove them, hold the component still until the solder cools and the joint solidifies.

7. **Place the soldering iron securely in its stand.**

   Never place a hot soldering iron on your work surface.

8. **Use your diagonal cutters to clip excess component leads as close to the solder joint as possible.**

Be careful to use just the right amount of solder (which means applying solder for just the right amount of time): Use too little, and you form a weak connection; use too much, and the solder may form globs that can cause short circuits.

You can damage many electronic components if you expose them to prolonged or excessive heat, so take care to apply the soldering iron only long enough to heat a component lead for proper soldering — no more, no less.

To avoid damaging heat-sensitive components (such as transistors), attach an alligator clip to the lead between the intended solder joint and the body of the component. This way, any excess heat will be drawn away through the clip and will not damage the component. Be sure to let the clip cool before you handle it again.

Inspecting the joint

After soldering, you should inspect the joint visually to make sure it's strong and conductive. The cooled solder joint should be shiny, not dull, and should be able to withstand a gentle tug from one side. If you soldered a lead to a circuit board, you should be able to see a *fillet* (volcano-shaped raised area of solder) at the joint. If you see dull solder or jagged peaks, you know you have a *cold solder joint.* Cold joints are physically weaker than properly made joints, and they don't conduct electricity as well.

Cold solder joints can form if you move the component while the solder is still cooling, if the joint is dirty or oily, or if you failed to heat the solder properly. Resoldering without first desoldering often produces cold solder joints because the original solder isn't heated enough.

If you have a cold solder joint, it's best to remove the existing solder (as described in the next section, "Desoldering when necessary"), clean the surfaces with isopropyl alcohol, and reapply fresh solder.

Desoldering when necessary

At some point in your work with electronics, you're bound to run into a cold solder joint, a backward-oriented component, or some other soldering mishap. To correct these mistakes, you have to remove the solder at the joint, and then apply new solder. You can use a desoldering pump (also known as a solder sucker and described shortly), a solder wick, or both to remove solder from the joint.

Use *solder wick* (also called *solder braid*), which is a flat braid of copper, to remove hard-to-reach solder. You place it over unwanted solder and apply heat. When the solder reaches its melting point, it adheres to the copper wire, which you then remove and dispose.

Exercise care when using solder wick. If you touch the hot braid, you can get a serious burn — copper conducts heat extremely well.

A *desoldering pump,* also known as a *solder sucker,* uses a vacuum to suck up excess solder that you melt with your soldering iron. There are two types:

- **A spring-loaded plunger-style pump:** To use a spring-loaded pump, you depress the plunger and position the nozzle over the joint that you want to remove. Then you carefully position the soldering iron tip into the joint to heat the solder, avoiding contact with the end of the pump. As the solder begins to flow, you release the plunger to suck up the solder. Finally, you expel the solder from the pump (into a waste receptacle) by depressing the plunger one more time. Repeat these steps as needed to remove as much of the old solder as possible.

  Don't store a desoldering pump with a cocked plunger. The rubber seal can become deformed, diminishing the vacuum to the extent that the pump may be unable to suck up any solder.

- **A bulb-style pump:** A bulb desoldering pump works a lot like the spring-loaded variety, except you squeeze the bulb to create the vacuum and release it to suck up the solder. You may find it difficult to use this pump unless you mount the bulb on the soldering iron. In fact, a device called a *desoldering iron* consists of a soldering iron with a vacuum bulb piggybacked alongside it.

Cooling after soldering

Make it a habit to unplug — not just shut off — your soldering iron when you finish your soldering work. Brush the tip of the still-warm iron against a damp sponge to clean off excess solder. After the iron has cooled, you can use tip-cleaner paste to remove stubborn dirt. Then you can finish with three good practices:

- Make sure the iron is completely cool before storing it.
- Place your solder reel in a plastic bag to keep it from getting dirty.
- Always wash your hands when you finish soldering, because most solder contains lead, which is poisonous.

Practicing safe soldering

Even if you plan to solder just one connection, you should take the appropriate precautions to protect yourself — and those around you. Remember, the iron will reach temperatures exceeding 700°F, and most solder contains poisonous lead. You (or a nearby friend or pet) may unwittingly be hit by popping, sputtering solder if you run across the occasional air pocket or other impurity in your reel. Just one small drop of solder hitting you in the eye or a tumbling soldering iron that lands on your bare foot can ruin a day, a body part — and a friendship.

Set up your work area — and yourself — with soldering safety in mind. Make sure the room is well ventilated, you have the iron placed snugly in its stand, and you position the electrical cord to avoid snags. Wear shoes (no flip-flops!), safety glasses, and an antistatic wristband when soldering. Avoid bringing your face too close to hot solder, which can irritate your respiratory system and possibly sputter. Keep your face to one side, and use a magnifying glass, if needed, to see tiny components that you're soldering.

Never solder a circuit with power flowing through it! Make sure the battery or other power supply is disconnected before you apply your soldering iron to components. If your iron has an adjustable temperature control, dial up the recommended setting for the solder you're using. And if your soldering iron accidentally goes belly-up, *stand back* and let it fall. If you try to grab it, Murphy's Law says you'll grab the hot end.

Finally, always unplug your soldering iron when you're done and promptly wash your hands.

Creating a Permanent Circuit

So you've perfected the world's greatest circuit, and you want to make it permanent. The most common way to transfer your circuit and make lasting connections is to use a perforated board (perfboard). A *perfboard* is a simple type of printed circuit board (PCB) designed for prototyping.

In this section, I explain what a PCB is and discuss different types of perfboards.

Exploring a printed circuit board

Most printed circuit boards consist of a nonconductive (usually plastic) layer, known as the *substrate,* with copper interconnections on one or both sides. (Many industrial-strength PCBs contain multiple layers, with components embedded in the substrate.) There are two types of copper interconnections: *pads,* which are small copper circles to which you solder components leads, and *traces,* which are short copper paths that, like wires, create connections between pads.

PCBs come in two basic varieties:

- **Through hole:** These PCBs contain predrilled holes surrounded by copper pads on one or both sides of the board. The holes are spaced 0.1 inch apart to accommodate both ICs and discrete components. You mount discrete components, such as resistors, diodes, and capacitors, to one side of the board by inserting their leads into the holes, soldering the leads to the copper pads on the other side of the board, and then trimming the leads. You can mount ICs directly to the board or you can mount *IC sockets* to the board and then insert the IC into the socket.
- **Surface mount:** These PCBs are solid (no holes). Specially designed surface-mount components — which look nothing like their through-hole counterparts — are mounted and attached to one side of the board. Surface-mount technology (SMT), which includes the boards and components, is geared for high-density circuits and large-scale automated circuit assembly processes.

Although you can find SMT components and SMT-compatible prototyping breadboards, they are challenging to work with because the components are tiny, their leads are even tinier, and it's difficult to keep the components in place while soldering. That's why I recommend you stick to perfboards, which are relatively inexpensive through-hole boards specifically designed for prototyping.

Making circuit boards with plug-and-play ICs

When you build circuit boards that include integrated circuits, use an IC socket instead of soldering the IC directly onto the board. You solder the socket onto the board, and then you plug in the IC and hit the switch.

IC sockets come in different shapes and sizes to match the integrated circuits they're meant to work with. For example, if you have a 16-pin integrated circuit, choose a 16-pin socket.

Here are some good reasons for using sockets:

- **Soldering a circuit board can generate static.** By soldering to the socket rather than to the actual IC, you can avoid ruining CMOS or other static-sensitive ICs.
- **ICs are often among the first things to go bad when you're experimenting with electronics.** Having the option to pull out a chip that you suspect is bad so you can replace it with a working one makes troubleshooting a whole lot easier.
- **You can share an expensive IC, such as a microcontroller, among several circuits.** Just pull the part out of one socket and plug it into another.

Sockets are available in all sizes to match the different pin arrangements of integrated circuits. They don't cost much — just a few pennies for each socket.

Relocating your circuit to a perfboard

Perfboards are circuit boards with predrilled holes arranged in grids so they can be used for prototyping. Most perfboards contain copper pads and traces. You can find a variety of single- and double-sided perfboards at your local or online electronics supplier. Figure 15-8 shows a small sampling of perfboards. Other shapes and sizes are readily available, including round preprinted boards with different diameters and bare perfboards with no copper connections.

Some perfboards are laid out just like a solderless breadboard, with multiple holes connected in terminal strips and bus strips. For instance, the board shown in the lower left of Figure 15-8 has the same geometry as a 550-hole solderless breadboard.

Sometimes called *solderable breadboards,* these perfboards enable you to transfer your circuit from a solderless breadboard to a perfboard easily because you don't have to alter the layout. You simply pick the parts off your solderless breadboard, insert them in the corresponding perfboard holes, solder their leads to the copper pads, and trim the leads. You use insulated wires in the same way you did in the original solderless breadboard: to connect components that aren't already electrically connected by the metal strips of the perfboard, soldering them into place. The drawback of using solderable breadboards is that they waste space.

Figure 15-8: A variety of perfboards are available for use in constructing permanent circuits. Just clean the board (if necessary) and add electronics components.

Other general-purpose perfboards consist of predrilled holes spaced at regular intervals in a square or rectangular grid. Different perfboard layouts are available. You choose the grid layout that best suits your needs. For instance, if your circuit uses many ICs, choose a model that has buses running up and down the board, such as the one shown in Figure 15-9. (Alternating the buses for the power supply and ground also helps to reduce undesirable inductive and capacitive effects.) If you need to conserve space, construct your circuit using a bare perfboard and *point-to-point wiring,* in which you pass components leads through the holes and solder components directly to each other.

If you design a small circuit, you can use just half a perfboard. Before transferring the components, cut the board with a hacksaw while wearing a protective mask to avoid breathing in the dust produced by the saw. Clean the portion of the board that you want to use and solder away. Some preprinted PCBs are scored so you can simply snap the board to create two or four smaller boards.

Figure 15-9: Several buses run up and down this perfboard, which is well suited for circuits that use multiple ICs.

Many perfboards come with mounting holes at the corners so that you can secure the board inside whatever enclosure your project provides (such as the chassis of a robot). If your perfboard does not have mounting holes, you may want to leave space at the edges of the board so you can drill holes. Alternatively, you can secure the board to a frame or within an enclosure by using double-sided foam tape. The tape cushions the board and prevents breakage, and the thickness of the foam prevents the underside of the board from touching the chassis.

Making a custom circuit board

After you become experienced in the ways of designing and building electronics projects, you may want to graduate to the big time and create your own custom circuit board, geared for a particular circuit design. You can make (yes, make) your very own custom PCB — just like the electronics manufacturers do. PCBs are reliable, rugged, allow for higher-density circuits, and enable you to include non-standard size components that may not fit in other types of circuit boards.

Making a printed circuit board is a fairly involved process — and beyond the scope of this book — but here's the lowdown on some of the steps involved:

1. You make a blank PCB by gluing or laminating a thin sheet of copper, known as *cladding,* onto the surface of a plastic, epoxy, or phenolic base. This forms a sort of "blank canvas" for the creation of a circuit.
2. You prepare a mask of your circuit layout, transfer it onto clear transparent film, and use the mask to expose a sheet of sensitized copper to strong ultraviolet light.
3. You dip the sensitized, exposed sheet into a developer chemical, producing a pattern (called a *resist pattern*) of the circuit board layout.
4. You form the circuit layout by etching away the portions of the copper not protected by the resistor — leaving behind just the printed circuit design, which consists of *pads* (contact points for components) and *traces* (interconnects).
5. You drill holes into the center of each pad so you can mount components on the top of the board with leads poking through the holes.
6. Finally, you solder each component lead to the board's pads.

To find out exactly how to make your own PCB, you can search the Internet with the following keywords: *make printed circuit board* — and you'll find tutorials, illustrations, and even videos that explain the process in great detail.

16

Mastering Your Multimeter to Measure Circuits

In This Chapter

- Introducing your new best friend: your multimeter
- Using a multimeter to measure all kinds of things
- Setting up and calibrating your multimeter
- Making sure electronic components are working properly
- Probing around your circuits
- Identifying the cause of circuit problems

Your excitement builds as you put the finishing touches on your circuit. With close friends standing beside you, eager to witness the first of your ingenious electronics exploits, you hold your breath as you flip the power switch, and . . .

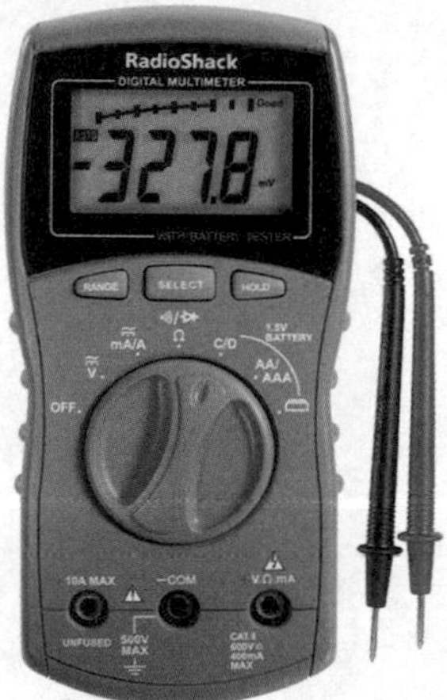

Nothing. At least, nothing at first. Then, disappointment, disillusionment, and disbelief as your friends — and your confidence — slowly retreat from the scene.

You ask yourself, "What could possibly be wrong?" Then you notice it: Smoke is emanating from what used to be a resistor. And then you realize that you used a 10 Ω resistor instead of a 10 kΩ resistor, trusting your weary eyes and mind to read and interpret resistor stripes properly. Oops!

In this chapter, you find out how to use a versatile tool — the multimeter — to perform important face-saving checks on electronic circuits and components. These tests help you determine whether everything is A-OK before you start showing off your circuitry to friends and family. When you've finished reading this chapter, you will realize that your multimeter is as important to you as an oxygen tank is to a scuba diver: You can both get along okay on your own for a while, but sooner or later, you're bound to suffer unless you get some help.

Multitasking with a Multimeter

A *multimeter* is an inexpensive handheld testing device that can measure voltage, current, and resistance. Some can also test diodes, capacitors, and transistors. With this one handy tool, you can verify proper voltages, test whether you have a short circuit, determine whether there's a break in a wire or connection, and much more. Make friends with your multimeter, because it can help you make sure your circuits work properly and is an invaluable tool for scouting out circuit problems.

Figure 16-1 shows a typical midpriced multimeter. You turn a dial to select the type of measurement you want to make. You then apply the metal tips of the two test leads (one red, one black) to a component or some part of your circuit, and the multimeter displays the resulting measurement.

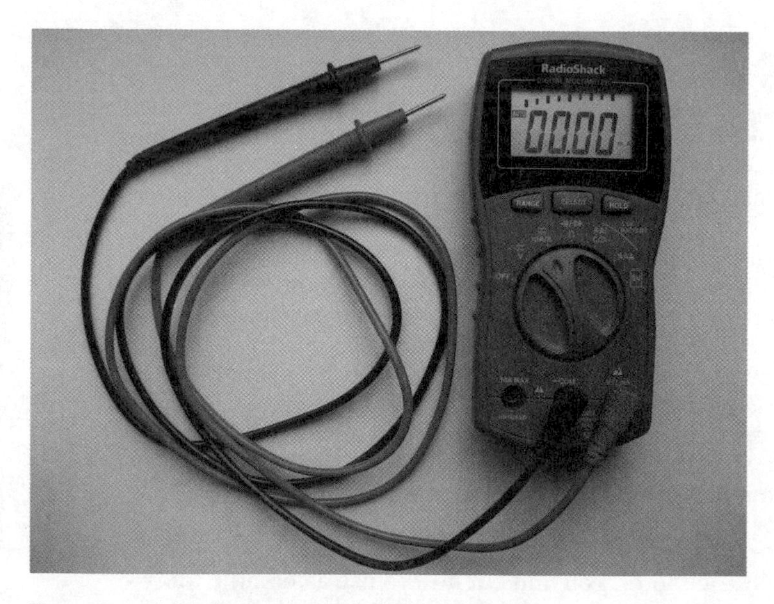

Figure 16-1: Multimeters measure voltage, resistance, current, and continuity.

Multimeter test leads have conical tips that you hold in contact with the component you're testing. You can purchase special spring-loaded test clips that slip over the tips, making it easier for you to attach the test leads to the component leads or other wires. (See Figure 16-2.) These insulated test clips ensure a good connection between the test leads and whatever it is you're testing, while preventing accidental contact with another part of the circuit.

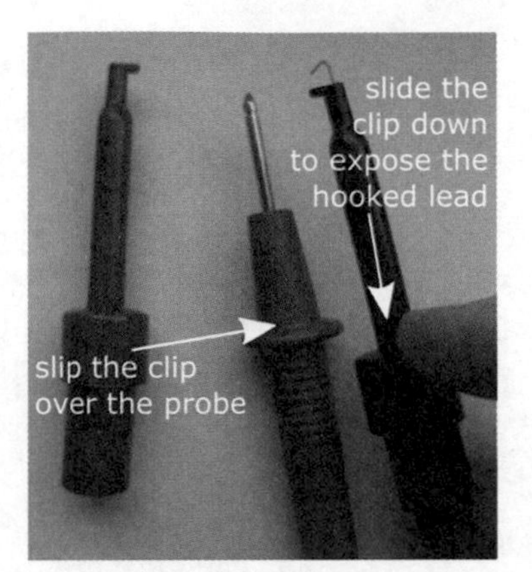

Figure 16-2: Spring-loaded test clips prevent accidental contact.

It's a voltmeter!

Multimeters can measure both DC and AC voltages, providing a variety of voltage measurement ranges, from 0 volts to a maximum voltage. A typical set of DC voltage ranges is 0–0.25 V, 0–2.5 V, 0–10 V, 0–50 V, and 0–250 V.

Using your multimeter as a *voltmeter,* you can measure the voltage of a battery outside a circuit or *under load* (meaning when it is providing power in a circuit). You can also use your multimeter with your circuit powered up if you want to test voltages dropped across circuit elements and (for that matter) voltages at various points in your circuit with respect to ground.

You can often pinpoint the location of a problem in your circuit by using your multimeter. It can verify whether the proper voltage is supplied to a component, such as a light-emitting diode (LED) or a switch. You use multimeter tests to narrow the field of suspects until you find the culprit causing all your headaches.

Voltmeters are so important in electronics, they have their own circuit symbol, which is shown in Figure 16-3, left. You may see this symbol with leads touching points in a circuit you read about on a website or in an electronics book. It tells you to take a voltage measurement across the two indicated points.

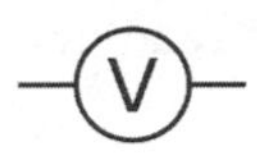

Ω

voltmeter ammeter ohmmeter

Figure 16-3: Circuit symbols for common test instruments.

It's an ammeter!

Your multimeter also functions as an *ammeter,* a device that measures the electric current going through a circuit. You use this function of the multimeter to determine whether a circuit or component is drawing too much current. If your circuit has more current going through it than it's designed to handle, the components may get overheated and damage your circuit permanently.

The circuit symbol for an ammeter is shown in Figure 16-3, center.

Ohm my! It's an ohmmeter, too!

You can measure the resistance of an individual component or an entire circuit (measured in ohms, as detailed in Chapter 5) with your multimeter functioning as an *ohmmeter.* You use this function to check up on wires, resistors, motors, and many other components. You always test resistance with the circuit *unpowered.* If the circuit is powered, current flowing through it can invalidate the resistance readings — or damage the meter.

The circuit symbol for an ohmmeter is shown in Figure 16-3, right.

If you're measuring the resistance of an individual component, take it out of the circuit before you test it. If you test a resistor when it's wired into a circuit, you'll get the equivalent resistance between two points, which is not necessarily the resistance of just your resistor. (See Chapter 5 for more on equivalent resistance.)

Because resistance or (for that matter) lack of resistance can reveal short circuits and open circuits, you can use your ohmmeter to sniff out problems such as breaks in wires and hidden shorts between components. A short circuit generates an ohmmeter reading of zero (or virtually zero) resistance; an open circuit generates a readout of infinite resistance. If you test the resistance from one end of a wire to the other and you get an infinite readout, you know there must be a break somewhere along the length of the wire. Such tests are known as *continuity tests.*

By measuring resistance, you can tell whether the following circuit elements and connections are working properly:

- **Fuses:** A blown fuse generates an infinite resistance reading, indicating an open circuit.
- **Switches:** An on switch should generate a zero (or low) resistance reading; an off switch should generate an infinite reading.

- **Circuit board traces:** A bad copper trace (line) on a printed circuit board acts like a broken wire and generates an infinite resistance reading.
- **Solder joints:** A bad solder joint may generate an infinite resistance reading.

Many multimeters include an audible continuity-testing feature. By turning the meter's selector to continuity or tone, you can hear a beep whenever the meter detects continuity in a wire or connection. If the wire or connection doesn't have continuity, the meter stays silent. The tone gives you a convenient way to check an entire circuit without having to keep your eye on the meter.

Exploring Multimeters

Multimeters range from bare-bones handheld models that cost less than $10 to feature-rich hobbyist models that cost from $30 to over $100, to sophisticated industrial bench-top models that cost more than $1000.

Even a low-end multimeter can help you understand what's going on in low-voltage circuits. However, unless you're cash-strapped, it's a good idea to spend a little more on a multimeter to get more features; you're sure to find them useful as you expand your electronics horizons.

Choosing a style: analog or digital

Most multimeters today, including the one shown in Figure 16-1, are *digital multimeters,* which provide readouts on a digital (numeric) display. You may also find some older-style *analog multimeters,* which use a needle to point to a set of graduated scales. You can see an analog multimeter in Figure 16-4.

Figure 16-4: This circa-1980s analog multimeter uses a needle to indicate voltage, current, and other values.

Using an analog multimeter can be a bit challenging. After selecting the type of testing (voltage, current, or resistance) and the range, you must correlate the results by using the appropriate scale on the meter face, and estimate the reading as the needle swings into action. It's easy to get an erroneous reading — due to misinterpreted scale divisions, mental arithmetic errors, or a compromised view of where the needle is pointing. In addition, resistance measurements are imprecise because the measurement scale is compressed at high resistance values.

Digital multimeters display each measurement result as a precise number, taking the guesswork out of the reading process. Most handheld digital multimeters are accurate to within 0.8% for DC voltages; the pricey bench-top varieties are over 50 times more accurate. Many digital multimeters also include an *autoranging* feature, which means that the meter automatically adjusts itself to display the most accurate result possible. Some have special testing features for checking diodes, capacitors, and transistors.

Analog multimeters out-perform digital multimeters when it comes to detecting *changing* readings. But if you don't have much of a need for that feature, your best bet is to get a digital multimeter because of its ease of use and more accurate readings.

Taking a closer look at a digital multimeter

All digital multimeters perform the basic voltage, resistance, and current measurements. Where they differ is in the range of values they can measure, the additional measurements they can perform, the resolution and sensitivity of their measurements, and the extra bells and whistles they come with.

Be sure to read through the manual for the multimeter you purchase. It contains a description of the features and specifications for your meter, as well as important safety precautions.

Here's what you'll find when you explore a digital multimeter:

- **Power switch/battery/fuse:** The on/off switch connects and disconnects the battery that powers the multimeter. Many multimeters use standard-size batteries, such as a 9-volt or AAA cell, but pocket-size meters use a coin-type battery. Most multimeters use an internal fuse to protect themselves against excessive current or voltage; some come with a spare fuse (if yours doesn't, buy one).

  Avoid using rechargeable batteries in a multimeter; they may produce erroneous results for some models.

- **Function selector:** Dial this knob to choose a test to perform (voltage, current, resistance, or something else) and, on some models, the range setting you want to use. Some multimeters are more "multi" than others, and include one or more of the following categories: AC amperes, capacitance, transistor gain (h_{FE}), and diode test. Many multimeters further divide some measurement categories into three to six different ranges; the smaller the range, the greater the sensitivity of the reading. Figure 16-5 shows close-ups of two function selector dials.

Figure 16-5: Digital multimeters provide a wide variety of measurement options.

- **Test leads and receptacles:** Inexpensive multimeters come with basic test leads, but you can purchase higher-quality coiled leads that stretch out to several feet and recoil to a manageable length when not in use. You may also want to purchase spring-loaded clip leads (refer to Figure 16-2). Some multimeters with removable test leads provide more than two receptacles for the leads. You insert the black test lead into the receptacle labeled GROUND or COM, but the red lead may be inserted into a different receptacle depending on what function and range you dial up. Many meters provide additional input sockets for testing capacitors and transistors, as shown in the photo on the left in Figure 16-5.
- **Digital display:** The readout is given in units specified by the range you've dialed up. For instance, a reading of 15.2 means 15.2 V if you've dialed a 20 V range, or 15.2 millivolts (mV) if you've dialed a 200 mV range. Most digital multimeters designed for hobbyists have what's called a *3½ digit display:* Its readout contains three or four digits, where each of the three right-most digits can be any digit from 0 to 9, but the optional fourth digit (that is, the left-most — most significant — digit) is limited to 0 or 1. For instance, if set to a 200 V range, such a multimeter can give readouts ranging from 00.0 V to 199.9 V.

Some multimeter models don't have a dedicated on/off switch; instead, the function selector knob has an off setting. Make sure you move the knob to the off setting after you finish taking measurements so you don't run down the battery. If you inserted the red test lead into a different receptacle to measure current, be sure to move the lead back to the receptacle that allows you to measure voltage and resistance; if you forget to do this, you may smoke your meter.

Homing in on the range

Many digital multimeters (and most analog multimeters) require that you select the range before the meter can make an accurate measurement. For example, if you're measuring the voltage of a 9 V battery, you set the range to the setting closest to (but still above) 9 volts. For most meters, this means you select the 20 V or 50 V range.

If you select too big a range, the reading you get won't be as accurate. (For instance, on a 20 V range setting, your 9 V battery may produce a reading of 8.27 V, but on a 200 V range setting, the same battery produces a reading of 8.3 V. You often need as much precision in your readings as possible.)

If you select too small a range, a digital multimeter typically displays an *overrange indicator* (for instance, a flashing 1 or OL or OF), whereas the needle on an analog meter shoots off the scale, possibly damaging the precision needle movement (so make sure you start with a large range and dial it down, if necessary). If you see an overrange indicator when testing continuity, it means the resistance is so high that the meter can't register it; it's fairly safe to assume that's an open circuit.

The autoranging feature found on many digital multimeters makes it even easier to get a precise reading. For instance, when you want to measure voltage, you set the meter function to volts (either DC or AC) and take the measurement. The meter automatically selects the range that produces the most precise reading. If you see an overrange indicator, that's telling you the value is too high to be measured by the meter. Autoranging meters don't require range settings, so their dials are a lot simpler.

Some autoranging multimeters have a manual range override feature. The multimeter shown in the photo on the right in Figure 16-5 contains a button labeled Range which, when pressed, steps through five manual range options. You have to carefully interpret the display to determine which range you've selected.

There's a limit to what a multimeter can test. You call that limit its *maximum range*. Most consumer multimeters have roughly the same maximum range for voltage, current, and resistance. For your hobby electronics, any meter with the following maximum ranges (or better) should work just fine:

DC volts: 1,000 V

AC volts: 500 V

DC current: 200 mA (milliamperes)

Resistance: 2 MΩ (two megohms, or 2 million ohms)

What if you need to test higher currents?

Most digital multimeters limit current measurements to less than one amp. The typical digital multimeter has a maximum range of 200 milliamperes. Attempting to measure substantially higher currents may cause the fuse in the meter to blow. Many analog meters, especially older ones, support current readings of 5 or 10 amps, maximum.

You may find analog meters that can tolerate a high-ampere input handy if you're testing motors and circuits that draw a lot of current. If you have only a digital meter with a limited-milliampere input, you can still measure higher currents indirectly by using a low-resistance, high-wattage resistor. To do this, you place a 1 Ω, 10-watt resistor in series with your circuit so that the current you want to measure passes through this test resistor. Then you use your multimeter as a voltmeter, measuring the voltage dropped across the 1 Ω resistor. Finally, you apply Ohm's Law to calculate the current flowing through the test resistor as follows:

$$current = \frac{voltage}{resistance} = \frac{V}{1\,\Omega}$$

Because the nominal value of the resistor is 1 Ω, the current (in amps) through the resistor has roughly the same value as the voltage (in volts) you measure across the resistor. Note that the resistor value will not be exactly 1 Ω in practice, so your reading may be off as much as 5%–10%, depending on the tolerance of your resistor and the accuracy of your meter. To get a more accurate result for your current measurement, first measure the *actual* resistance of your 1 Ω resistor, and then use that resistance in your calculation for current. (You can get a refresher on Ohm's Law in Chapter 6.)

Setting Up Your Multimeter

Before testing your circuits, you must make sure that your meter is working properly. Any malfunction gives you incorrect testing results — and you may not even realize it. To test your multimeter, follow these steps:

1. **Make sure that the test probes at the end of the test leads are clean and screwed in all the way.**

   Dirty or corroded test probes can cause inaccurate results. Use electronic contact cleaner to clean both ends of the test probes and, if necessary, the connectors on the meter.

2. **Turn on the meter and dial it to the ohms (Ω) setting.**

   If the meter isn't autoranging, set it to low ohms.

3. **Plug both test probes into the proper connectors of the meter and then touch the ends of the two probes together (see Figure 16-6).**

   Avoid touching the ends of the metal test probes with your fingers while you're performing the meter test. The natural resistance of your body can throw off the accuracy of the meter.

4. **The meter should read 0 (zero) ohms or very close to it.**

   If your meter doesn't have an autozero feature, look for an adjust (or zero adjust) button to press. On analog meters, rotate the meter's zero adjust knob until the needle reads 0 (zero). Keep the test probes in contact and wait a second or two for the meter to set itself to zero.

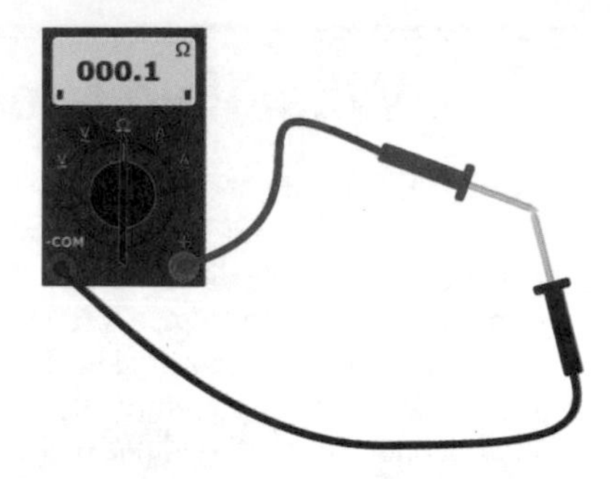

Figure 16-6: Touch the test probes of the meter together and verify a zero ohms reading to make sure the meter is working properly.

5. **If you don't get any response from the meter when you touch the test probes together, recheck the dial setting of the meter.**

   Nothing happens if you have the meter set to register voltage or current. If you make sure that the meter has the right settings and it still doesn't respond, you may have faulty test leads. If necessary, repair or replace any bad test leads with a new set.

You can consider the meter *calibrated* when it reads zero ohms with the test probes *shorted together* (held together so they're touching each other). Do this test each time you use your meter, especially if you turn off the meter between tests.

If your meter has a continuity setting, don't use it to zero-adjust (calibrate) the meter. The tone may sound when the meter reads a few ohms, so it doesn't give you the accuracy you need. Recalibrate the multimeter using the ohms setting, and not the continuity setting, to ensure proper operation.

Operating Your Multimeter

When you use your multimeter to test and analyze circuits, you must consider what settings to dial up, whether you're testing components individually or as part of a circuit, whether the circuit should be powered up or not, and where you place the test leads (in series or in parallel with whatever you're testing).

Think of your multimeter as an electronic component in your circuit (because in a way, it is). If you want to measure voltage, your meter must be placed *in parallel* with the section of the circuit you're measuring because voltages across parallel branches of a circuit are the same. If you want to measure current, your meter must be placed *in series* with the section of the circuit you are measuring because components in a series circuit carry the same current. (You can read about series and parallel connections in Chapter 4.)

In the following sections, I explain how to use your multimeter to measure voltage, current, and resistance in the simple resistor-LED (light-emitting diode) circuit that you see in Figure 16-7.

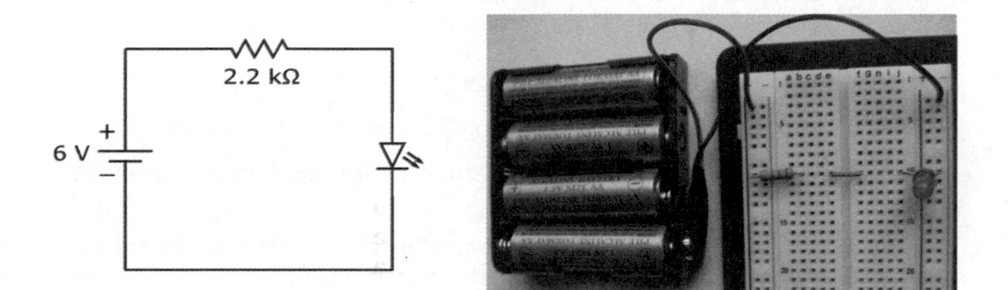

Figure 16-7: A simple resistor-LED circuit.

If you choose to build the resistor-LED circuit and try out your multimeter, here are the parts you need:

- Four 1.5-volt AA batteries
- One four-battery holder (for AA batteries)
- One battery clip
- One 2.2 kΩ 1/4-watt (minimum) resistor (red-red-red)
- One red LED (any size; I use 5 mm)
- One solderless breadboard
- One short jumper wire

Refer to the photo in Figure 16-7 as you build the circuit. Make sure you orient the LED so that its shorter lead is connected to the negative side of your battery pack. You can find the details about constructing circuits using a solderless breadboard in Chapter 15.

The most important thing to remember about solderless breadboards is that they have internal electrical connections. For instance, the five holes in row 10, columns a through e are connected to each other; the five holes in row 10, columns f through j are connected to each other. And all the holes in each individual *power rail* (those columns labeled + and –) are connected to each other, but the four power rails are not connected to each other.

Measuring voltage

To examine voltage levels — that is, the voltage drop from a point in your circuit to ground — throughout your circuit using a multimeter, your circuit must be powered up. You can test the voltage at almost any point in a circuit. Here's how to measure voltages:

1. **Select the type of voltage (AC or DC).**

   In the case of our resistor-LED circuit, select DC volts.

2. **If your multimeter isn't autoranging, choose the range that gives you the most sensitivity.**

   If you're not sure what range to select, start with the broadest range and dial it down later if the measurement falls in a lower range. In the resistor-LED circuit (refer to Figure 16-7), the maximum voltage you can expect to measure is equal to the supply voltage, which is nominally 6 V. Select a range of 10 V or 20 V (meaning 0–10 V or 0–20 V), depending on what your multimeter offers.

3. **Measure a voltage level.**

   Attach the black lead to the ground connection in the circuit and attach the red lead to the point in the circuit that you want to measure. This action places your multimeter in parallel with the voltage drop from a point in your circuit to ground.

4. **Measure the voltage drop across a circuit component (for instance, a resistor or LED).**

   Attach the black lead to one side of the component and attach the red lead to the other side of the component. This action places your multimeter in parallel with the voltage drop you want to measure.

   If you attach the black lead to the more negative side of the component (that is, the side that is closer to the negative side of the battery pack) and the red lead to the more positive side of the component, your multimeter will register positive volts. If you attach your multimeter leads the other way, your multimeter will register negative volts.

Refer to Figure 16-8 for examples of using a multimeter to measure two different voltage drops in the resistor-LED circuit. In the left image, the meter is measuring the voltage that powers the entire circuit, so the multimeter reads 6.4 V. (Note that the fresh batteries I use are supplying more than their nominal voltages.) In the right image, the meter is measuring the voltage drop across the LED, which is 1.7 V.

In some circuits, such as those that handle audio signals, voltages may change so rapidly that your multimeter may not be able to keep up with the *voltage swings*. To test fast-changing signals, you need a logic probe (for digital signals only) or an oscilloscope.

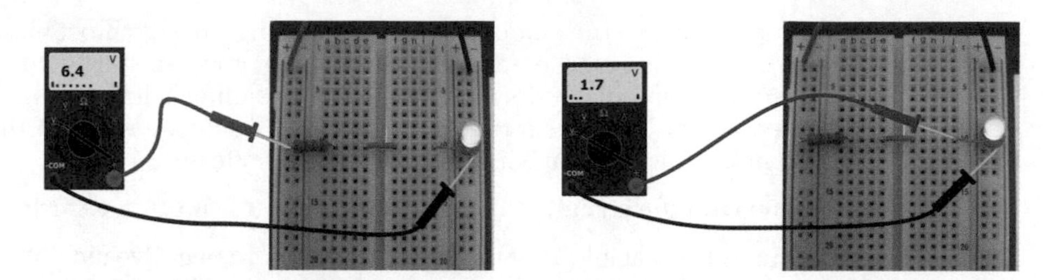

Figure 16-8: Measuring voltage drops in a resistor-LED circuit.

Measuring current

To measure current, you connect your multimeter in a way that ensures that the same current you want to measure passes through the meter when the circuit is powered up. In other words, you must insert your multimeter in the circuit *in series with* the component you want to measure current through. This setup, as shown in Figure 16-9, is different from the voltmeter configuration.

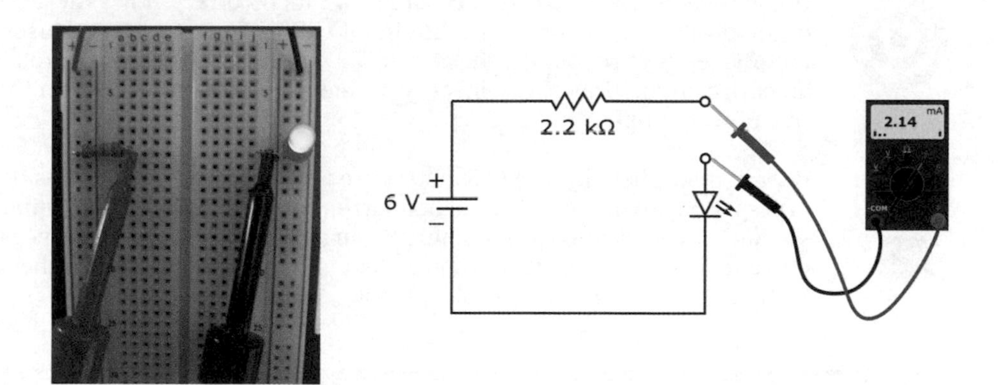

Figure 16-9: To measure current, connect the meter in series with the circuit or component.

These are the steps you follow to measure current:

1. **Select the type of current (AC or DC).**

   In the case of our resistor-LED circuit, select DC amps.

2. **If your multimeter isn't autoranging, choose the range that gives you the most sensitivity.**

   If you're not sure what range of current to select, start with the broadest range, and dial it down later if the measurement falls within a lower range. Because the amount of current passing through most electronic circuits can be measured in milliamps (mA), you can start with a range of 200 mA and dial it down to 20 mA if the reading is less than 20 mA. For the resistor-LED circuit (refer to Figure 16-9), select a range of 10 mA or 20 mA, depending on what your multimeter offers.

3. **Interrupt the circuit at the point where you want to measure current.**

   Attach the black lead of the meter to the more negative side of the circuit and the red lead to the more positive side. This action places your multimeter in series with the component you want to measure current through.

In the resistor-LED circuit (refer to Figure 16-7), current has only one path to flow through, so you can interrupt the circuit between any two components. One way to interrupt the circuit is to remove the jumper that connects the resistor to the LED. Then to measure the current, connect the black multimeter lead to the LED and the red multimeter lead to the resistor, as shown in Figure 16-9.

In a circuit with parallel branches, current splits at each *node* (the connection between branches). To measure how much current flows through one of the branches, you interrupt the circuit within that branch and insert your meter's leads to reconnect the circuit. To measure how much overall current an entire circuit draws, you insert your meter's leads in series with the positive power supply.

Bear in mind that many digital meters are limited to testing currents of 200 mA or less. Be careful: Don't test higher current if your meter isn't equipped to do so. And never leave your multimeter in an ammeter position after measuring current. You can damage the meter. Get in the habit of turning off the meter immediately after running a current test.

Don't blow your fuse!

Many analog and digital meters provide a separate input (test lead receptacle) for testing current, usually marked as A (for amps) or mA (for milliamps). Some multimeters provide an additional input for testing higher currents, typically up to 10 or 20 amps. The multimeter shown on the left in Figure 16-5 has two inputs for testing current, labeled mA and 20A.

Be sure to select the appropriate input *before* making any current measurement. Forgetting to do this step may either blow a fuse (if you're lucky) or damage your meter (if you're unlucky).

Measuring resistance

You can run lots of different tests by using your multimeter as an ohmmeter that measures resistance. Obviously, you can test resistors to check their values or see whether they've been damaged, but you can also examine capacitors, transistors, diodes, switches, wires, and other components by using your ohmmeter. Before you measure resistance, however, be sure to calibrate your ohmmeter (as described in the earlier section "Setting Up Your Multimeter").

If your multimeter has specific features for testing capacitors, diodes, or transistors, I recommend that you use those features rather than the methods that I give you in the following sections. But if you have a bare-bones multimeter without those features, these methods can help you.

Testing resistors

Resistors are components that limit current through a circuit. (Refer to Chapter 5 for details.) Sometimes you need to verify that the nominal resistance value marked on the body of a resistor is accurate, or you may want to investigate whether a suspicious-looking resistor with a bulging center and third-degree burns has gone bad.

To test a resistor with a multimeter, follow these steps:

1. **Turn off the power before you touch your circuit, and then disconnect the resistor you want to test.**
2. **Set your multimeter to read ohms.**

   If you don't have an autoranging meter, start at a high range and dial down the range as needed.
3. **Position the test leads on either side of the resistor.**

   It doesn't matter which test lead is placed on what side of the resistor, because resistors don't care about polarity.

   Take care not to let your fingers touch the metal tips of the test leads or the leads of the resistor. If you touch them, you add the resistance of your body into the reading, producing an inaccurate result.

The resistance reading should fall within the tolerance range of the nominal value marked on the resistor. For instance, if you test a resistor with a nominal value of 1 kΩ and a tolerance of 10%, your test reading should fall in the range of 900 to 1,100 Ω. A bad resistor may be completely open inside (in which case you may get a reading of infinite Ω), may be shorted out (in which case you get a reading of zero Ω), or may have a resistance value outside its stated tolerance range.

Testing potentiometers

As with a resistor, you can test a *potentiometer,* or *pot* — which is a variable resistor — using the ohms setting on your multimeter. (For more about pots, refer to Chapter 5.)

Here's how to do the test:

1. **Turn off the power before you touch your circuit, and then remove the potentiometer.**
2. **Set your multimeter to read ohms.**

   If you don't have an autoranging meter, start at a high range and dial down the range as needed.
3. **Position the test leads on two of the potentiometer's leads.**

   Depending on where you put the leads, you can expect one of these results:

   - With the meter leads applied to one fixed end (point 1) and the *wiper,* or variable lead (point 2), as shown in Figure 16-10, turning the dial shaft in one direction increases the resistance, and turning the dial shaft in the other direction decreases the resistance.
   - With the meter leads applied to the wiper (point 2) and the other fixed end (point 3), the opposite resistance variation happens.
   - If you connect the meter leads to both fixed ends (points 1 and 3), the reading that you get should be the maximum resistance of the pot, no matter how you turn the dial shaft.

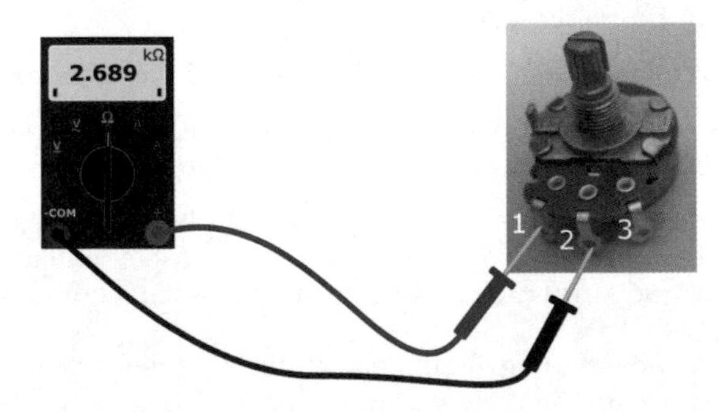

Figure 16-10: Connect the test leads to the first and center, center and third, and first and third terminals of the pot.

As you turn the shaft of the potentiometer, take note of any sudden changes in resistance, which may indicate a fault inside the pot. Should you find such a fault, replace the pot with a new one.

Testing capacitors

You use a *capacitor* to store electrical energy for a short period of time. (For coverage of capacitors, see Chapter 7.) If your multimeter doesn't have a capacitor-testing feature, you can still use the multimeter in the ohmmeter setting to help you decide whether to replace a capacitor.

Here's how to test a capacitor:

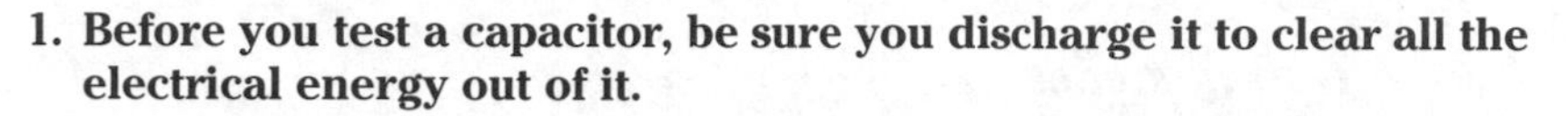

1. **Before you test a capacitor, be sure you discharge it to clear all the electrical energy out of it.**

   Large capacitors can retain a charge for long periods of time — even after you remove power.

   To discharge a capacitor, you short out its terminals through an insulated *bleeder jumper* (as shown in Figure 16-11), which is simply a wire with a large (1 MΩ or 2 MΩ) resistor attached. The resistor prevents the capacitor from shorting out, which would make it unusable.

2. **Remove the bleeder jumper.**

3. **Dial your multimeter to ohms and touch the test leads to the capacitor leads.**

   Unpolarized capacitors don't care which way you connect the leads, but if you're testing a polarized capacitor, connect the black lead to the negative terminal of the capacitor, and the red lead to the positive terminal. (Chapter 7 explains how to determine capacitor polarity.)

4. **Wait a second or two, and then note the reading.**

   You'll get one of these results:

   - A good capacitor shows a reading of infinity when you perform this step.
   - A zero reading may mean that the capacitor is shorted out.
   - A reading of between zero and infinity could be indicative of a leaky capacitor, one that is losing its capability to hold a charge.

This test doesn't tell you the capacitance value or whether the capacitor is open, which can happen if the component becomes structurally damaged inside or if its *dielectric* (insulating material) dries out or leaks. An open capacitor reads infinite ohms, just like a good capacitor. For a conclusive test, use a multimeter with a capacitor-testing function.

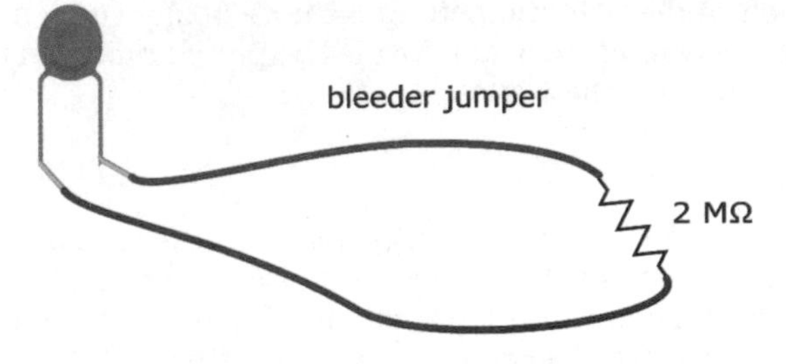

Figure 16-11: Purchase or make a bleeder jumper, used for draining excess charge from a capacitor.

Testing diodes

A *diode* is a semiconductor component that acts like a one-way value for current. (To get the details on diodes, see Chapter 9.) If your multimeter doesn't have a diode-check setting, you can use the ohms setting to test most types of diodes.

To test a diode, follow these steps:

1. **Set your meter to a low-value resistance range.**
2. **Connect the black lead to the diode's cathode (negative side, with a stripe) and the red lead to its anode (positive side).**

   The multimeter should display a low resistance.
3. **Reverse the leads, and you should get an infinite resistance reading.**

If you're not sure which end is up in a diode you've got on hand, you can use your multimeter to identify the anode and the cathode. Run resistance tests with leads connected one way, and then with the leads connected the other way. For the lower of the two resistance readings, the red lead is connected to the anode and the black lead is connected to the cathode.

Testing transistors

A *bipolar transistor* is essentially two diodes in one package, as illustrated in Figure 16-12. (For a PNP transistor, both diodes are reversed.) If your multimeter has neither a transistor-checking feature nor a diode-checking feature, you can use the ohms setting to test most bipolar transistors, in much the same way you test diodes: You set your meter to a low-value resistance range, and test each diode within your transistor in turn.

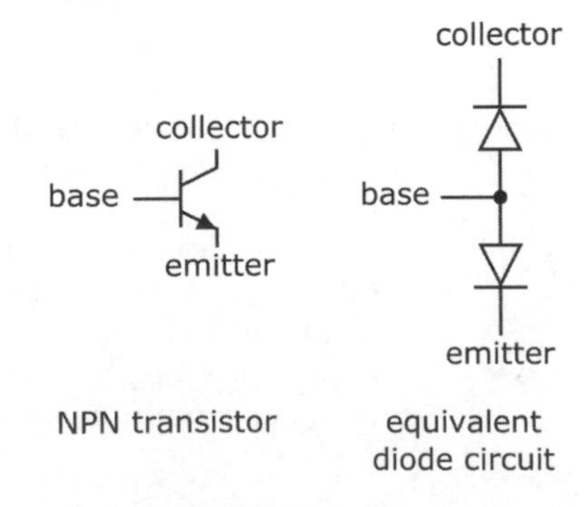

Figure 16-12: A bipolar transistor is like two diodes in one package.

Use the following test *only* with bipolar transistors. Testing with a multimeter can permanently damage some types of transistors, especially field-effect transistors (FETs). If you're not sure what type of transistor you have, look it up in a datasheet before you test it. You can often find the datasheet by searching the Internet for the component-identification number; for example, you might search for *2N3906 datasheet).*

If you're testing an NPN transistor (such as the one shown in Figure 16-12), follow these steps:

1. **Set your meter to a low-value resistance range.**
2. **Connect the black lead to the collector of the transistor and the red lead to the base.**

   The multimeter should display a low resistance.
3. **Reverse the leads.**

   You should get an infinite resistance reading.
4. **Connect the black lead to the emitter and the red lead to the base.**

   The meter should display a low resistance.
5. **Reverse the leads.**

   The meter should display infinite resistance.

For a PNP transistor, the readings should be the opposite of what they are for an NPN transistor.

Testing wires and cables

You can use your multimeter as an ohmmeter to run continuity tests on wires and cables. You may want to do this so you can sniff out breaks inside wires and *short circuits,* or unintended continuity, between two wires in a cable.

Even wire resists the flow of electrons

Why don't you *always* get 0 Ω when you test a wire, especially a long wire? All electrical circuits exhibit some resistance to the flow of current; the ohms measurement tests this resistance. Even short lengths of wire have resistance, but it's usually well below 1 Ω, so it's not an important test subject for continuity or shorts. However, the longer the wire, the greater the resistance — especially if the wire has a small diameter. Usually, the larger the wire, the lower its resistance per foot. Even though the meter doesn't read exactly 0 Ω, you can assume proper continuity in this instance if you get a low ohms reading.

To test for continuity in a single wire, connect the multimeter test leads to either end of the wire and dial up a low-range ohms setting. You should get a reading of 0 Ω or a very low number of ohms. A reading of more than just a few ohms indicates a possible break in the wire, causing an *open circuit.*

To test for a short between different wires that shouldn't be electrically connected, you set the meter to measure ohms, and then connect one of the test leads to an exposed end of one wire and the other test lead to an exposed end of the other wire. If you get a reading of 0 Ω or a low number of ohms, you may have a short circuit between the wires. A higher reading usually means your wires are not shorted together. (Note that you may get a reading other than infinite ohms if the wires are still connected to your circuit when you make the measurement. Rest assured that your wires are not shorted unless your reading is very low or zero.)

Testing switches

Mechanical switches can get dirty and worn, or sometimes even break, making them unreliable or completely unable to pass electrical current. Chapter 4 describes four common types of switches: single-pole, single-throw (SPST); single-pole, double-throw (SPDT); double-pole, single-throw (DPST); and double-pole, double-throw (DPDT). Depending on the switch, there may be zero, one, or two off positions, and there may be one or two on positions.

You can use your multimeter set to ohms to test any of these switches. Be sure to familiarize yourself with the on/off position(s) and the terminal connections of the particular switch you're testing — and run tests for each possibility. With your test leads connected across the terminals of any input/output combination placed in the off position, the meter should read infinite ohms; the on position should give you a reading of 0 Ω.

You can most easily test switches by taking them out of a circuit. If the switch is still wired into a circuit, the meter may not show infinite ohms when you place the switch in the off position. If, instead, you get a reading of some value other than 0 Ω, you can assume the switch is operating properly as an open circuit when it's in the off position.

Testing fuses

Fuses are designed to protect electronic circuitry from damage caused by excessive current flow and, more importantly, to prevent a fire if a circuit overheats. A blown fuse is an open circuit that's no longer providing protection, so it has to be replaced. Before you test a fuse, make sure you either safely remove it from the circuit or power down the circuit. Then, set your multimeter to the ohms setting and touch one test lead to either end of the fuse. If the meter reads infinite ohms, it means you have a burned-out fuse.

Running other multimeter tests

Many digital multimeters include extra functions that test specific components, such as capacitors, diodes, and transistors. These tests provide more definitive results than the resistance measurements I discuss earlier in this section.

If your multimeter has a capacitor-testing feature, it will display the value of the capacitor. This can come in handy because not all capacitors follow the industry standard identification scheme. Refer to your multimeter manual for the exact procedure because the specifics vary from model to model. Be sure to observe the proper polarity when connecting the capacitor to the test points on the meter.

If your multimeter has a diode-checking feature, you can test a diode by attaching the red test lead to the anode (positive terminal) of the diode, and the black lead to the cathode (negative terminal). You should get a fairly low, but not zero, reading (for instance, 0.5). Then reverse the leads and you should get an overrange reading. If you get two zero readings or two overrange readings, chances are your diode is bad (that is, shorted out or open).

You can use the diode-checking feature to test bipolar junction transistors, treating them as two individual diodes (refer to Figure 16-12).

If your multimeter has a transistor-checking feature, follow the procedure outlined in the manual, which varies from one model to another.

Using a Multimeter to Check Your Circuits

One of the top benefits of a multimeter is that it can help you analyze the rights and wrongs of your circuits. By using the various test settings, you can verify the viability of individual components and confirm that voltages and currents are what they should be. Sooner or later, you hook up a circuit that doesn't work right away — but your multimeter can help you sniff out the problem if you can't resolve the problem by physically checking all your connections.

To troubleshoot your circuit, you should first mark up your circuit diagram with component values, estimated voltage levels at various points in the circuit, and expected current levels in each branch of the circuit. (Often the process of marking up the diagram uncovers a math error or two.) Then use your multimeter to probe around.

Here's a quick list of items to check as you troubleshoot your circuit:

- Power supply voltages
- Individual component functionality and actual values (out of the circuit)
- Continuity of wiring
- Voltage levels at various points in the circuit
- Current levels through part of the circuit (without exceeding the current capabilities of your multimeter)

Using a step-by-step procedure, you can test various components and parts of your circuit and narrow the list of suspects until you either uncover the cause of your circuit problem or admit you need professional help — from your friendly neighborhood electronics guru.

17

Putting Projects Together

In This Chapter

- Creating unique light blinkers and flashers
- Rigging an alarm
- Constructing an adaptable siren
- Sounding off with your own amplifier
- Designing a traffic signal
- Making beautiful music

Getting up to speed on electronics really pays off when you get to the point where you can actually build a project or two. In this chapter, you get to play with several fun, entertaining, and educational electronics gadgets that you can build in half an hour or less. I selected the projects for their high cool factor and their simplicity. I've kept parts to a minimum, and the most expensive project costs under $15 or so to build.

I've given you some detailed procedures for the first project, so work through that one first. Then, you should be able to follow the circuit schematics and build the rest of the projects on your own. Check Chapter 11 if you need a refresher on schematics, and browse through Chapter 3 if you'd like to review basic circuit concepts. And if the projects don't seem to work as advertised (it happens to the best of us), review Chapter 16, arm yourself with a multimeter, and start troubleshooting!

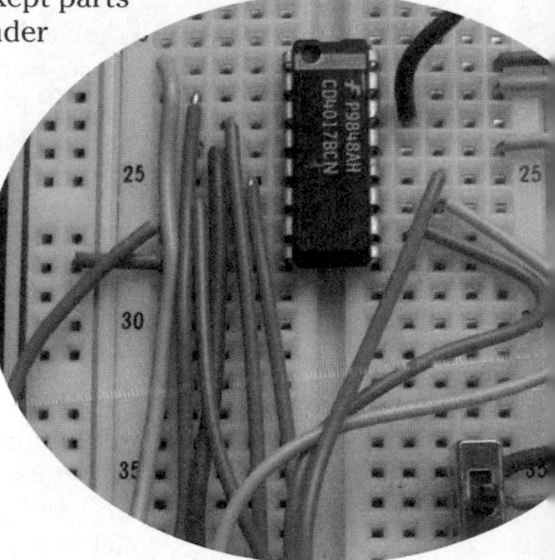

Getting What You Need Right Off the Bat

You can build all the projects in this chapter on a solderless breadboard. Of course, feel free to build any of the projects on a regular soldered circuit board, if you want to keep them around. There's more detail about breadboarding and building circuits in Chapter 15. If you get stuck on any of these projects, hop to that chapter to help you through.

You can find all the parts that you need to construct the projects in this chapter at any electronics store or online electronics retailer. If you don't have a well-stocked electronics outlet near you, check out Chapter 19 for some mail-order electronics parts suppliers.

Unless otherwise noted, use these guidelines when selecting components:

- All resistors are rated for 1/4 W or 1/8 W and 5% or 10% tolerance. I include the color code for each resistor value in the parts list for each project.
- All capacitors are rated at a minimum of 25 V. I note the type of capacitor that you need (for instance, disc or electrolytic) in the parts list for each project.

If you want to understand the ins and outs of one or more of the electronic components you use in these projects, review the material in Chapters 4–7 and 9–12. You'll find information on switches in Chapter 4, details about resistors and Ohm's Law in Chapters 5 and 6 respectively, and a treatise on capacitors in Chapter 7. Chapter 9 explains diodes, Chapter 10 discusses transistors, and two of the integrated circuits (ICs) used in these projects are covered in Chapter 11. Wires, power sources, and other parts (for instance, sensors, speakers, and buzzers) are discussed in Chapter 12.

You can learn a lot about a particular IC by looking up its datasheet on the Internet or doing an Internet search for application notes about the part. These information sources tell you much more than just the pinout and power requirements for the ICs; they often provide sample circuits and tips on getting the best performance from the chip. Projects in this chapter use the 555 timer chip, the 4017 decade counter IC, and the LM386 audio amplifier chip.

Creating an LED Flasher Circuit

Your first mission — should you choose to accept it — is to build a circuit containing a single light-emitting diode (LED) that blinks on and off at a rate that you can vary. This may sound simple (and, thanks to the 555 timer IC, it is), but getting the LED to blink means you must successfully build a complete circuit, limit the current in your circuit so that it doesn't fry your LED, and set up a timer to switch the current on and off so that the light blinks. After you accomplish your mission, you modify the circuit to create a multi-LED flasher that you can mount to the back of your bike to alert motorists to your presence.

If you've already built some circuits on a solderless breadboard (say, by following the instructions in previous chapters), you may be somewhat of a pro by now. You may not need the detailed circuit-building instructions that follow in this section, but it's still a good idea to read along as I explain how to select components for this first circuit.

Exploring a 555 flasher

You can see the schematic of the single LED flasher in Figure 17-1. (If you need a quick refresher course on reading schematics, flip to Chapter 14.) Here are the parts you need to build this circuit:

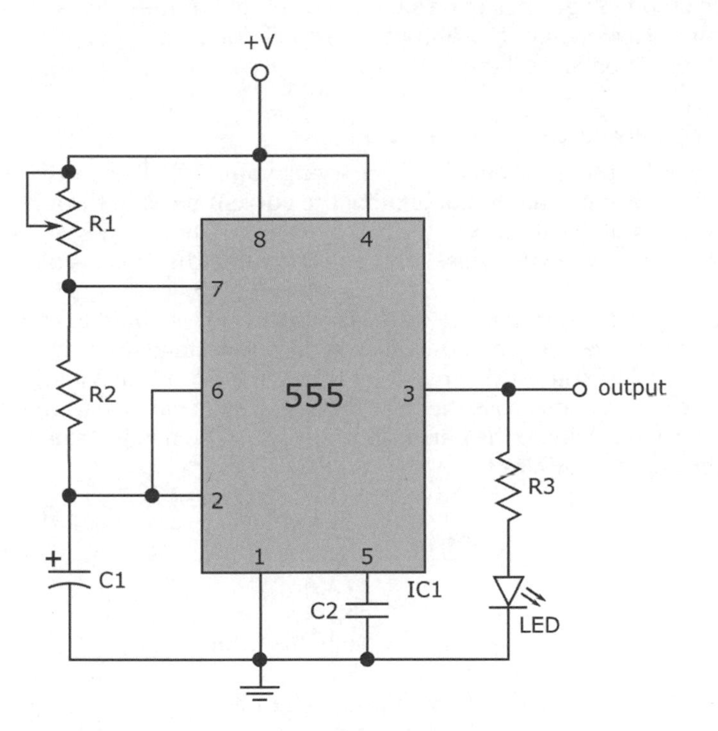

Figure 17-1: Schematic of the single LED flasher circuit.

- 9-volt battery (with battery clip)
- IC1: LM555 timer IC
- R1: 1 MΩ potentiometer
- R2: 47 kΩ resistor (yellow-violet-orange)
- R3: 330 Ω resistor (orange-orange-brown)

- C1: 4.7 μF electrolytic (polarized) capacitor
- C2: 0.01 μF disc (nonpolarized) capacitor
- LED: Light-emitting diode (any size and any color)

Before you build the LED flasher, you might want to do a quick analysis to understand exactly how it works.

The cornerstone of the LED flasher (as well as other projects in this chapter) is the 555 timer IC. You can use this versatile part in a variety of ways, as explained in Chapter 11. For this project, the 555 timer is configured as an *astable multivibrator* (which is a fancy way of saying it's an oscillator) generating an ongoing series of on/off pulses at regular intervals, sort of like an electronic metronome. The output of the 555 timer IC, at pin 3, is what you use to switch on and off the LED current.

Limiting current through the LED

Resistor *R3* is there to keep you from frying your LED. This lowly resistor performs the important job of limiting the current passing through the LED. The output voltage at pin 3 of the 555 timer varies between 9 V (the positive power supply) when the pulse is on and 0 V when the pulse is off.

Assuming the forward voltage drop across the LED is about 2.0 V (a typical value), you know that when the pulse is on, the voltage drop across resistor *R3* is about 7 V. You get this result by taking the 9 V at pin 3 and subtracting the 2 V dropped across the LED. From that, you can use Ohm's Law (see Chapter 6) to calculate the current through *R3*, which is the same as the current through the LED, as follows:

$$\begin{aligned} current &= \frac{voltage}{resistance} \\ &= \frac{7\ \text{V}}{330\ \Omega} \\ &\approx 0.021\ \text{A} = 21\ \text{mA} \end{aligned}$$

Now, that's a current your LED can safely handle!

Controlling the timing of the pulse

Resistors *R1* and *R2* and capacitor *C1* control both the width and the on/off timing interval of the pulse generated by the 555 timer IC (see Chapter 11 for details). This project uses a potentiometer to vary *R1* so you can change the rate of the blinking light from slow waltz to fast samba.

The *time period,* T, is the total time it takes for one up-and-down pulse:

$$\text{T} = 0.693 \times (R1 + 2R2) \times C1$$

To figure out the range of time periods that the 555 timer generates, first use 47,000 for *R2* (47 kΩ) and 0.0000047 for *C1* (4.7 μF) in the equation for T. Then calculate the low end of the timing interval by using 0 for *R1* in the equation, and calculate the high end of the timing interval by using 1,000,000 for *R1* in the equation. You should expect the timing interval to range from roughly 0.3 seconds to 3.6 seconds as you dial the pot from 0 Ω to 1 MΩ.

Building the LED flasher circuit

To see whether the light blinks at about the rate your calculations say it should, build the LED flasher circuit and try it out! Use Figure 17-2 as your guide. If this is the first circuit you're building, you may want to follow the detailed instructions in this section.

You can make it easier to use battery power when you build a circuit on a breadboard using just a switch and a few wires, as shown in Figure 17-2. Use a single-pole, double-throw (SPDT) switch to connect the positive battery terminal to the top positive supply rail (in Chapter 4, I explain this process in detail). Then connect the positive supply rails on top and bottom using a red wire, and the negative supply rails on top and bottom using a black wire, so that power is available on both the top and the bottom of the board.

Figure 17-2: An LED flasher with parts mounted on a solderless board. (555 timer IC pin labels have been added.)

Here are the steps to build the LED flasher circuit:

1. **Collect all the components you need for the project.**

   See the parts list in the "Exploring a 555 flasher" section for a rundown of what you need. Nothing is worse than starting a project, only to have to stop halfway through because you don't have everything at hand!

2. **Carefully insert the 555 timer chip into the middle of the board.**

   It's common practice to insert an IC so that it straddles the empty middle row of the breadboard and the *clocking notch* (that little indentation or dimple on one end of the chip) faces the left of the board.

3. **Insert the two fixed resistors, *R2* and *R3*, into the board, following the schematic and the sample breadboard in Figure 17-2.**

4. **Insert the two capacitors, *C1* and *C2*, into the board, following the schematic and the sample breadboard in Figure 17-2.**

   As noted in Chapter 11, the pins on IC chips are numbered counterclockwise, starting at the clocking notch. If you've placed the 555 timer IC with the clocking notch facing the left side of the board, the pin connections are as shown in Figure 17-2.

   Make sure you orient the polarized capacitor properly, with its negative side connected to ground.

5. **Solder wires to the potentiometer (*R1*) to connect it into the breadboard.**

   Use 22-gauge solid-strand hookup wire. The color doesn't matter. Note that the potentiometer has three connections to it. One connection (at either end) goes to pin 7 of the 555; the other two connections (at the other end and the center) are joined, or bridged, and attached to the positive side of the power supply.

6. **Connect the LED as shown in the schematic and the photo.**

   Observe proper orientation when inserting the LED: Connect the cathode (negative side, with the shorter lead) of the LED to ground. Check the packaging that came with your LED to make sure you get it right. (If you don't, and you insert the LED backwards, nothing bad will happen, but the LED won't light. Simply remove the LED and reinsert it, the other way around.)

7. **Use 22-gauge single-strand wire, preferably already precut and trimmed for use with a solderless breadboard, to finish making the connections.**

   Use the sample breadboard shown in Figure 17-2 as a guide to making these jumper-wire connections.

8. ***Before* applying power, double-check your work. Verify all the proper connections by cross-checking your wiring against the schematic.**
9. **Finally, attach the 9 V battery to the positive supply and ground rails of the breadboard.**

   It's easier if you use a 9 V battery clip, which contains prestripped leads. You may want to solder 22-gauge solid hookup wire to the ends of the leads from the clip; this makes it easier to insert the wires into the solderless breadboard. Remember: The red lead from the battery clip is the positive terminal of the battery, and the black lead is the negative terminal, or ground.

Checking your handiwork

When you apply power to the circuit, the LED should flash. Rotate the *R1* knob to change the speed of the flashing. Does the LED blink at the rate you expect it to? If your circuit doesn't work, disconnect the 9 V battery and check the connections again.

Here are some common mistakes to look for:

- **555 IC inserted backward:** This mistake can damage the chip, so if this happens, you might want to try another 555.
- **LED inserted backward:** Pull it out and reverse the leads.
- **Connection wires and component leads not pressed firmly enough into the breadboard sockets:** Be sure that each wire fits snugly into the breadboard, so there are no loose connections.
- **Wrong component values:** Double-check, just in case!
- **Dead battery:** Try a new one.
- **Circuit wired wrong:** Have a friend take a look. Fresh eyes can catch mistakes that you might not notice.

You can use your multimeter to test voltages, currents, and resistances in your circuit. As described in Chapter 16, such tests can help you identify the cause of circuit problems. Your multimeter can tell you whether your battery has enough juice, whether your diode is still a diode, and much, much more.

If you're building a circuit that's new to you, it's good electronics practice to build it on a solderless breadboard first. That's because you often need to tweak a circuit to get it working right. When you have it working to your satisfaction, you can make the circuit permanent if you like. Just take your time — and remember to double- (and even triple-) check your work. Don't worry — you'll be an old hand in no time.

Creating an LED Bike Flasher

You can expand the simple LED flasher circuit to create an inexpensive multi-LED flasher that you can use to increase your safety when you go for out for a ride on your bike or a run in the park. Or you can just wear it on your shirt to impress your friends.

Look at the circuit in Figure 17-3. Other than the additional LEDs at the output of the 555 timer IC and the use of a fixed resistor instead of a potentiometer for *R1,* this circuit seems identical to the simple LED flasher circuit from the previous section. And it is. Well, except for the values of *R1, R2,* and *C1,* which are the components that determine the pulse rate that controls the blinking of the LEDs.

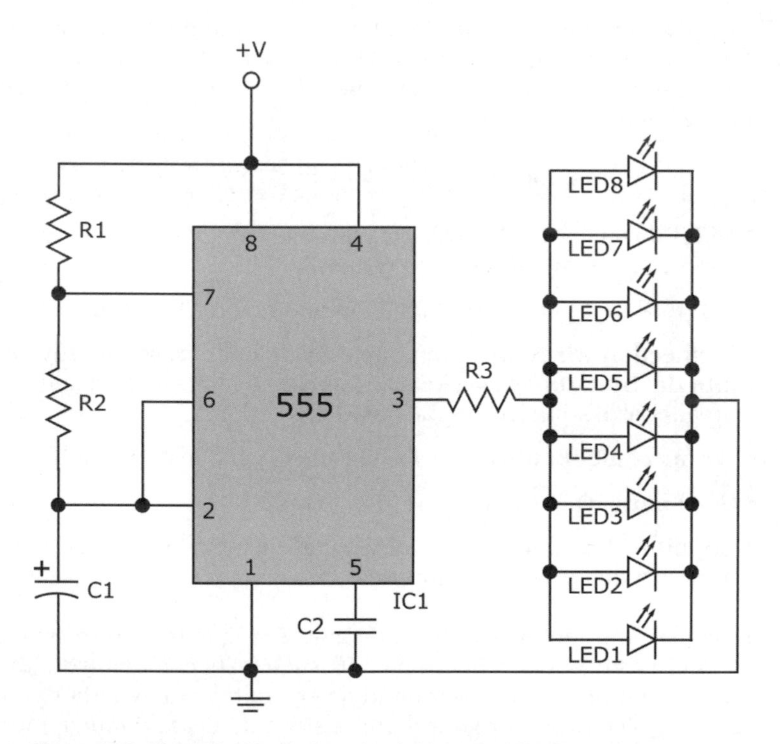

Figure 17-3: The LED bike flasher circuit. Values of *R1, R2,* and *C1* are selected to create a rapid pulse train that controls the LED flashing action.

For a bike flasher, you want the LEDs to flash at a rapid clip, but not so fast that you can't distinguish one blink from another. The values shown next for *R1, R2,* and *C1* generate a timing interval of roughly two pulses per second (2 Hz). I also suggest that you use ultrabright LEDs, which are similar to standard LEDs except that they have clear plastic cases so that the light appears to be brighter.

Here is the parts list for the LED bike flasher:

- 9-volt battery (with battery clip)
- IC1: LM555 timer IC
- R1, R2: 1 kΩ resistor (brown-black-red)
- R3: 330 Ω resistor (orange-orange-brown)
- C1: 220 μF electrolytic (polarized) capacitor
- C2: 0.01 μF disc (nonpolarized) capacitor
- LED1-8: Ultrabright light-emitting diode (5 mm, any color)

If you'd like to change the flash rate, try using different values of *R1* (or *R2* or *C1*). For instance, using 220 Ω resistors (red-red-brown) for both *R1* and *R2* produces a flash rate of about 10 pulses per second (10 Hz). And remember to add an on/off switch for the battery if you make this circuit permanent.

Catching Intruders with a Light-Sensing Alarm

Figure 17-4 shows you a schematic of a light-sensing alarm. The idea of this project is simple: If a light comes on, the alarm goes off.

You build the alarm around a 555 timer chip, which acts as a tone generator. The 555 timer is configured (once again) as an oscillator, and the values of *R3, R4,* and *C1* are selected to create an output pulse train (on pin 3) at a frequency in the audible range (20 Hz to 20 kHz).

Note that the reset pin (pin 4) of the 555 timer is not tied to the positive power supply, as it has been in 555 oscillator circuits described previously in this chapter. This fact is significant because if the reset pin is tied high, the 555 timer will just keep on oscillating (and the speaker will keep producing a tone) as long as power is applied. If a low voltage is applied to the reset pin, however, the 555 timer's internal timing circuit is reset, the output (pin 3) goes low, and the speaker is quiet.

So, to make the 555 timer sound the alarm only when light is present, you need to create a light-sensitive switch and use it to control the reset pin on the 555 timer. The left side of the circuit in Figure 17-4 provides the light-sensitive switch in the form of a photoresistor-transistor combination.

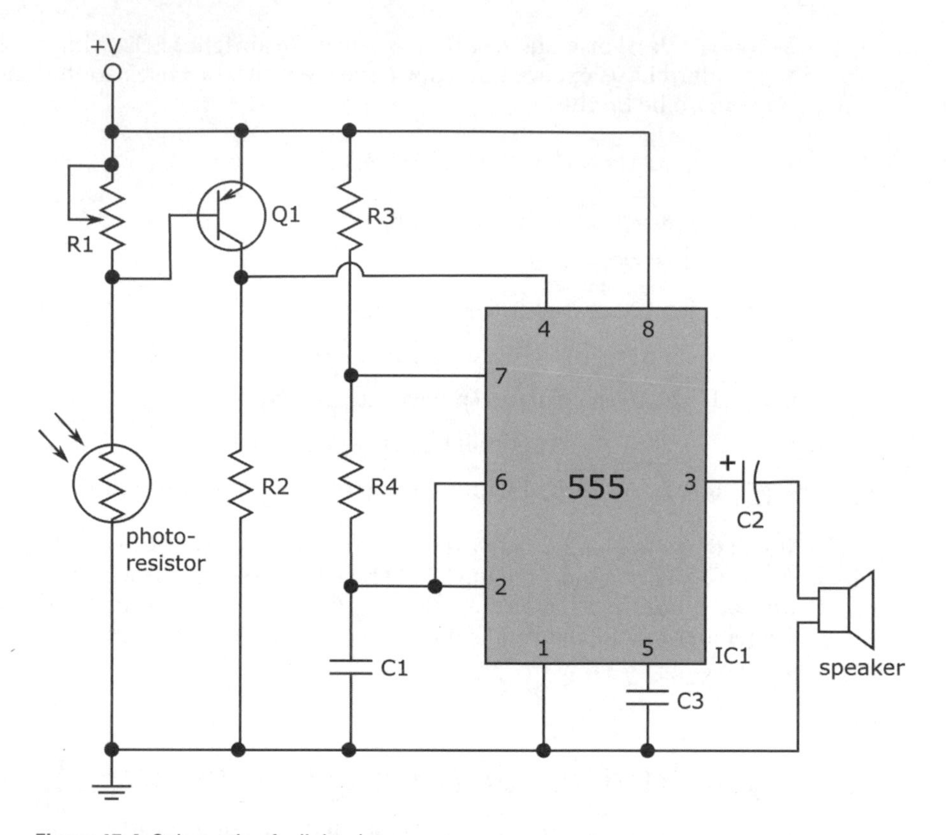

Figure 17-4: Schematic of a light alarm.

Transistor *Q1* plays the role of the switch, sometimes conducting current and sometimes not conducting current. (You find out what controls *Q1* soon.) Transistor *Q1* controls the 555 timer reset pin as follows:

- **When the transistor is not conducting current, the voltage on pin 4 (reset) of the 555 timer goes low.**

  If the transistor is not conducting current, no current is flowing through resistor *R2*, so the voltage drop across *R2* is 0, and the voltage where the collector of transistor *Q1* (that is, the terminal on the lower right side of the transistor in Figure 17-4) meets pin 4 of the 555 timer is 0.

- **When the transistor is conducting current, the voltage at pin 4 (reset) of the 555 timer goes high.**

  In Chapter 10, you see that when a transistor is fully conducting, the voltage drop from the collector to the emitter is nearly 0, so in this circuit, the voltage at the collector is nearly equal to the 9 V power supply voltage.

Whether the transistor is on (conducting) or off (nonconducting) depends on what's happening at the base (left terminal) of the transistor. The base voltage is controlled by a voltage divider (see Chapter 6 for details), which consists of the potentiometer (*R1*) and the photoresistor. To turn *Q1* on, the voltage at its base must go low (see Chapter 11 for details of PNP transistor operation). With little or no light present, the resistance of the photoresistor is very high, so the voltage at the base of transistor *Q1* is high and *Q1* is off. When light hits the photoresistor, its resistance decreases, the voltage at the base of *Q1* goes low, and the transistor is on.

The bottom line is that the light alarm has two possible states:

- **Darkness:** The resistance of the photoresistor is very high, so the base voltage of transistor *Q1* is high, causing *Q1* to turn off. With no current flowing through *R2*, reset pin 4 of the 555 timer is low. As a result, the 555 timer is not oscillating (no alarm).
- **Light:** The resistance of the photoresistor is low, so the base voltage of transistor *Q1* is low, causing *Q1* to turn on and conduct current through *R2*, which then raises the voltage on reset pin 4 of the 555 timer. As a result, the 555 timer oscillates and the alarm sounds.

You can adjust the sensitivity of the alarm by turning the pot *(R1)*, which simply alters the voltage divider ratio, so more (or less) light is needed to turn transistor *Q1* on. Decide for yourself whether you want to sense a change from total darkness to low light or from low light to bright light.

Assembling the light alarm parts list

Here's the shopping list for the light alarm project:

- 9-volt battery (with battery clip)
- IC1: LM555 timer IC
- Q1: 2N3906 PNP transistor
- R1: 100 kΩ potentiometer
- R2: 3.9 kΩ resistor (orange-white-red)
- R3: 10 kΩ resistor (brown-black-orange)
- R4: 47 kΩ resistor (yellow-violet-orange)
- C1, C3: 0.01 μF disc (nonpolarized) capacitor
- C2: 4.7 μF electrolytic or tantalum (polarized) capacitor
- Speaker: 8 Ω, 0.5 W speaker
- Photoresistor: Experiment with different sizes; for example, a larger photoresistor will make the circuit a little more sensitive

Making your alarm work for you

You can apply this light alarm in several practical ways. Here are a few ideas:

- Put the light alarm inside a pantry so it goes off whenever someone raids the chocolate chip cookies. Keep your significant other out of your stash — or keep yourself on that diet! When the pantry door opens, light comes in and the alarm goes off.
- Do you have a complex electronics project in progress in the garage that you don't want anybody to disturb? Place the alarm inside the garage, near the door. If someone opens the garage door during the day, light comes through and the alarm sounds.
- Build your own electronic rooster that wakes you up at daybreak. (Who needs an alarm clock?)
- Create a room-in-use alert system — but don't use a noisy speaker. Instead, replace *C2* and the speaker with a 330 Ω resistor and an LED. The LED will light up when the photoresistor detects light. Wire the circuit so that most of the parts are in a box in the room and the LED is mounted outside the door to the room.

Playing the C-Major Scale

Figure 17-5 shows a schematic for a primitive electronic keyboard. The circuit may look complicated, but it really is fairly simple if you understand how the 555 timer IC operates as an oscillator.

Here are the parts you need to build the C-major scale circuit:

- 9-volt battery (with battery clip)
- IC1: LM555 timer IC
- R1, R7: 2.2 kΩ resistor (red-red-red)
- R2: 10 kΩ resistor (brown-black-orange)
- R3: 10 kΩ potentiometer
- R4: 820 Ω resistor (grey-red-brown)
- R5, R6: 1.8 kΩ resistor (brown-grey-red)
- R8: 1.2 kΩ resistor (brown-red-red)
- R9: 2.7 kΩ resistor (red-violet-red)
- R10: 3 kΩ resistor (orange-black-red)

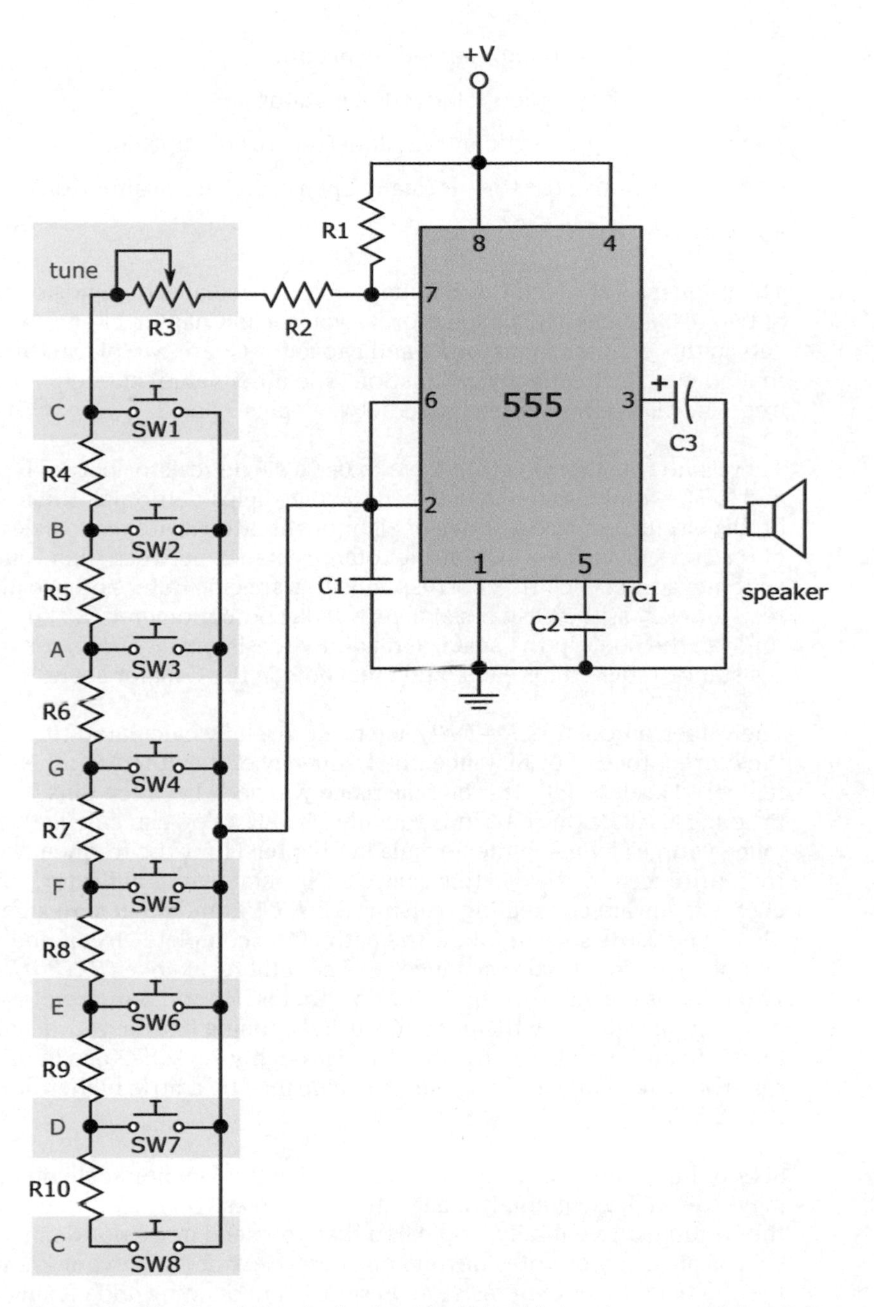

Figure 17-5: The C-major scale circuit, with labels added to show you which switch controls each note and how to tune the scale.

- C1: 0.1 μF disc (nonpolarized) capacitor
- C2: 0.01 μF disc (nonpolarized) capacitor
- C3: 4.7 μF electrolytic or tantalum (polarized) capacitor
- SW1-8: momentary on, normally open switches (pushbutton)
- Speaker: 8 Ω, 0.5 W speaker

The frequency at which the 555 timer output oscillates depends on the values of two resistances and a capacitor, as you see in Chapter 11 and in other projects in this chapter. Resistor *R1* and capacitor *C1* are two of the three values that go into the frequency calculation. The other value that helps determine frequency is the resistance found between pins 7 and 2.

There's no rule that says you have to use a single resistor between pins 7 and 2. The total resistance between the pins helps determine the frequency. In this circuit, you use a series of eight pushbutton switches to select a series of resistors in such a way that the total resistance between pins 7 and 2 helps generate a frequency that corresponds to a specific note. You use a 10 kΩ resistor *(R2)* as the base resistance, a 10 kΩ potentiometer *(R3)* to tweak, or tune, all the notes in the scale, and additive resistors *(R4–R10)* for the total resistance required for each individual note in the C-major scale.

The values of resistors *R4–R10* have been carefully calculated to produce the correct tones. For instance, the frequency of the note A on the equal-tempered scale is 440 Hz. The resistance you need between pins 7 and 2 to produce a 440 Hz pulse train is roughly 15.1 kΩ. (You can calculate this resistance yourself by using the formula in Chapter 11 for the frequency of a pulse train produced by the 555 timer used as an astable multivibrator.) By pressing SW3, you are connecting resistors *R2, R3, R4,* and *R5* in series between pins 7 and 2. (Be sure to follow the path of the complete circuit and see for yourself what the total resistance is.) The total resistance *(R2+R3+R4+R5)* is 12.6 kΩ plus the value of the 10 kΩ pot (that is, *R3*). If your circuit is tuned properly (by adjusting the pot while using a tuning fork or your accurately tuned piano, if you like), the pot value is roughly 2.5 kΩ. (Keep in mind that resistor values can vary, so your pot value may be a little higher or lower than 2.5 kΩ.)

Set up the circuit and try it out! You can play the C-major scale on it, and maybe even the beginning of a few tunes, such as "Do Re Mi" and "America the Beautiful." Eventually, you'll find that you need more notes, such as sharps and flats, or notes beyond one octave. Armed with your knowledge of the 555 timer IC, adding resistors in series, and opening and closing circuits with switches, you can build out this C-major circuit to create more interesting tunes.

Scaring Off the Bad Guys with a Siren

Unless you carry a badge (a real one, not the one in your toy box), you can't arrest any bad guys when you set off the warbling siren that you build in this project, shown in Figure 17-6. But the siren sounds cool, and you can use it as an alarm to notify you if somebody's getting at your secret stash of baseball cards, vintage Frank Sinatra records, a signed copy of *Mister Spock's Music from Outer Space* record, or whatever.

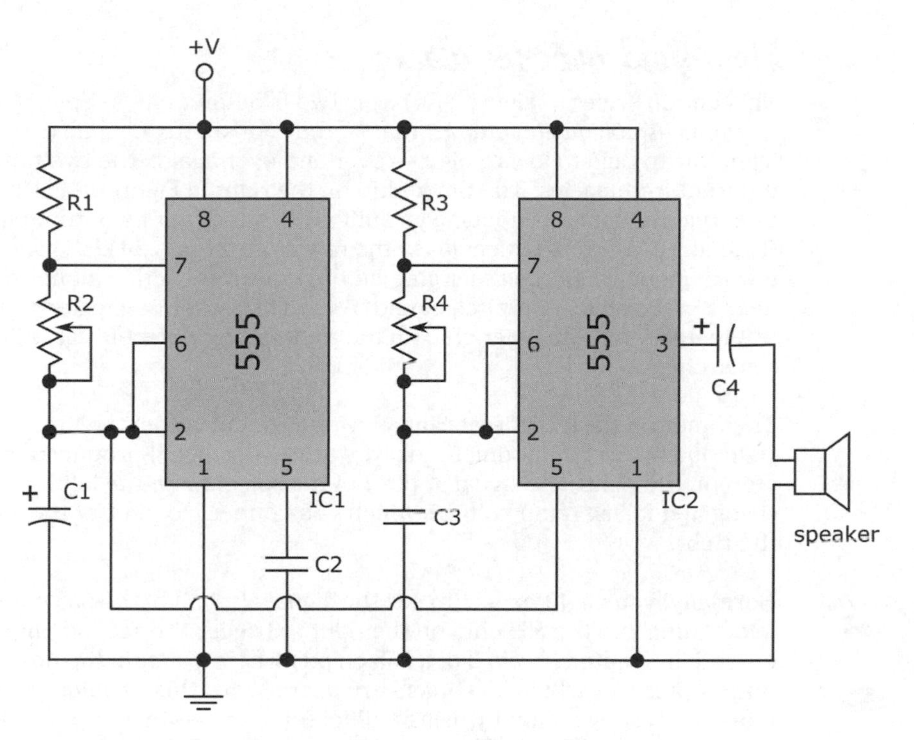

Figure 17-6: A police-type siren made from two 555 timer ICs.

Scoping out the 555 siren parts list

To start alarming your friends, gather these parts to build the circuit:

- 9-volt battery (with battery clip)
- IC1, IC2: LM555 timer IC
- R1, R3: 2.2 kΩ resistors (red-red-red)
- R2: 50 kΩ potentiometer

- R4: 100 kΩ potentiometer
- C1: 47 μF electrolytic (polarized) capacitor
- C2: 0.01 μF disc (nonpolarized) capacitor
- C3: 0.1 μF disc (nonpolarized) capacitor
- C4: 4.7 μF electrolytic or tantalum (polarized) capacitor
- Speaker: 8Ω, 0.5 W speaker

How your warbler works

This circuit (refer to Figure 17-6) uses two 555 timer chips. You rig both chips to act as *astable multivibrators;* that is, they constantly change their output from low to high to low to high — over and over again. The two timers run at different frequencies. The timer chip on the right in Figure 17-6 is configured as a *tone generator,* producing an audible frequency at its output pin, pin 3. (Humans can hear frequencies in the range of 20 Hz to 20 kHz, give or take a few frequencies.) If the timer chip on the right were acting alone, you would hear a steady, medium-pitch sound from the speaker connected to its output. But instead, the 555 timer chip on the right is acting in concert with the 555 timer chip on the left.

The timer on the left operates at a lower frequency than the timer on the right and is used to modulate (okay, warble) the signal produced by the timer chip on the right. The signal at pin 2 of the 555 chip on the left is a slowly rising and falling ramp voltage, which you connect to pin 5 of the 555 chip on the right.

Normally, you might expect to see the signal at pin 3 of the 555 chip on the left feeding into the 555 chip on the right to trigger the second chip. As discussed in Chapter 11, pin 3 of a 555 chip is where you find the up-and-down output pulse for which 555 timers are famous. For this warbler, you get a more interesting sound by using a different signal — that at pin 2 — to trigger the second 555 chip. The signal at pin 2 of the 555 chip on the left rises and falls slowly as capacitor *C1* charges and discharges. (Chapter 7 explains just how a capacitor charges and discharges; this rising and falling capacitor voltage triggers the up-and-down pulse waveform that the 555 timer outputs at pin 3, which you're not using.) By feeding this capacitor voltage (at pin 2 of the 555 chip on the left) into the control pin (pin 5) of the 555 chip on the right, you override the internal trigger circuitry of the second chip, using a varying trigger signal instead, which helps make your warbler warble.

By adjusting the two potentiometers, *R2* and *R4,* you change the pitch and speed of the siren. You can produce all sorts of siren and other weird sound effects by adjusting these two potentiometers. You can operate this circuit at any voltage between 5 V and about 15 V. To power the gadget, use an easy-to-find 9 V battery (included in the parts list in the preceding section).

Building an Audio Amp with Volume Control

Give your electronics projects a big mouth with a little amplifier designed around parts that are inexpensive and easy to find at most electronics suppliers, such as the LM386 power amplifier IC. This amp boosts the volume from microphones, tone generators, and many other signal sources.

Figure 17-7 shows the schematic for this project, which consists of just 10 parts and a battery. You can operate the amplifier at voltages between 5 V and about 15 V. A 9 V battery does the trick.

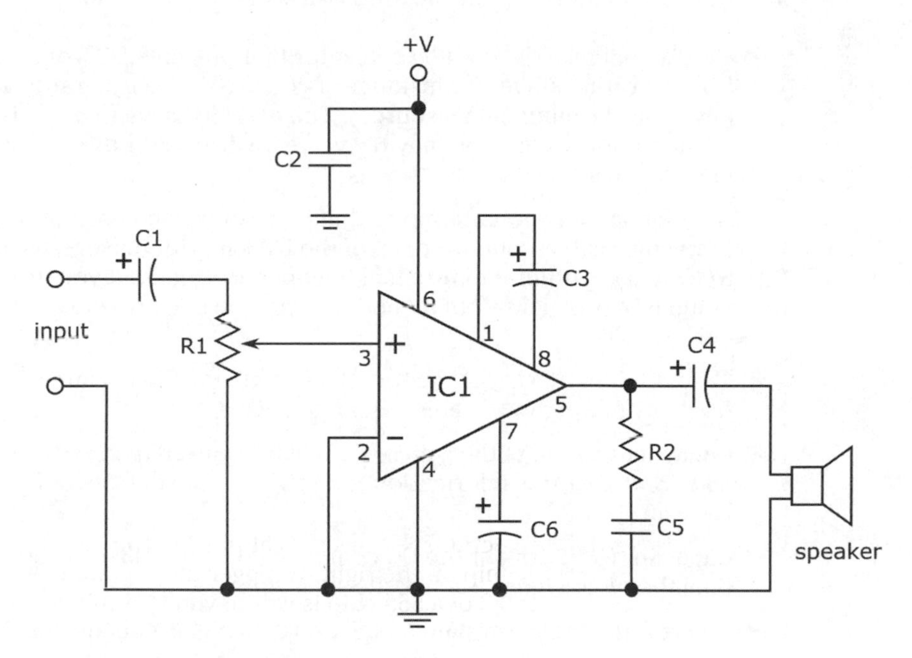

Figure 17-7: Schematic of the audio amplifier.

Here's a rundown of the parts you need for this project:

- 9-volt battery (with battery clip)
- IC1: LM386 power amplifier
- R1: 10 kΩ potentiometer (optional)
- R2: 10 Ω resistor (brown-black-orange)
- C1: 10 µF electrolytic (polarized) capacitor (optional)
- C2: 0.1 µF disc (nonpolarized) capacitor

- C3, C6: 10 µF electrolytic (polarized) capacitor
- C4: 220 µF electrolytic (polarized) capacitor
- C5: 0.047 µF disc (nonpolarized) capacitor
- Speaker: 8 Ω, 0.5 W speaker

Just connect a signal source (for instance, a RadioShack condenser microphone, part 270-092, which requires a DC power source) across the inputs, making sure to connect the ground of the signal source to the common ground of the amplifier circuit. The LM386 does most of the work for you in this little circuit. Here's what the other parts in this circuit do:

- *C1* is an optional decoupling capacitor that prevents DC from passing from an earlier stage (for instance, if you use a tone generator or other device as the input signal source). The parts list says to use a 10 µF capacitor for *C1*, but you may try values as low as 0.1 µF (or try eliminating *C1* and observe what happens).
- *R1* is an optional potentiometer that you can use to control the volume. (You connect the wiper to pin 3 of the LM386, with one end terminal to *C1* and the other end terminal to common ground. If you don't want volume control, leave out *R1* and connect the negative side of *C1* to pin 3 of the LM386.)
- *C2* and *C6* are bypass capacitors that isolate the LM386 internal circuitry from any power supply noise, hum, or spikes.
- Capacitor *C3* boosts the gain of the LM386 from 20 (without *C3*) to about 200. (Note that this information is straight off the datasheet for the LM386.)
- Capacitor *C4* filters out the DC component of the LM386 output so that only the audio signal reaches the speaker.
- The resistor-capacitor pair, *R2-C5*, prevents high-frequency oscillations.

This simple circuit puts out a whole lotta sound in a small and portable package. And the better the microphone and speaker, the better the sound!

Try connecting your portable music player to the input of this audio amp. Get an old set of earbuds and do the following: Cut off one of the earbuds, strip the insulation off the wires, and identify the signal and ground wires. Then connect signal and ground across the amplifier inputs, insert the earbud plug into your music player's jack, and enjoy the sound!

If you want to witness the effects of noise on a sensitive circuit, remove capacitors *C2* and *C6* (but do not replace them with wires or anything else) and try out the amplifier. You will probably hear lots of crackling.

Creating Light Chasers

If you were a fan of the *Knight Rider* television series that aired in the '80s, you remember the sequential light chaser that the KITT Car sported in front. In this section, I show you two versions of the light chaser, each of which uses just two inexpensive ICs and a handful of other parts. You may want to choose one circuit or the other to build. Light Chaser 1 is a little easier to understand. Light Chaser 2 adds a layer of complexity to increase the cool factor.

The parts lists for the two circuits are nearly the same. The main parts list, which includes labels that reference the schematics shown in Figures 17-8 and 17-9, for both circuits is as follows:

- 9-volt battery (with battery clip)
- IC1: LM555 timer IC
- IC2: 4017 CMOS decade counter IC
- R1: 1 MΩ potentiometer
- R2: 47 kΩ resistor (yellow-violet-orange)
- R3: 330 Ω resistor (orange-orange-brown)
- C1: 0.1 μF disc (nonpolarized) capacitor
- C2: 0.01 μF disc (nonpolarized) capacitor

In addition to the main parts list, Light Chaser 1 uses these parts:

- LED1–LED10: LED (any size and any color)

In addition to the main parts list, Light Chaser 2 uses these parts:

- LED1–LED6: LED (any size and any color)
- D1–D8: 1N4148 diode

Building Light Chaser 1

The schematic for Light Chaser 1 is shown in Figure 17-8. For this design, each of 10 LEDs will light up in succession (that is, following the pattern 1-2-3-4-5-6-7-8-9-10), and the pattern will repeat itself continuously as long as the circuit has power.

The 4017 decade counter and other CMOS chips are very sensitive to static electricity, and you can easily fry the part if you aren't careful. Make sure you take special precautions, such as wearing an antistatic wrist strap (as described in Chapter 13), before handling the 4017 CMOS IC.

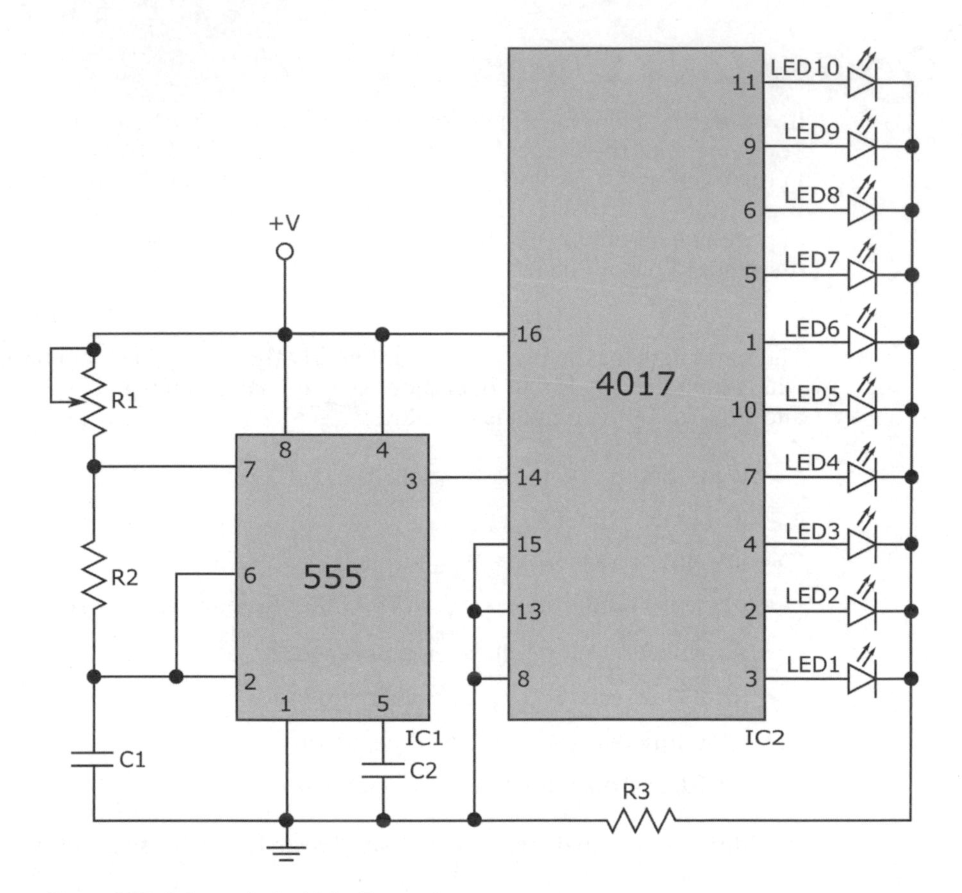

Figure 17-8: Schematic for Light Chaser 1.

Controlling the lights

The circuit for Light Chaser 1 in Figure 17-8 has two sections:

- **The brains:** A 555 timer IC makes up the first section, on the left of the schematic. You wire this chip to function as an astable multivibrator (see Chapter 11 for details). The 555 produces a series of pulses on its output pin (pin 3); you determine the speed of the pulses by dialing potentiometer *R1*.
- **The body:** The second section, on the right of the schematic, contains a 4017 CMOS decade counter chip with LEDs connected to each of its 10 output pins. As I explain in Chapter 11, pins 1–7 and pins 9–11 on the 4017 chip go from low to high one at a time (but not in the same order as the pin numbers) when a trigger signal is connected to pin 14. Resistor *R3* limits the current passing through whichever LED is activated at any given time.

- **The connection:** The LEDs are switched when the 4017 receives a pulse (on pin 14) from the 555 output (pin 3). You wire the 4017 so it repeats the 1-to-10 sequence, over and over again, for as long as the circuit has power. By adjusting the pot *(R1),* you can change the speed of the lighting sequence.

Arranging the LEDs

You can build Light Chaser 1 on a solderless breadboard to try it out. If you plan to make it into a permanent circuit, give some thought to the arrangement of the ten LEDs. For example, to achieve different lighting effects, you can try the following:

- **Put all the LEDs in a row, in sequence:** The lights chase each other up (or down) over and over again.
- **Put all the LEDs in a row, but alternate the sequence left and right:** Wire the LEDs so the sequence starts from the outside and works its way inward.
- **Place the LEDs in a circle so the LEDs sequence clockwise or counter-clockwise:** This light pattern looks like a roulette wheel.
- **Arrange the LEDs in a heart shape:** You can use this arrangement to make a unique Valentine's Day present.

Building Light Chaser 2

Figure 17-9 shows another way to build a light chaser. The left side of Light Chaser 2 is the same as the left side of Light Chaser 1, so the brains of both circuits operate in the same way. The right side of Light Chaser 2 is set up so that the LEDs light up sequentially from *LED1* through *LED6* and then back down to *LED1.* The lighting sequence follows this repeating pattern: 1-2-3-4-5-6-5-4-3-2. By adjusting the pot *(R1),* you can change the speed of this bidirectional lighting sequence.

Note that each of the middle LEDs (*LED2–LED5*) is connected to two output pins on the 4017 decade counter IC, with a diode between each output pin and the LED. By connecting two output pins to one LED, you enable that LED to light up twice during each count from 0 to 9. You need to include a diode at each output pin to prevent current from flowing back into the 4017 chip. (Chapter 9 explains that diodes operate like one-way valves for current.) For instance, when 4017 pin 5 goes high, current flows from pin 5 through *D8* and *LED5* (lighting it up), but diode *D7* prevents any of that current from flowing back into the 4017 chip through pin 10.

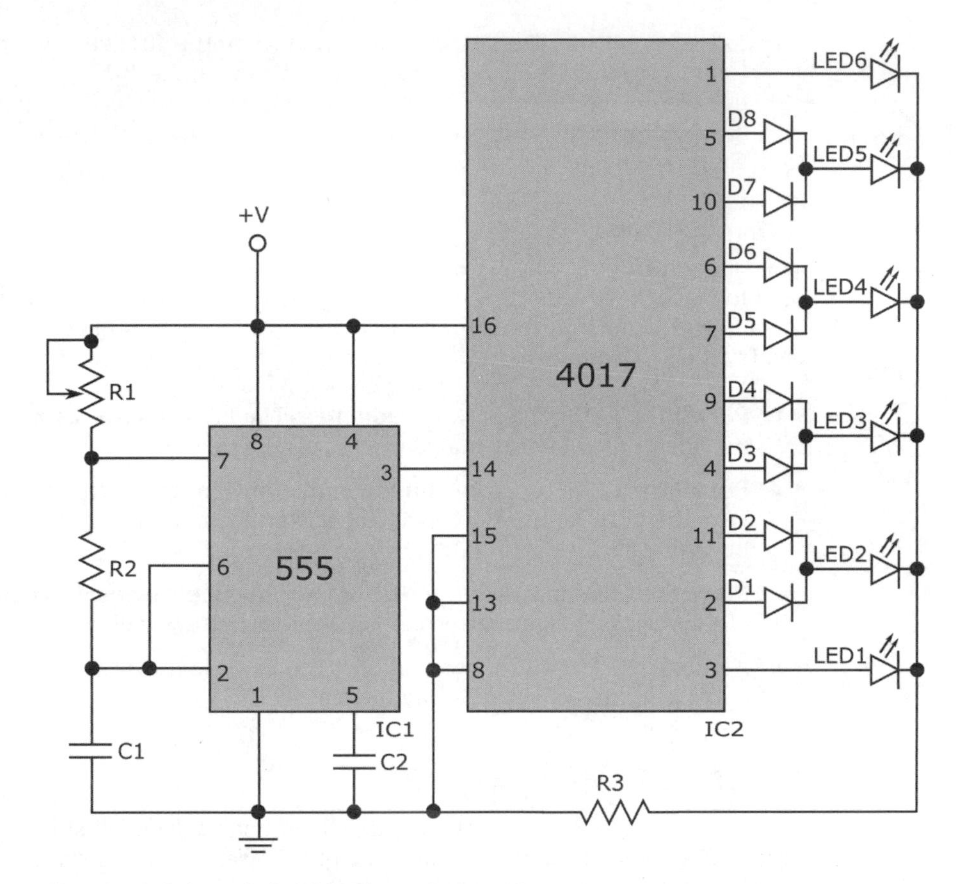

Figure 17-9: Schematic for Light Chaser 2.

Red Light, Green Light, 1-2-3!

In this section, I show you how to use a 555 timer chip and the 4017 decade counter (once again) to build a simulated green-yellow-red traffic signal. The schematic for the traffic signal is shown in Figure 17-10. Here are the parts you need to build this circuit:

- 9-volt battery (with battery clip)
- IC1: LM555 timer IC
- IC2: 4017 CMOS decade counter IC
- R1: 100 kΩ potentiometer
- R2: 22 kΩ resistor (red-red-orange)
- R3: 330 Ω resistor orange-orange-brown)
- C1: 100 μF electrolytic (polarized) capacitor

- C2: 0.01 μF disc (nonpolarized) capacitor
- LED1: green LED (any size)
- LED2: yellow LED (any size)
- LED3: red LED (any size)
- D1–D10: 1N4148 diode

Diode *D5* is not necessary in the circuit shown in Figure 17-10. However, you need diode *D5* for some of the tweaks to the design of this circuit I describe later, so you might as well include *D5* from the get-go.

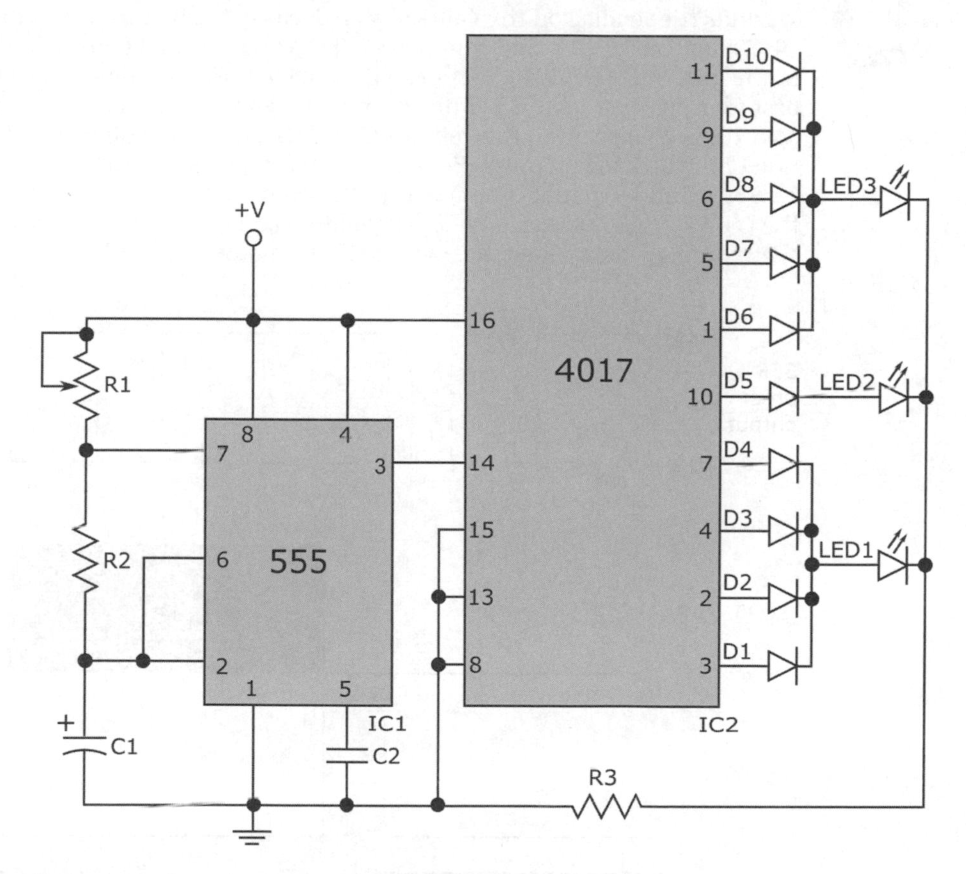

Figure 17-10: Schematic for a three-light traffic signal.

The 555 timer chip is used in astable mode to generate a low-frequency square-wave pulse on output pin 3. Note that the value of capacitor *C1* is 100 μF — much larger than the *C1* value used to control the light chaser circuits in the previous section. The larger the capacitance, the longer it takes to charge the capacitor, and the longer it takes to trigger the 555 chip via pin 2. So the 555 timer output (pin 3) oscillates at a much slower rate than in the light chaser circuits.

By varying the resistance of the potentiometer *(R1),* you control the timing cycle, but because this pot is smaller than the pot used in the light chaser circuits, you can't vary the timing quite as much. The full duration of the timing cycle (that is, the time it takes to complete one up-and-down pulse on pin 3 of the 555 timer IC) is designed to range from about 3 seconds to about 10 seconds.

Each timing pulse from the 555 chip triggers the decade counter to count, so the 10 outputs of the decade counter go high one at a time every 3 to 10 seconds (depending on the value of *R1*). Because the first four outputs of the 4017 chip (pins 3, 2, 4, and 7, in that order) are connected (through diodes) to the green LED (*LED1*), the green LED will light up and remain lit during the first four pulses from the 555 timer chip. The fifth output of the 4017 chip (pin 10) is connected to the yellow LED (*LED2*), so it will light up for the duration of the fifth timing pulse. Because the sixth through tenth outputs (pins 1, 5, 6, 9, and 11, in that order) of the 4017 chip are tied (through diodes) to the red LED *(LED3),* the red LED will light up and stay lit for the sixth through tenth timing pulses. Then, the cycle will repeat. (See Figure 17-11.)

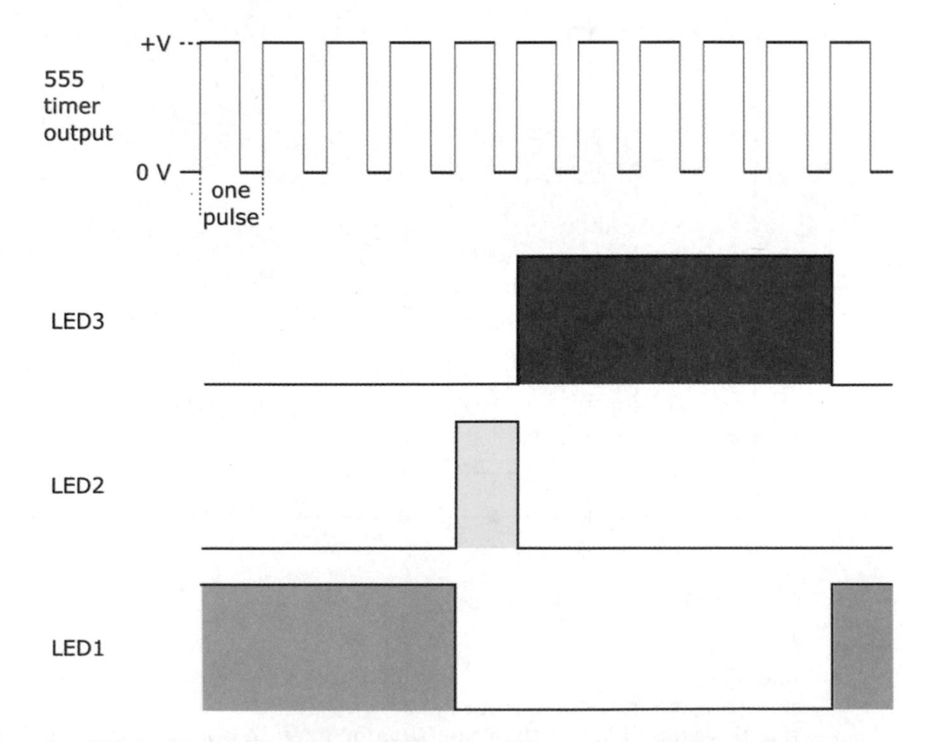

Figure 17-11: The pulse train from the 555 timer chip triggers the 4017 decade counter to activate the green, yellow, and red LEDs according to a designed timing sequence.

Here are some ways in which you can tweak your traffic light design:

- **Change the duration of the yellow and red lights:** Disconnect the cathode (negative side) of diode *D6* from *LED3,* and then connect it (that is, the cathode of diode *D6*) to *LED2*. Now the yellow LED will stay lit for two (instead of one) pulses, and the red LED will stay lit for four (instead of five) pulses.
- **Add another yellow state:** In the UK, traffic lights transition from green to yellow to red to yellow to green. To create this sequence, disconnect the cathode of diode *D10* from *LED3,* and then connect that cathode to *LED2*. Now the last pulse in the 10-pulse train will activate the yellow LED, creating the timing sequence used in the UK.
- **Change the speed of the overall timing sequence:** Replace *C1* with a 47 μF capacitor. The entire timing sequence should be roughly half of what it used to be. Or try using different values for *R2* or a different potentiometer for *R1*. (Refer to Chapter 11 for timing equations.)
- **Create a blinking red light (these are popular in New Jersey):** Remove the yellow and green LEDs (*LED1* and *LED2*) so that you're essentially disconnecting 4017 output pins 2, 3, 4, 7, and 10. Replace *C1* with a 4.7 μF capacitor. Your red light blinker should cycle on and off every 0.5 to 5 seconds (depending on the value of the pot, *R1*). (Of course, this is overkill. You don't need the 4017 decade counter to do this; you just need a 555 timer IC and some resistors and capacitors.)
- **Replace the pot (*R1*) with a fixed resistor:** If you're happy with a specific timing sequence and want to build a permanent circuit, there's no need to use a bulky potentiometer.

Maybe you may know some kids who would love to have a traffic signal to use while they're playing with toy cars and trucks, riding their Big Wheels in your driveway, or playing "Red Light, Green Light, 1-2-3!" with a bunch of friends. You can test the circuit on a solderless breadboard, tweak the design to suit the needs of your young customers, and then build a permanent circuit and enclose it in a shiny box that has three holes for the LEDs and a hook or stand for mounting. (If you do create such a project, remember to include an on/off switch for your battery.)

Part IV
The Part of Tens

Check out www.dummies.com/extras/electronics for a look at ten amazing individuals whose discoveries and inventions helped shape modern electronics.

In this part . . .

- Discovering ways to expand your knowledge of electronics
- Adding advanced testing tools to your electronics arsenal
- Finding out where to buy your stash of electronic components and tools

Chapter 18

Ten Ways to Explore Electronics Further

In This Chapter

- Trying your hand at ready-made project kits and other hobbyists' circuit designs
- Arming yourself with testing and simulation tools
- Discovering the fundamentals of computer architecture
- Delving into programmable microcontrollers and single-board computers

Ready to build on your newfound knowledge of electronics? Want to expand your horizons and create programmable electronics projects? This chapter provides you with a list of ideas for enhancing your electronics experience.

Surfing for Circuits

Thousands of project ideas are available on the Internet. Use your favorite search engine to find projects in topics or specific parts that interest you. For instance, search for *simple audio circuits* or *555 timer circuits* to get loads of ideas — some with complete explanations, schematics, and photographs of a breadboarded circuit. Or choose an idea for a circuit and see if one is out there already. A search for *door alarm circuit,* for example, turns up many simple circuit ideas and even YouTube videos.

Getting a Jumpstart with Hobby Kits

If you want to make some cool things happen but don't want to start from scratch, you can purchase one or more electronics hobby kits. These kits include everything you need to build a functional circuit: all the electronic components, wire, circuit board, and detailed instructions for putting the circuit together. Some even include an explanation of how the circuit works.

You'll find kits for light-sensitive alarms, simulated traffic signals, electronic combination locks, adjustable timers, decorative light displays, and much more. Many of the parts sources mentioned in Chapter 19 provide ready-made kits at reasonable prices. You can practice your circuit-building and analysis skills using these kits, and then move on to designing, building, and testing your own circuits from scratch.

Simulating Circuit Operation

If you have a complicated circuit design or just want to understand more about how a particular circuit will behave when powered up, you can use a *circuit simulator.* This software program uses computer-based models of circuit components to predict the behavior of real circuits. You tell it what components and power supplies you're using and how they should be wired, and the software tells you whatever you want to know about the operation of the circuit: the current through any component, voltage drops across components, circuit response across various frequencies, and so forth.

Many circuit simulators are based on an industry-standard algorithm called SPICE (Simulation Program with Integrated Circuit Emphasis); you can use them to simulate and analyze various circuits — analog, digital, and *mixed signal* (that is, incorporating both analog and digital). You can download a free evaluation copy of one such simulator, Multisym, and try it out yourself by visiting `www.ni.com/multisim/try/`.

Scoping Out Signals

An *oscilloscope* is a piece of test equipment that displays how a voltage varies with time as a trace across a cathode-ray tube (CRT) or other display that contains a calibrated grid. You use a scope to visualize what is happening to rapidly varying voltages in your circuits. If you're interested in building audio amplifiers and other circuits that have time-varying signals, such as sound, an oscilloscope can come in handy and help you understand circuit operation and pinpoint errors. A good scope costs a few hundred dollars but you may be able to find some great deals on eBay or Craigslist.

Counting Up Those Megahertz

You can use a *frequency counter* (or frequency meter) to help you determine whether your AC circuit is operating properly. By touching the leads of this test device to a signal point in a circuit, you can measure the frequency of that signal. For example, suppose you create an infrared transmitter and the light from that transmitter is supposed to pulse at 40,000 cycles per second (also known as 40 kHz). If you connect a frequency counter to the output of the circuit, you can verify that the circuit is indeed producing pulses at 40 kHz — not 32 kHz, 110 kHz, or some other Hz.

Generating a Variety of Signals

To test a circuit's operation, it often helps to apply a known signal input to the circuit, and observe how the circuit behaves. You can use a *function generator* to create repeating-signal AC waveforms in a variety of shapes and sizes — and apply the generated waveform to the input of the circuit you're testing. Most function generators develop three kinds of waveforms: sine, triangle, and square. You can adjust the frequency of the waveforms from a low of between 0.2 Hz and 1 Hz to a high of between 2 MHz and 20 MHz. Some function generators come with a built-in frequency counter so you can accurately time the waveforms you generate. You can also use a stand-alone frequency counter to fine-tune the output of your function generator.

Exploring Basic Computer Architectures

In Chapter 11, you find out how *logic gates* (AND, OR, NAND, and so on) process *bits* (1s and 0s) of data and that you can purchase specialized integrated circuits (ICs) that contain logic gates. Chapter 11 also shows you a basic circuit diagram of a half-adder circuit that uses just two logic gates. Circuits such as the half-adder form the foundation of computer architectures. By connecting multiple logic gates in just the right way, you can create circuits that compute, store, and control information (series of 1s and 0s organized into groups of 8 called *bytes*). Start your journey into the fascinating field of computer architecture by building digital logic circuits that use LEDs as output indicators. (Check out `www.doctronics.co.uk/4008.htm` for a detailed description of how to build a 4-bit binary full adder.)

Microcontrolling Your Environment

In Chapter 11, you get a peek at a *microcontroller,* which is a tiny computer on a chip. You create a program on your computer, and download the program onto the chip. Then when you power up the chip, it follows the instructions in your program. The BASIC Stamp and PICAXE microcontrollers are inexpensive alternatives that use the easy-to-learn BASIC programming language. However, the beginner-friendly Arduino microcontroller system, which uses a C-like programming language, has exploded in popularity in recent years due to its simplified integrated development environment (IDE), versatility, affordability, and enormous online user community.

You can purchase a complete Arduino Starter Kit — including a microcontroller, IDE, project book, breadboard, cables, servomotor, photoresistor, tilt sensor, temperature sensor, and other discrete components — for less than $85 on `https://store.arduino.cc`. (And eBay lists some knock-off Arduino kits with gobs of cool extras for around $50.) These comprehensive microcontroller kits enable you to program circuits to interact with your environment, take readings from sensors, make decisions based on those readings, and execute actions based on those decisions. Arduino can provide you with an entrée into the field of robotics. Check out *Arduino For Dummies* by John Nussey (Wiley Publishing, Inc.), and keep an eye on the growing field of user-friendly microcontrollers, because new features (such as integrated Wi-Fi) and competitive products are appearing on the market.

Getting a Taste of Raspberry Pi

The Raspberry Pi is a series of single-board computers that you connect to your TV or monitor and a standard keyboard. The original Pi runs the Linux operating system, but the second generation Pi runs both Linux and a version of Windows 10. You program the Pi using Python or any one of a number of IDEs. Although it's not as beginner-friendly as Arduino, the Raspberry Pi is as inexpensive (roughly $35) and has a large online user community. Intended as an educational tool for teaching kids programming skills, the Raspberry Pi is enjoying tremendous popularity and finding itself at the heart of many computer-embedded hobby projects, such as an Internet radio and an infrared bird box. You'll learn more about programming rather than down-and-dirty electronics with the Raspberry Pi, but you certainly can come up with ideas that integrate the two skill sets. Check out *Raspberry Pi For Dummies,* 2nd Edition by Sean McManus (Wiley Publishing, Inc.), or visit `www.raspberrypi.org`.

Try, Fry, and Try Again

Perhaps the best way to expand your knowledge of electronics is to develop your own ideas, design some circuits, build and test them, and then go back and tweak your design. Sometimes the only way to find out what the limitations of various parts or designs are is to fry a few LEDs, toast a couple of ICs, or stay up all night probing the depths of a nonworking circuit until you figure out what's wrong with it. To quote a popular science teacher, Valerie Frizzle, "Take chances, make mistakes, get messy!" (But, please, take safety precautions no matter what else you do!)

Ten Great Electronics Parts Sources

In This Chapter

- Parts sources from around the globe
- Avoiding hazardous substances
- Understanding the pros and cons of surplus parts

Looking for some great sources for your electronic parts? This chapter gives you some perennial favorites, both inside and outside North America. This list is by no means exhaustive; you can find literally thousands of specialty outlets for new and used electronics. Plus, Amazon and eBay provide virtual marketplaces for all sorts of sellers — from established retailers to individuals selling parts and components out of their homes. But the sources I list here are among the more established in the field, and all have web pages for online ordering. (Some also offer print catalogs.)

North America

Check out these online resources if you're shopping in the United States or Canada. Most of these outlets ship worldwide, so if you live in a different country, you can still consider buying from these stores. Just remember that shipping costs may be higher, and you may have to pay an import duty, depending on your country's regulations.

All Electronics

www.allelectronics.com

All Electronics runs a retail store in the Los Angeles area and sends orders worldwide. Most of its stock is *new surplus,* meaning the merchandise is brand-new but has been overstocked by the company. All Electronics has a printed catalog, which is also available in PDF format on its website. Stock changes frequently, and the latest updates are available only on the website.

Allied Electronics

www.alliedelec.com

Allied Electronics is what's known as a *stocking distributor.* It offers goods from a variety of manufacturers, and most parts are available for immediate shipping from Allied's warehouse in Fort Worth, Texas. Allied is geared toward the electronics professional, but it welcomes hobbyists, too. The Allied catalog is *huge,* and it's available on the website as well as in print. You can also find parts using the search feature on Allied's website.

Digi-Key

www.digikey.com

If you want it, Digi-Key probably has it. Like Allied Electronics, Digi-Key is a stocking distributor, carrying thousands upon thousands of items. Digi-Key welcomes small orders and offers reasonable USPS shipping rates. Its online ordering system includes detailed product information, price, available stock levels, and even links to product datasheets. The site offers a handy search engine so you can quickly locate what you're looking for, as well as an online interactive catalog (with magnification capabilities you'll need). Digi-Key will also send a free printed catalog, but to read the tiny print, you have to get out your glasses. The text has to be teeny-weeny to fit everything in.

Electronic Goldmine

www.goldmine-elec.com

Electronic Goldmine sells new and surplus parts, from the lowly resistor to exotic lasers. Its website is organized by category, which makes ordering easy. (One category, labeled "Rare and Esoteric," may be where Doc Brown got the flux capacitor that made time travel possible in *Back to the Future*.) Listings for most parts include a color picture and a short description. Be sure to check out the nice selection of project kits.

Jameco Electronics

www.jameco.com

Jameco sells components, kits, tools, and more, offering both convenient online and catalog ordering. You can browse the website by category, or if you know the part number you're interested in — such as a 2N2222 transistor — you can find it by entering the part number in a search box. You can also use the search feature for categories of parts, such as motors, batteries, or capacitors. Just type the category term, press Enter, and off you go.

Mouser Electronics

www.mouser.com

Similar to Allied and Digi-Key, Mouser is a stocking distributor with tens of thousands of parts on hand. If you can't find it at Mouser, it probably doesn't exist. Mouser carries more than 165,000 resistors alone, listed in the "Passive Components" category. You can order from its online store or its humongous print catalog. Feel free to request a printed catalog to keep on your nightstand, or let your mouse do the walking through Mouser's online catalog.

Parts Express

www.parts-express.com

Parts Express specializes in electronic parts and other equipment for audiovisual aficionados. You'll find an ample selection of individual components, complete with user reviews, as well as project kits and other do-it-yourself resources on the Parts Express website. Check out the project showcase, comprehensive list of formulas (including Ohm's Law), and technical glossary, and don't forget to review the electronics safety information. You'll also find Parts Express on eBay, Facebook, and Twitter!

RadioShack

www.radioshack.com

RadioShack is perhaps the world's most recognized source for hobby electronics, but alas, "The Shack" filed for Chapter 11 bankruptcy in early 2015 and the future of its brand is uncertain. As of mid-2015, roughly half of RadioShack's 4,000 retail stores were in the process of closing, but it's likely you can still find a location within reasonable driving distance of your home. If you need a resistor, capacitor, or transistor right away, you definitely won't find it at Wal-Mart or BestBuy, but you may find it at your local RadioShack. RadioShack also has an online store, RadioShack.com, but I received a lot of “out of stock online” messages when searching for components multiple times in mid-2015. Fingers crossed that the local stores survive and thrive!

Outside North America

Electronics is popular all over the globe! Here are some websites you can visit if you live in places such as Australia or the UK. As with North American online retailers, most of these folks also ship worldwide. Check their ordering pages for details.

Premier Farnell (UK)

`www.farnell.com`

Based in the UK, but with operations in 24 countries across Europe, Asia Pacific, and the Americas, Premier Farnell stocks some 500,000 products. It operates under several company names, including Farnell element14 (Europe), element14 (Asia Pacific), Newark element14 (North America), and Farnell Newark (Brazil). To order products, start at `www.farnell.com`, select your country from the extensive list on the home page, and you’ll have access to a vast array of products.

Maplin (UK)

`www.maplin.co.uk`

Maplin provides convenient online ordering for shoppers in the UK and the Republic of Ireland. The company also supports dozens of retail stores throughout the UK and Ireland.

What’s RoHS Compliance?

When you’re shopping around for parts, you may see the term *RoHS compliant* next to some of the parts. The term *RoHS* (pronounced “ROW-haas”) refers to the Restriction of Hazardous Substances directive adopted in 2003 by the European Union (E.U.). The RoHS directive, which took effect in 2006, restricts the placement on the E.U. market of new electrical and electronic devices that contain more than a specified level of lead and five other hazardous substances. Companies producing consumer and industrial electronics need to worry about RoHS compliance if they want to sell products in E.U. countries (and China, which has its own RoHS specification), but if you are just tinkering around with electronics in your house, you need not worry about using lead-free solder and other RoHS-compliant parts. Just don’t let your cat munch on your solder.

New or Surplus?

Surplus is a loaded word. To some, it means junk that just fills up the garage, like musty canvas tents or funky fold-up shovels that the US Army used back in the 1950s. To the true electronics buff, surplus has a different meaning: affordable components that help stretch the electronics-building dollar.

Surplus just means that the original maker or buyer of the goods doesn't need it any more. It's simply excess stock for resale. In the case of electronics, surplus seldom means used, as it might for other surplus components, such as motors or mechanical devices that have been reconditioned. Except for hard-to-find components — such as older amateur radio gear — surplus electronics are typically brand-new, and someone still actively manufactures much of this equipment. In this case, surplus simply means extra.

The main benefit of shopping at the surplus electronics retailer is cost: Even new components are generally lower priced than at the general electronics retailers. On the downside, you may have limited selection — whatever components the store was able to purchase. Don't expect to find every value and size of resistor or capacitor, for example.

Remember that when you buy surplus, you don't get a manufacturer's warranty. Sometimes that's because the manufacturer is no longer in business. Although most surplus sellers accept returns if an item is defective (unless it says something different in their catalogs), you should always consider surplus stuff as-is, with no warranty implied or intended (and all that other lawyer talk).

Glossary

Here are many of the terms you'll run into throughout your electronics life. Knowing these terms will help you become fluent in electronics-speak.

alkaline battery: A type of nonrechargeable battery. *See also* battery.

alternating current (AC): Current characterized by a change in direction of the flow of electrons. *See also* direct current (DC).

ampere: The standard unit of electric current, commonly referred to as amps. One ampere is the strength of an electric current when 6.241×10^{18} electrically charged particles move past the same point within a second. *See also* current, I.

amplitude: The magnitude of an electrical signal, such as voltage or current.

analog circuit: A circuit that processes analog signals. *See also* digital circuit.

analog signal: A varying voltage or current that constitutes a one-to-one mapping of a physical quantity, such as sound or displacement.

anode: The terminal of a device into which conventional current (hypothetical positive charge) flows. In power-consuming devices, such as diodes, the anode is the positive terminal; in power-releasing devices, such as batteries, the anode is the negative terminal. *See also* cathode.

antistatic wrist strap: A device used to prevent the buildup of static electricity on individuals working on sensitive electronic equipment.

autoranging: A feature of some multimeters that automatically sets the test range.

AWG (American Wire Gauge): *See* wire gauge.

battery: A power source that uses an electrochemical reaction to produce a positive voltage at one terminal and a negative voltage at the other terminal. This process involves placing two different types of metal in a certain type of chemical. *See also* alkaline battery, lithium battery, nickel-cadmium (NiMH) battery, nickel-metal hydride (NiCd) battery and, zinc-carbon battery.

biasing: Applying a small amount of voltage to a diode or to the base of a transistor to establish a desired operating point.

bipolar transistor: A common type of transistor consisting of two fused pn-junctions. *See also* transistor.

bit: Short for *binary digit*. A digit that has a value of 0 or 1.

breadboard: Also known as *prototyping board* or *solderless breadboard;* a rectangular plastic board (available in a variety of sizes) that contains groups of electrically interconnected contact holes. You plug in components — resistors, capacitors, diodes, transistors, and integrated circuits, for example — and then string wires to build a circuit. *See also* solder breadboard.

bus strip: *See* power rail.

byte: A grouping of eight bits used as a basic unit of information for storage in computer systems.

cable: A group of two or more wires protected by an outer layer of insulation, such as a common power cord.

capacitance: The capability to store energy in an electric field, measured in farads. *See also* capacitor.

capacitor: A component that provides the property of capacitance in a circuit. *See also* capacitance.

cathode: The terminal of a device from which conventional current (hypothetical positive charge) flows. In power-consuming devices, such as diodes, the cathode is the negative terminal; in power-releasing devices, such as batteries, the cathode is the positive terminal. *See also* anode.

circuit: A complete path that allows electric current to flow.

cladding: An extremely thin sheet of copper that you glue over a base made of plastic, epoxy, or phenol to make a printed circuit board.

closed circuit: An uninterrupted circuit through which current can flow. *See also* open circuit.

closed position: The position of a switch that allows current to flow. *See also* open position.

cold solder joint: A defective joint that occurs when solder doesn't properly flow around the metal parts.

commutator: A device used to change the direction of electric current in a motor or generator.

component: A part used in a circuit, such as a battery or a diode.

conductor: A substance through which electric current can move freely.

connector: A metal or plastic receptacle on a piece of equipment (a phone jack in your wall, for example) that a cable end fits into.

continuity: A type of test you perform with a multimeter to establish whether a circuit is intact between two points.

conventional current: The flow of hypothetical positive charge from positive to negative voltage; the reverse of real current. *See also* real current.

current: The flow of electrically charged particles. *See also* ampere, I.

desoldering pump: *See* solder sucker.

digital circuit: A circuit that processes digital signals. *See also* analog circuit.

digital signal: A pattern consisting of just two voltage or current levels representing binary digital data.

diode: A semiconductor electronic component consisting of a pn-junction that allows electric current to flow one way more easily than the other way. Diodes are commonly used to convert alternating current to direct current by limiting the flow of current to one direction.

direct current (DC): A type of current in which the electrons move in only one direction, such as the electric current generated by a battery.

double-pole, double-throw (DPDT) switch: A type of switch that has two input contacts and four output contacts. It is a dual on/on switch that behaves like two SPDT switches acting in sync.

double-pole, single-throw (DPST) switch: A type of switch that has two input contacts and two output contacts. It is a dual on/off switch that behaves like two SPST switches acting in sync.

DPDT: *See* double-pole, double-throw (DPDT) switch.

DPST: *See* double-pole, single-throw (DPST) switch.

earth ground: A direct electrical connection to the earth. *See also* ground.

electric current: *See* current.

electrical signal: The pattern over time of an electrical current. Often, the way an electrical signal changes its shape conveys information about something physical, such as the intensity of light, heat, or sound, or the position of an object, such as the diaphragm in a microphone or the shaft of a motor.

electricity: The displacement of electrons along a conductor.

electromagnet: A temporary magnet consisting of a coiled wire around a piece of metal (typically, an iron bar). When you run current through the wire, the metal becomes magnetized. When you shut off the current, the metal loses its magnetic quality.

electromotive force: An attractive force between positive and negative charges, measured in volts.

electron: A negatively charged subatomic particle. *See also* proton.

ESD (electrostatic discharge): *See* static electricity.

fillet: A raised area formed by solder.

floating ground: A circuit ground that isn't connected to earth ground.

flux: A wax-like substance that helps molten solder flow around components and wire and ensures a good joint.

frequency: A measurement of how often an AC signal repeats, measured in cycles per second, or hertz (Hz). The symbol for frequency is *f*. *See also* hertz (Hz).

gain: The amount that a signal is amplified (the voltage of the signal coming out divided by the voltage of the signal coming in).

gauge: *See* wire gauge.

ground: A connection in a circuit used as a reference (0 volts) for a circuit. *See also* earth ground.

heat sink: A piece of metal that you attach securely to the component that you want to protect. The sink draws off heat and helps prevent the heat from destroying the component.

helping hands clamp: Also called a third-hand clamp; adjustable clips that hold small parts while you're working on projects.

hertz (Hz): The measurement of the number of cycles per second in alternating current. *See also* frequency.

high signal: In digital electronics, a signal at or near 5 volts (typically 3–5 V) that represents one of two binary states.

I: The symbol for conventional current, measured in amperes (amps). *See also* ampere, current.

IC: *See* integrated circuit (IC).

inductance: The capability to store energy in a magnetic field (measured in henrys). *See also* inductor.

inductor: A component that provides the property of inductance to a circuit. *See also* inductance.

infrared temperature sensor: A kind of temperature sensor that measures temperature electrically.

insulator: A substance through which electric current is unable to move freely.

integrated circuit (IC): Also known as a chip; a component that contains several miniaturized components, such as resistors, transistors, and diodes, connected in a circuit that performs a designated function.

inverter: Also known as a NOT gate; a single-input logic gate that inverts the input signal. A low input produces a high output, and a high input produces a low output. *See also* logic gate.

inverting mode: A process by which an op amp flips an input signal to produce the output signal.

jack: A type of connector. *See also* connector.

joule: A unit of energy.

lithium battery: A lightweight disposable battery that generates about 3 volts and has a higher capacity than does an alkaline battery. *See also* battery.

live circuit: A circuit to which you've applied voltage.

logic gate: A digital circuit that accepts input values and determines which output value to produce based on a set of rules.

low signal: In digital electronics, a signal at or near 0 volts (typically 0–2 V) that represents one of two binary states.

microcontroller: A programmable integrated circuit.

multimeter: An electronics testing device used to measure such factors as voltage, resistance, and current.

negative temperature coefficient (NTC) thermistor: A resistor whose resistance decreases with a rise in temperature. *See also* resistor, thermistor.

nickel-cadmium (NiCd) battery: The most popular type of rechargeable battery. Some NiCad batteries exhibit the memory effect, requiring that they be fully discharged before they can be recharged to full capacity. See *also* battery.

nickel-metal hydride (NiMH) battery: A type of rechargeable battery that offers higher energy density than does a NiCd rechargeable battery. *See also* battery.

nominal value: The stated value of a resistor or other component. The real value can vary up or down from the nominal value based on the tolerance of the device. *See also* tolerance.

N-type semiconductor: A semiconductor doped with impurities so that it has more free electrons than a pure semiconductor.

ohm: A unit of resistance; its symbol is Ω. *See also* R, resistance.

Ohm's Law: An equation that defines the relationship between voltage, current, and resistance in an electrical circuit.

open circuit: A type of circuit in which a wire or component is disconnected, preventing current from flowing. *See also* closed circuit.

open position: The position of a switch that prevents current from flowing. *See also* closed position.

operational amplifier (op amp): An integrated circuit containing several transistors and other components. In many applications, it performs much better than an amplifier made from a single transistor. For example, an op amp can provide uniform amplification over a much wider range of frequencies than can a single-transistor amplifier.

oscillator: A circuit that generates a repeating electronic signal.

oscilloscope: An electronic device that measures voltage, frequency, and various other parameters for varying waveforms.

pad: A contact point on a printed circuit board used for connecting components.

Phillips: Both a screw with a plus-sign-shaped (+) slot in its head and the screwdriver used with it.

piezoelectric effect: The capability of certain crystals, such as quartz or topaz, to expand or contract when you apply voltage to them or to produce voltage when you squeeze or move them.

pn-junction: The point of contact between a P-type semiconductor, such as silicon doped with boron, and an N-type semiconductor, such as silicon doped with phosphorus. The pn-junction is the foundation for diodes and bipolar transistors. *See also* bipolar transistor, diode.

positive temperature coefficient (PTC) thermistor: A device whose resistance increases with a rise in temperature. *See also* resistance, thermistor.

potentiometer: A variable resistor that allows for the continual adjustment of resistance from virtually zero ohms to a maximum value.

power: The measure of the amount of work that electric current does while running through an electrical component, measured in watts.

power rail: A series of interconnected electrical contact holes in a column of a solderless breadboard that is intended to be used for power distribution. Also known as bus strip.

precision resistor: A type of resistor with low tolerance (the allowable deviation from its stated, or nominal, value). *See also* nominal value, tolerance.

proton: A positively charged subatomic particle. *See also* electron.

prototyping board: *See* breadboard.

P-type semiconductor: A semiconductor doped with impurities so that it has fewer free electrons than a pure semiconductor.

pulse: A burst of current or voltage, usually beginning with an abrupt rise and ending with an abrupt fall.

pulse-width modulation: A method of controlling the speed of a motor that turns voltage on and off in quick pulses. The longer the on intervals, the faster the motor.

R: The symbol for resistance. *See also* ohm, resistance.

RC time constant: A calculation of the product of resistance and capacitance that defines the length of time it takes to charge a capacitor to two-thirds of its maximum voltage or to discharge it to one-third of its maximum voltage.

real current: The flow of electrons from a negative to a positive voltage. *See also* conventional current.

relay: A device that acts like a switch in that it closes or opens a circuit depending on the voltage supplied to it.

resistance: A measure of a component's opposition to the flow of electric current, measured in ohms. *See also* ohm, R.

resistor: A component with a fixed amount of resistance that you can add to a circuit to restrict the flow of current. *See also* resistance.

rosin flux remover: A detergent used after soldering to clean any remaining flux to prevent it from oxidizing a circuit. Available in a bottle or spray can.

schematic: A drawing showing how components in a circuit are connected.

semiconductor: A material, such as silicon, that has some of the properties of both conductors and insulators.

semiconductor temperature sensor: A kind of temperature sensor that measures temperature electrically.

sensor: An electronic component that senses a condition or an effect, such as heat or light, and converts it into an electrical signal.

series circuit: A circuit in which the current runs through each component sequentially.

short circuit: An accidental connection between two wires or components allowing current to pass through them rather than through the intended circuit.

single-pole, double-throw (SPDT) switch: A type of switch that has one input contact and two output contacts. It switches the input between two choices of outputs. It is also known as an on/on or changeover switch.

single-pole, single-throw (SPST) switch: A type of switch that has one input contact and one output contact. It is also known as an on/off switch.

60/40 rosin core: Solder containing 60% tin and 40% lead (the exact ratio can vary a few percentage points) with a core of rosin flux. This type of solder is ideal for working with electronics. *See also* solder, soldering.

slide switch: A type of switch where you slide the switch forward or backward to turn something (such as a flashlight) on or off.

solar cell: A type of semiconductor that generates a current when exposed to light.

solder: A metal alloy that is heated and applied to two metal wires or leads and allowed to cool, forming a conductive joint. *See also* 60/40 rosin core, soldering.

solder breadboard: A breadboard on which you solder components in place. *See also* breadboard.

soldering: The method you use in your electronics projects to assemble components on a circuit board to build a permanent electrical circuit. Rather than using glue to hold things together, you use small globs of molten metal, or solder. *See also* solder.

soldering iron: A wand-like tool that consists of an insulated handle, a heating element, and a polished metal tip used to apply solder.

soldering pencil: *See* soldering iron.

solderless breadboard: *See* breadboard.

solder sucker: Also known as a desoldering pump; a tool consisting of a spring-loaded vacuum used for removing excess solder.

solder wick: Also known as solder braid; a device used to remove hard-to-reach solder. The solder wick is a flat braid of copper that works because the copper absorbs solder more easily than the tin plating of most components and printed circuit boards.

solid wire: A wire consisting of only a single strand.

SPDT: *See* single-pole, double-throw (SPDT) switch.

SPST: *See* single-pole, single-throw (SPST) switch.

static electricity: Charge that builds on or within an object and remains stationary until a path is provided for the charge to flow. Lightning is a form of static electricity.

strain relief: A device that clamps around a wire and prevents you from tugging the wire out of the enclosure.

stranded wire: A metal wire consisting of several fine wires twisted together into a bundle that is wrapped in insulation.

stray capacitance: Energy that's stored in a circuit unintentionally when electric fields occur between wires or leads that are placed too close together.

terminal: A piece of metal to which you hook up wires (as with a battery terminal).

thermistor: A resistor whose resistance value varies with changes in temperature.

thermocouple: A type of sensor that measures temperature electrically.

third-hand clamp: Also called helping hands clamp; a small, weighted clamp that holds parts while you solder.

tinning: The process of heating up a soldering tool and applying a small amount of solder to the tip to prevent solder from sticking to the tip.

tolerance: The variation from the nominal value of a component due to the manufacturing process expressed as a percentage. *See also* nominal value.

trace: A wire on a circuit board that runs between the pads to electrically connect the components.

transistor: A semiconductor device that's commonly used to switch and amplify electrical signals.

V: The symbol for voltage, also represented by E. *See also* voltage.

variable capacitor: A capacitor whose capacitance can be dynamically altered mechanically or electrically. *See also* capacitance, capacitor.

variable inductor: A coil of wire surrounding a movable metal slug. By turning the slug, you change the inductance of the coil.

variable resistor: *See* potentiometer.

voltage: An attractive force between positive and negative charges.

voltage divider: A circuit that uses voltage drops to produce voltage lower than the supply voltage at specific points in the circuit.

voltage drop: The resulting lowering of voltage when voltage pulls electrons through a resistor (or other component), and the resistor absorbs some of the electrical energy.

wire: A long strand of metal, usually made of copper, that you use in electronics projects to conduct electric current.

wire gauge: A system of measurement of the diameter of a wire.

zinc-carbon battery: A low-quality, disposable battery. *See also* battery.

Index

• Numbers •

• A •

• B •

• D •

• E •

• F •

• G •

• H •

• I •

• M •

• N •

• O •

• Q •

• R •

• S •

• T •

• W •

• X •

• Z •

About the Author

Cathleen Shamieh is an electrical engineer and high-tech writer with extensive engineering and consulting experience in the fields of medical electronics, speech processing, and telecommunications.

Dedication

To my family, in heaven and on earth, and to Julie, whose perseverance in the face of great difficulty is an inspiration to me.

Author's Acknowledgments

I am grateful to the entire Wiley team for their hard work, support, and professionalism. Special thanks to my top-notch editor, Susan Pink, for her attention to detail, sense of humor, and friendly exchanges about topics ranging from Buffalo to bees. The phrase, "this what, please?" is forever engrained in my brain! Thanks, too, to Kirk Kleinschmidt, for unearthing technical inaccuracies and sharing insights based on a wealth of experience, and to Katie Mohr for shaping this project and allowing my deadlines to slip a bit.

I'd also like to acknowledge the developers of Inkscape for providing the user-friendly, open-source vector graphics program I relied on to create and edit the figures in this book, and to the many unnamed users in the Inkscape community who created and publicly shared clip art of everything from rulers to resistors.

Finally, I'd like to thank Bill, Kevin, Peter, Brendan, and Patrick for their steadfast love and support.

Publisher's Acknowledgments

Acquisitions Editor: Katie Mohr

Project Editor: Susan Pink

Copy Editor: Susan Pink

Technical Editor: Kirk Kleinschmidt

Editorial Assistant: Claire Brock

Sr. Editorial Assistant: Cherie Case

Project Coordinator: Kumar Chellapa

Cover Image: ©Daniel Schweinert/shutterstock

Investing For Dummies
Eric Tyson
Windows 8 For Dummies
Andy Rathbone
500 Spanish Verbs For Dummies
Cecie Kraynak, MA
iPhone For Dummies
Edward C. Baig
Bob "Dr. Mac" LeVitus

Math & Science

Algebra I For Dummies, 2nd Edition
978-0-470-55964-2

Anatomy and Physiology For Dummies, 2nd Edition
978-0-470-92326-9

Astronomy For Dummies, 3rd Edition
978-1-118-37697-3

Biology For Dummies, 2nd Edition
978-0-470-59875-7

Chemistry For Dummies, 2nd Edition
978-1-118-00730-3

1001 Algebra II Practice Problems For Dummies
978-1-118-44662-1

Microsoft Office

Excel 2013 For Dummies
978-1-118-51012-4

Office 2013 All-in-One For Dummies
978-1-118-51636-2

PowerPoint 2013 For Dummies
978-1-118-50253-2

Word 2013 For Dummies
978-1-118-49123-2

Music

Blues Harmonica For Dummies
978-1-118-25269-7

Guitar For Dummies, 3rd Edition
978-1-118-11554-1

iPod & iTunes For Dummies, 10th Edition
978-1-118-50864-0

Programming

Beginning Programming with C For Dummies
978-1-118-73763-7

Excel VBA Programming For Dummies, 3rd Edition
978-1-118-49037-2

Java For Dummies, 6th Edition
978-1-118-40780-6

Religion & Inspiration

The Bible For Dummies
978-0-7645-5296-0

Buddhism For Dummies, 2nd Edition
978-1-118-02379-2

Catholicism For Dummies, 2nd Edition
978-1-118-07778-8

Self-Help & Relationships

Beating Sugar Addiction For Dummies
978-1-118-54645-1

Meditation For Dummies, 3rd Edition
978-1-118-29144-3

Seniors

Laptops For Seniors For Dummies, 3rd Edition
978-1-118-71105-7

Computers For Seniors For Dummies, 3rd Edition
978-1-118-11553-4

iPad For Seniors For Dummies, 6th Edition
978-1-118-72826-0

Social Security For Dummies
978-1-118-20573-0

Smartphones & Tablets

Android Phones For Dummies, 2nd Edition
978-1-118-72030-1

Nexus Tablets For Dummies
978-1-118-77243-0

Samsung Galaxy S 4 For Dummies
978-1-118-64222-1

Samsung Galaxy Tabs For Dummies
978-1-118-77294-2

Test Prep

ACT For Dummies, 5th Edition
978-1-118-01259-8

ASVAB For Dummies, 3rd Edition
978-0-470-63760-9

GRE For Dummies, 7th Edition
978-0-470-88921-3

Officer Candidate Tests For Dummies
978-0-470-59876-4

Physician's Assistant Exam For Dummies
978-1-118-11556-5

Series 7 Exam For Dummies
978-0-470-09932-2

Windows 8

Windows 8.1 All-in-One For Dummies
978-1-118-82087-2

Windows 8.1 For Dummies
978-1-118-82121-3

Windows 8.1 For Dummies, Book + DVD Bundle
978-1-118-82107-7

Available in print and e-book formats.

Available wherever books are sold. **For more information or to order direct visit www.dummies.com**

of For Dummies

Custom Publishing

Reach a global audience in any language by creating a solution that will differentiate you from competitors, amplify your message, and encourage customers to make a buying decision.

Apps • Books • eBooks • Video • Audio • Webinars

Brand Licensing & Content

Leverage the strength of the world's most popular reference brand to reach new audiences and channels of distribution.

For more information, visit www.Dummies.com/biz